D1195080

Astronomical Spectrographs and their History

Astronomical spectrographs analyse light emitted by the Sun, stars, galaxies and other objects in the Universe, and have been used in astronomy since the early nineteenth century. This book provides a comprehensive account of spectrographs from an historical perspective, from their theory and development over the last 200 years, to the recent advances of the early twenty-first century.

The author combines the theoretical principles behind astronomical spectrograph design with their historical development. Spectrographs of all types are considered, with prism, grating or grism dispersing elements. Included are Cassegrain, coudé, prime focus, échelle, fibre-fed, ultraviolet, nebular, objective prism, multi-object instruments and those which are ground-based, on rockets and balloons, or in space.

The book contains several tables listing the most significant instruments, around 900 references, and over 150 images, making it an indispensable reference for professional astronomers, graduate students, advanced amateur astronomers, and historians of science.

JOHN HEARNSHAW is Professor of Astronomy in the Department of Physics and Astronomy at the University of Canterbury, New Zealand. He has won the Mechaelis Prize for astronomy in New Zealand, and has twice been awarded the Alexander von Humboldt Fellowship in Germany. He chairs the International Astronomical Union Program Group for the Worldwide Development of Astronomy.

Astronomical Spectrographs and their History

John Hearnshaw
University of Canterbury, New Zealand

CAMBRIDGE UNIVERSITY PRESS
Cambridge, New York, Melbourne, Madrid, Cape Town, Singapore, São Paulo, Delhi

Cambridge University Press
The Edinburgh Building, Cambridge CB2 8RU, UK

Published in the United States of America by Cambridge University Press, New York

www.cambridge.org
Information on this title: www.cambridge.org/9780521882576

First published 2009

Printed in the United Kingdom at the University Press, Cambridge

A catalogue record for this publication is available from the British Library

Library of Congress Cataloging-in-Publication Data
Hearnshaw, J. B.
Astronomical spectrographs and their history / John Hearnshaw.
    p.   cm.
ISBN 978-0-521-88257-6
1. Spectrograph – History.   2. Spectrum analysis – History.   I. Title.
QB873.H43   2009
522′.67–dc22

                                               2008045408

ISBN 978-0-521-88257-6 hardback

Contents

Preface

Few astronomers would dispute the pivotal rôle that the astronomical spectrograph has played in the development of astrophysics. Of all astronomical instruments other than the telescope itself, none other can compete with the spectrograph for the range of new astronomical knowledge it has provided, and for the insights it has given on the physical nature of the celestial bodies in the Universe. Together with the predecessor of the spectrograph, the visual spectroscope, these instruments have revolutionized our knowledge of the Sun, the planets, stars, gaseous nebulae, the interstellar medium, galaxies and quasars.

Without the spectrograph, we would know nothing of solar or stellar composition, nothing about stellar rotation rates, and much less than we do on stellar space motions and binary stars. Even the real nature of the stars themselves would be a matter of conjecture and debate. And we would have rudimentary knowledge of the conditions prevailing in gaseous and planetary nebulae and of the nature of external galaxies beyond the Milky Way. There would be no Hubble's law, and hence no direct knowledge of the expansion of the Universe other than indirect inference based on Olbers' paradox or on theoretical prediction. Quasars would not be easily distinguished from stars, and the study of radio galaxies and active galactic nuclei would be limited to their morphological properties in optical or radio images. In short, optical spectrographs have underpinned almost every branch of astrophysics in the past century and a half.

This monograph is concerned with the astronomical spectrograph and its predecessor, the spectroscope. Only optical spectrographs are considered, that is, those using visible or ultraviolet light, except for a brief discussion of near infrared solar spectroscopy. A chapter on infrared spectrometers would have been desirable, but neither time nor space permitted its inclusion. And only those employing prism or grating dispersing elements (including grisms) are included in the discussion.

Two aspects of astronomical spectrographs are considered, both their historical development and the theory underpinning their design. I believe each half of the story presented here complements the other; to understand the history of this subject, one needs a good grounding in the theory. Likewise, present-day designers and observers should never forget the history of their subject and the rich rewards it can confer.

In the first chapter, the historical development of the instrument from the earliest experiments of Fraunhofer to the present day are described in some detail. Secondly, the basic principles of spectrograph design are reviewed, with an emphasis on the principles of achieving the desired resolving power and the maximum light throughput. Properties of dispersing elements, be they prisms or gratings, are also discussed. The third and following chapters give further details of the history, theory and development of several important types of spectrograph, namely the coudé and échelle spectrographs, solar spectrographs, the objective prism spectrograph, ultraviolet and nebular spectrographs and multi-object spectrographs. A comprehensive list of references cited is given after each chapter.

The final chapter discusses ten pioneering spectrographs of the late twentieth and early twenty-first centuries. In the past decade or so, spectrograph design has made substantial advances. Notable are the development of high dispersion échelle spectrographs and of multi-object spectrographs with optical fibre feeds. Detector developments, especially the charge-coupled device (or CCD), have revolutionized the practice of astronomical spectroscopy, and these advances are amongst those reviewed.

Also in these pages, credit will be given to the people who have designed, built and used spectrographs in astronomy. This is because this is not a textbook, but a synthesis about the history, design and applications of astronomical spectrographs, as well as about spectroscopists.

Astronomical Spectrographs and their History should be seen as a natural sequel to my earlier book, *The Analysis of Starlight* (Hearnshaw, Cambridge University Press, 1986), which discussed the history of stellar spectroscopy, but which only briefly discussed instrumental history and eschewed theory altogether. The two volumes together give a comprehensive account of the development of this science over the past two centuries.

Much of this monograph was researched while I was on two sabbatical leaves from the University of Canterbury, New Zealand. The first occasion was from mid 1996 to mid 1997. I spent six months from July 1996 at the South African Astronomical Observatory in Cape Town, where the outstanding astronomical library was ideal for researching material for the first two chapters. In early 1997 I visited the Astrophysikalisches Institut Potsdam (AIP) for four

months, and continued working in the Babelsberg library of that institution. Finally I spent the last two months of the sabbatical year at the Dominion Astrophysical Observatory in Victoria, British Columbia. I am grateful to all three institutions for access to their excellent library resources.

The second sabbatical was from September 2003 when I spent three months in the library at the Vatican Observatory in Castelgandolfo. From March 2004 I continued this sabbatical with four months in the library of Lund Observatory in southern Sweden. All these institutions have outstanding astronomical libraries.

Further work was undertaken in the library of the University of Canterbury in New Zealand and on a brief visit in 1997 to the US Naval Observatory library in Washington, DC.

Acknowledgements

I gratefully acknowledge the people who kindly made it possible for me to visit some of the great astronomical libraries of the world where I researched material for this book. In Cape Town, the late Professor Bob Stobie was director at the South African Astronomical Observatory in 1996 during my six months there, and he did everything possible to make my stay there as comfortable and productive as it was. The SAAO librarian, Ethleen Lastovica, did much to introduce me to her library and help me with locating materials in it. While in Cape Town, I read the pre-publication manuscript of Ian Glass' book *Victorian Telescope Makers* on Thomas and Howard Grubb. I am grateful to him for making available Fig. 1.9 showing a Grubb automatic prism spectroscope.

At the Astrophysical Institute Potsdam I am grateful for the support of the Alexander von Humboldt Stiftung during my four months in the Babelsberg branch of that institution in 1997. During this time, the late Dr Gerhard Scholz was my host, and I am grateful to him for his hospitality.

Also in 1997 I visited the Dominion Astrophysical Observatory in Victoria, BC for two months, and I thank the director, Dr Jim Hesser, for allowing me to work there. The late Dr Bev Oke kindly introduced me to the Keck low resolution imaging spectrometer (LRIS) during my stay in Victoria. Professor Colin Scarfe at the University of Victoria did much to facilitate my stay in that city.

In 2003 I spent three months working in the library at La Specola Vaticana, the Vatican Observatory, in Castelgandolfo. Father George Coyne was at that time director of the Specola, and he did everything to make my stay most comfortable and enjoyable. Father Juan Casanovas was in charge of the excellent library at La Specola, and he helped me on numerous occasions with locating material and discussing details of solar spectrographs. The chapter on solar spectrographs was written during my time there. Brother Guy Consolmagno assisted with the high resolution scanner used for some of the illustrations.

In 2004 I spent four months at Lund Observatory in Sweden. Professor Lennart Lindegren was at the time director and he and Professor Dainis Dravins did everything possible to facilitate my stay in the excellent Lund library.

I thank the late Professor Donald Osterbrock for information on the life and work of Frank Wadsworth, which I have included in Section 1.4. Professor David Gray commented on the theory of shadowing in échelle gratings. Drs Stephen Vogt and Harland Epps kindly received me at the University of California Santa Cruz, where I learnt more about the HIRES instrument at Keck. I also thank Drs Bob Tull and Phillip Macqueen (Phillip was my former Ph.D. student in New Zealand), both at Austin, University of Texas, for helpful discussions on spectrograph design. Another former graduate student, Dr Stuart Barnes, guided me in many aspects of the design of the Hercules spectrograph at Mt John, and this assistance will also have indirectly helped in the writing of this book. He also kindly provided Fig. 3.8.

I must also thank Dr David Latham at the Harvard-Smithsonian Center for Astrophysics, who first introduced me to the marvels of échelle spectrographs in 1974, long before I started to write this book. Without him, this work might never have been written. An even earlier mentor on coudé spectrographs was the late Dr Ted Dunham, who worked to develop the Mt Wilson coudé spectrograph, and who influenced me profoundly while I was a graduate student using his coudé spectrograph at Mt Stromlo in Canberra in the late 1960s and early 1970s.

Numerous other astronomers have kindly granted permission for me to use their diagrams and illustrations in this book. These are acknowledged in the table of figure sources at the end of the book.

1 · The historical development of astronomical spectroscopes and spectrographs

1.1 COLOUR, REFRANGIBILITY AND WAVELENGTH

The concepts of colour, refrangibility and wavelength have been crucial for the development of spectroscopes and spectroscopy. Indeed the first prism spectroscopes were built partly for measuring the refractive indices or refrangibility of different glasses, while diffraction gratings were used for early measurements of wavelength.

1.1.1 The refraction of light

The Dutch astronomer Willebrord Snell (1591–1626) is usually credited with the discovery of the law of refraction, in the early seventeenth century in Leiden. René Descartes (1596–1650) later included the law in his treatise *Dioptrics* of 1637 (without however acknowledging Snell), and he used it to account for the formation of a rainbow, but not explicitly for the colours that the rainbow produces.

Isaac Newton's (1642–1726) first paper, published by the Royal Society in 1672, but based on his optical experiments undertaken six years earlier, came to important conclusions on the relationship between refrangibility and colour [1]. Newton studied the refraction of the Sun's rays in a glass prism, and projected the resulting spectrum onto a wall of his room. He concluded: '...light consists of rays differently refrangible, which, without any respect to a difference in their incidence, were, according to their degrees of refrangibility, transmitted towards divers parts of the wall.' He found that the Sun's light was composed of rays of different primary colours, each with its different refrangibility, which gave rise to the dispersive power of the prism. He went on: 'As the rays of light differ in degrees of refrangibility, so they also differ in their disposition to exhibit this or that particular colour ... To the same degree of refrangibility ever belongs the same

colour, and to the same colour ever belongs the same degree of refrangibility.'

These experiments were the fundamental starting point for the science of spectroscopy. Newton himself referred to 'the celebrated Phaenomena of Colours', indicating that he was far from the first to study the solar spectrum (the word was his term for the 'colour-image'). His apparatus was a primitive spectroscope, and comprised a small hole in his window shutters, a glass prism, and a wall 22 feet distant from the prism used as a screen onto which the spectrum was projected. At first no lenses were employed, but later he inserted a lens after the prism and noted that a spectrum of greater purity could thereby be produced.

In his *Lectures opticae* of 1669 (but published posthumously in 1728) Newton used his theory of colours to account for the colours of the rainbow, which contained the primary 'Colours in order red, yellow, green, blue and purple, together with all the intermediate ones, that may be seen in the Rain-bow; whence will easily appear the Production of Colours in a Prism and the Rain-bow' [2].

It is interesting that Newton observed the same phenomenon of a continuous spectrum using the planet Venus, by allowing the planet's light to enter the eye directly from the prism. Even stars of first magnitude were observed in the same way, and he remarked that a telescope would both increase the quantity of light for stellar spectroscopy as well as reduce the undesirable effects of atmospheric scintillation.

Newton's experiments led him to believe that the dispersive power of prisms was related to the refractive index of the glass. He concluded: 'The denser the Matter of the Prism is, or the rarer the incompassing Medium is, *caeteris paribus*, the greater will be the Difference of Refraction, and hence the Appearance of the Colours will be more manifest' [2, see Proposition XXIV].

1.1.2   Wavelength, colour and spectral lines

Thomas Young (1773–1829) in 1801 was the first person to use a simple diffraction grating to demonstrate the wave nature of light and to show that the wavelength could be obtained from the groove spacing of the grating. His first gratings comprised a series of parallel grooves ruled on glass at the spacing of about 500 grooves per inch. Using sunlight incident at 45°, he found four bright orders due to the interference of the light. The sines of the angles of diffraction ($\sin \beta$) increased in accordance with the integers 1:2:3:4, and from this progression he was able to estimate the wavelength of sunlight [3]. According to Young, the visible spectrum covered a range of wavelength from 675 down to 424 nm, with yellow light corresponding to 576 nm.

Joseph Fraunhofer (1787–1826) continued experiments with diffraction gratings in the early 1820s. His first gratings were coarse transmission gratings, made by stretching fine parallel wires between the threads of two screws – typically with spacings of up to 325 wires per inch.[1] He was able to measure the angular positions and hence wavelengths of prominent absorption lines in the solar spectrum with such a grating, including a value of 5888 Å for the orange line that he labelled D [4].

In later experiments Fraunhofer produced gratings by ruling with a diamond directly onto glass. Thus in 1822 he produced one with 3340 grooves per inch and which gave a higher dispersion. He was able to confirm Young's finding that the diffraction orders had $\sin \beta \propto n$ (here n is the order number of the diffraction, an integer), instead of simply $\beta \propto n$ that he had used previously [5]. Gratings as finely ruled as 7790 grooves per inch were produced the following year [6].

One of Fraunhofer's main interests was the determination of refractive indices for the different glasses that were produced by the Benediktbeuern glass works in Bavaria. As refractive index varied with wavelength, he had the foresight to use different emission lines as wavelength standards at which refractive indices could be measured [4]. Thus the relationship between refractive index (measured by applying Snell's law of refraction to a prism spectroscope) and wavelength (from a diffraction grating of known groove spacing) could be explored. Such experiments laid the basis for the future of spectroscopy and of spectroscope design, as well as the design of telescopes employing achromatic doublets for the objective lenses.

Fraunhofer's early prism spectroscope comprised a 60° prism mounted in front of a small 25-mm aperture theodolite telescope with cross wires seen in the eyepiece. The prism in turn was 24 feet from a narrow opening in the window shutters. It was with this apparatus that he first recorded several hundred absorption lines in the solar spectrum [7, 8, 9].

In 1823, for his more extensive observations of stellar spectra, Fraunhofer used a larger telescope of aperture 10 cm, in front of which he mounted a prism of apex angle 37° 40′ to form an objective prism spectroscope [6]. With this instrument he was able to observe the absorption line spectra of six bright stars, as well as of Mars, Venus and the Moon.

None of Fraunhofer's spectroscopes employed a collimator. For his solar observations the light traversed a narrow slit of width 0.07 inches; at a distance of 24 feet from the prism, the slit subtended no more than an arc minute, giving a resolving power ($R = \lambda/\delta\lambda$) probably of about 2000, easily sufficient to observe the lines in the solar spectrum.

In England in 1802 William Wollaston (1766–1828) had earlier observed five lines (a term he introduced) in the solar spectrum, without realizing their true significance [10]. His apparatus was similar to Newton's, except that the crevice in his window shutters was just 0.05 inches wide, and his flint glass prism was placed at 10 to 12 feet. Thus the benefit of a narrow slit to improve resolving power was established by the pioneering spectroscopists in the early nineteenth century.

1.2   THE COLLIMATOR IN PRISM SPECTROSCOPES

A collimator was introduced into laboratory spectroscopes by several spectroscopists from 1839, in order to achieve a higher resolving power, and probably as the result of independent developments. The first such use was most likely in a spectroscope by Jacques Babinet (1794–1872), which was presented to the Académie des Sciences in 1839 by François Arago (1786–1853) [11],

---

[1] Fraunhofer quoted values in Paris inches. A Paris inch is 26.7 mm. They have been converted here to the more familiar imperial inch of 25.4 mm.

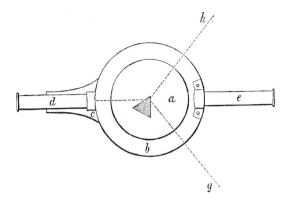

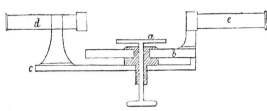

Figure 1.1. Prism spectroscope of William Simms, 1840, for solar spectroscopy. Here d is the collimator, e is the viewing telescope, a is the prism on the prism table and b is the graduated circle.

in which a lens was placed in front of the prism so as to render the rays parallel. The instrument was designed for measuring the refractive indices of glass.

A more detailed description of a slit and collimator spectroscope was presented by the famous optical instrument maker, William Simms (1793–1860) in 1840. His instrument is shown in Fig. 1.1. He introduced the word 'collimator', and wrote:

> There are two telescopes, e and d; one to be used for making the observation, and the other as a collimator. In general the former is attached to the graduated circle b, and the latter to the stage c; but their positions can be changed at pleasure. The observing telescope is fitted with eye-piece and cross wires, in the usual way; but the collimator has in the principal focus of the object-glass two metallic plates, the straight edges of which open parallel to each other, so as to form a line of any required breadth, and at the end of the tube a plane mirror reflects light from the sun or sky, through the slit of the collimator [12].

As Simms mentioned, the instrument was intended for solar spectroscopy. Yet his design also

became a model for many laboratory spectroscopes of the mid nineteenth century. In particular, the use of a graduated circle for the telescope and a rotating table to carry the prism permitted great ease of use when measuring angles of refracted rays.

In the following years collimators became increasingly common. William Swan (1818–94) included one in his astronomical spectroscope of 1856 [13]. By about 1860 the collimator was essentially a universal feature of all astronomical and laboratory spectroscopes.

1.3 SOME NOTABLE ASTRONOMICAL PRISM SPECTROSCOPES OF THE MID NINETEENTH CENTURY

The basic arrangement of a prism spectroscope devised by William Simms in 1840 was widely copied. It comprised a slit, a collimator, a circular prism platform carrying one or more prisms, and a viewing telescope with cross wires and an eyepiece. Both collimator and telescope tubes could rotate about the centre of the prism table with graduated scales.

One such instrument was made by C. A. von Steinheil (1801–70) in Munich for Gustav Kirchhoff (1824–87) in Heidelberg and was used for Kirchhoff's famous study of the solar spectrum, 1861–3 [14, 15]. There were four prisms in this instrument (see Fig. 4.5), three of 45° and one of 60°, which had to be repositioned manually on the table for different telescope angles when observing different spectral regions. The adjustable slit had a small reflecting prism over part of its length, allowing spark spectra to be viewed simultaneously in one half of the slit with a solar spectrum in the other half, thereby facilitating a direct comparison of the two for line coincidences.

At about the same time as Kirchhoff's solar spectroscopy, several astronomers almost simultaneously and independently embarked on a study of stellar spectra, by attaching prism spectroscopes to refracting telescopes. Giovanni Donati's (1826–73) spectroscope was one such instrument, which was attached to a lens of 41 cm aperture that comprised his refractor objective [16, 17]. The spectroscope itself had a single prism, a collimator and a small movable viewing telescope, but in place of a slit there was a cylindrical lens, which broadened the spectrum, making it easier to view.

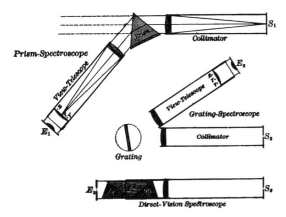

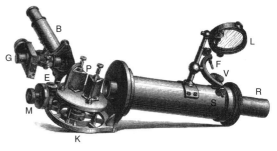

Figure 1.3. Two-prism Browning spectroscope used by Huggins and Miller on the 8-inch refractor at Tulse Hill Observatory, 1864.

Figure 1.2. Three different types of spectroscope that were depicted by C. A. Young in his *Text Book of General Astronomy* of 1888.

Von Steinheil also built a stellar spectroscope in 1862 which, like Donati's instrument, was slitless so as to admit the maximum amount of light [18]. The collimator and viewing telescope were in fixed positions, so as approximately to achieve minimum deviation.

Of the early stellar spectroscopists, one of the most successful was William Huggins (1824–1910). His stellar spectroscope was constructed with the assistance of William Miller (1817–70) and mounted on Huggins' 8-inch Clark refractor at Tulse Hill in London. The instrument was described in detail by them in their classic paper of 1864 [19]. The spectroscope comprised two flint glass prisms to achieve a high dispersion (about 430 Å/mm in the blue). They received light from a collimator and slit and a cylindrical lens ahead of the slit with its axis orthogonal to the slit height, broadened the spectrum. The viewing telescope had an aperture of 0.8 inches and was equipped with cross wires to enable accurate setting on spectral lines. The position of the viewing telescope was adjusted with a micrometer screw.

A comparison spectrum could be viewed simultaneously with this instrument, as with Kirchhoff's solar spectroscope, by means of a small reflecting prism directly in front of the slit, and by this means Huggins and Miller were able to carry out a qualitative analysis of the principal chemical elements to be found in the brightest stars [19].

Shortly afterwards, Huggins and Miller had a similar but less dispersive two-prism instrument constructed for them by John Browning (1835–1925), and

this second spectroscope was used for observations of the light of gaseous nebulae [20], though in this case the cylindrical lens, which was useful for stellar point sources, was generally dispensed with. Figure 1.3 shows the two-prism Browning spectroscope used by Huggins and Miller.

Huggins' illustrious contemporary in stellar spectroscopy research was the Jesuit priest Angelo Secchi (1818–78) in Rome. He chose a compact direct vision spectroscope for his initial investigations of stellar spectra. The instrument was made by the Parisian optician, August Hoffmann, who had made a similar spectroscope for Jules Janssen (1824–1907) [21]. The concept of the direct vision spectroscope was due to the Italian physicist Giovanni Battista Amici (1786–1863), in 1863. It comprised a flint glass prism in contact with two reversed crown glass prisms, so arranged as to give dispersion without deviation of the centre of the spectrum. In Hoffmann's arrangement, a higher dispersion was achieved with a train of two flint and three crown prisms – see Fig. 1.4.

Secchi used his instrument on the Merz 24-cm refractor at the Collegio Romano Observatory from December 1862. The instrument was described in papers in 1863 [22] and 1866 [23], though in the first of these a train of as many as nine prisms is shown. The spectroscope was donated to the Académie des Sciences in 1867 [24].

The circumstances in which Secchi first used a spectroscope for stellar spectroscopy were related in his book *The Stars* in 1879 [25]:

This spectroscope was known in Rome almost at the same time as the paper by Donati. Immediately I had the idea of using it on the stars, by mounting

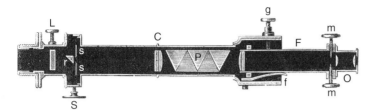

Figure 1.4. Angelo Secchi's direct vision Hoffmann spectroscope of 1862. P is the prism train comprising three crown glass and two flint glass prisms. s is the slit and C the collimator. The viewing telescope is F with eyepiece O. Light from a comparison source could be reflected off the small prism r in front of the slit.

it to our large refractor, in the hope of obtaining better results than those of Donati, thanks to the perfection of our Merz objective lens. The instrument was ordered at once, but did not arrive until December of the same year. Meanwhile Mons. Janssen had come to Rome to study the solar spectrum, and as he had with him one of these small instruments, I begged him to mount it on our refractor, so we could use it provisionally until my own arrived. He agreed, and we undertook together these first investigations which were presented to the Academy of Sciences in Paris.

These events, in which Janssen and Secchi mounted the Hoffmann spectroscope on the Merz refractor, were also reported by Secchi elsewhere [22]. When they did so, Secchi wrote, '...we were astonished by the magnificent results that were obtained at the first attempt'.

A few years later Secchi introduced an objective prism spectroscope to his stellar spectroscopy programme [26]. He described the new instrument thus:

The method that I have used is that which had been in a simple way adopted by Fraunhofer. It consists in putting a prism in front of the objective. For reasons of economy, I have had to limit the size of the prism to 6 inches (= 16 centimetres) diameter. Its refracting angle is 12 degrees and it is made from a very pure flint glass. It is supported in a suitable armature and placed in front of the objective of the large equatorial telescope and is capable of perfect adjustment in each coordinate. The aperture of the refractor remains thus reduced by more than half of its area. But, in spite of that, the light is so intense that it far exceeds that obtained with the use of direct vision prisms near the eyepiece [26].

Secchi used this objective prism instrument with two cylindrical lenses as an eyepiece, which broadened the spectra perpendicular to the dispersion. The objective prism and its mount were made by the firm of Merz in Munich, and it was the first thin prism of this type designed for a large refractor as a slitless spectroscope. It is shown in Fig. 5.1.

Lewis Rutherfurd (1816–92) was an amateur astronomer in New York state who also embarked on a programme of stellar spectroscopy in 1862, using his Fitz $11\frac{1}{4}$-inch refractor. He was a skilled instrumentalist and built his own slit spectroscope with a 60° flint glass prism and rotating viewing telescope [27]. A later paper discussed the simultaneous viewing of a comparison spectrum using a spirit lamp and he also experimented with carbon disulphide liquid prisms [28] (see also [29]). The carbon disulphide gave a high dispersion, especially useful at longer red wavelengths, and a high ultraviolet transparency, but suffered from a very high sensitivity of its refractive index to temperature (see [30] for details).

Rutherfurd criticized the spectroscopic instrumentation of his European contemporaries, finding a variety of deficiencies in the apparatus of Donati, Airy [31] and Secchi [32].

1.4 IMPROVEMENTS IN PRISM SPECTROSCOPE DESIGN IN THE LATER NINETEENTH CENTURY

In the second half of the nineteenth century numerous technical improvements in laboratory spectroscopes were devised, and some of these had a direct influence on the practice of astronomical spectroscopy. Some of these improvements resulted from the growth of firms specializing in optical and astronomical instruments.

Famous instrument makers of the later nineteenth century were mainly located in Britain and Germany, and included Troughton and Simms, Adam Hilger and John Browning (all in London), Howard Grubb (in Dublin), Schmidt and Hänsch, Hermann Wanschaff, Hans Heele, Otto Toepfer und Sohn (all in Berlin), G. Merz, C. A. von Steinheil (both in Munich) and August Hoffmann and Jules Duboscq (both in Paris). All these made prism spectroscopes for astronomy.[2]

Laboratory spectroscopy began developing strongly as a subject from about 1860, following the definitive study by Robert Bunsen (1811–99) and Gustav Kirchhoff on the flame and spark spectra of the different chemical elements [33, 34], which provided the basis for the chemical analysis of laboratory and celestial sources (see Fig. 1.5).

One of the improvements of the mid nineteenth century was the use of more than one prism to give higher dispersion and resolving power. The concept of a multiple-prism train was used by Hippolyte Fizeau (1819–96) and Léon Foucault (1819–68) in 1848, when they employed up to five prisms in their instrument, and hence obtained much superior resolving power, including for solar spectroscopy [35]. Such an arrangement was employed, for example, by William Huggins (and many others) for his laboratory study of the lines of the chemical elements, for which he used a spectroscope with six 45° prisms [36], as well as by Gustav Kirchhoff [14].

The London instrument maker John Browning, whose business flourished from about 1866 to 1905, built many multiple-prism spectroscopes and devised a mechanism for maintaining the circular arc of prisms always in the position of minimum deviation, regardless of the wavelength selected by the viewing telescope [37] – see Figs. 1.6 and 1.7.

His first instrument of this type with six prisms was built for John Peter Gassiot (1797–1877) at the Kew Observatory for solar work. It was criticized by Richard Proctor (1837–88) for having a fixed orientation of the first prism (relative to the collimator) [38], but subsequent improvements in the design overcame the problem alluded to [39, 40, 41]. A Browning spectroscope with seven flint glass prisms, though not of the automatic type, was used by Norman Lockyer for his pioneering work on the spectroscopy of the solar chromosphere [42, 43] – see Fig. 1.8. In Dublin, Howard Grubb included an automatic three-prism laboratory spectroscope in his firm's catalogue of 1885. An instrument of this type with compound prisms was supplied to the Royal Dublin Society.

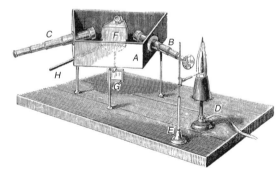

Figure 1.5. Early prism spectroscope used by Robert Bunsen for laboratory studies of flame spectra, circa 1860. A is the prism box, B the collimator, C the viewing telescope, D the burner, E a holder, F the prism.

---

[2] The Berlin instrument firm and clock-maker of Johann Bamberg acquired the Wanschaff company in 1919, Toepfer und Sohn also in 1919 and the Heele company in 1923, forming the company Carl Bamberg Friedenau. This eventually became part of Askania Werke AG in the early 1920s.

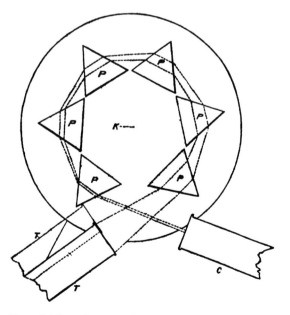

Figure 1.6. Browning automatic six-prism spectroscope, 1870. C is the collimator and T the movable viewing telescope.

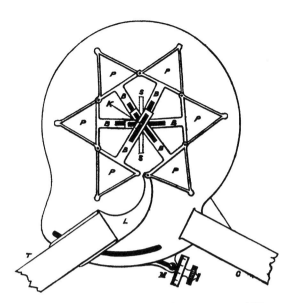

Figure 1.7. Browning automatic six-prism spectroscope, 1870, showing the mechanism for maintaining the prisms at minimum deviation when the position of the telescope T is adjusted by the micrometer M.

The idea of increasing the dispersion, not by adding more prisms, but by passing the light through one or more prisms twice after reflecting it back on itself, was another much used technique, which has been originally ascribed to the Paris optician Jules Duboscq (1817–86) in about 1860 (see [44, p. 511]). Such an arrangement was copied in various forms by many instrument makers. One of the better known was the automatic spectroscope (i.e. an instrument that always gave minimum deviation) built by Otto von Littrow (1843–64) in Vienna, which had four prisms, each used twice, thus giving the effective dispersion of eight prisms [45]. The light emerging from the collimator lens returned in the opposite direction through the same lens, which now served as the objective of the viewing telescope. This arrangement, in which the light returned back from the dispersing element along the same path, came to be known as the Littrow configuration, a term that has been widely used in spectroscope and spectrograph design, even for grating spectrographs, when the angles of incidence and refraction are arranged to be equal.

One method of achieving a double pass through the prisms was to fold the beam using a totally reflecting 90° prism at the end of the prism train. The

return beam then went back through the prisms, but in a higher level (the prisms had to be twice the height), which therefore enabled the separation of collimator and viewing telescope. Such an instrument was constructed by Howard Grubb (1844–1931) in 1870 for William Huggins [46]. Separating the collimator lens from the viewing telescope objective was advantageous, as it avoided back-scattered undispersed light from the collimator lens being seen as a bright background illumination superimposed on the spectrum.

Even more complicated designs were built or proposed. Alfred Cornu (1841–1902) had a single-prism spectroscope in which the prism was traversed four times [47]. Louis Thollon's (1829–87) spectroscope of 1878 had four prisms each traversed twice in upper and lower levels, which he used for observing the solar spectrum at high dispersion [48]. Secchi also acquired a high dispersion double-pass instrument of this type, which comprised four and a half prisms for dispersion in an automatic arrangement – see Fig. 1.10.

In Cambridge, H. F. Newall (1857–1944) built a single-prism double-pass spectroscope for stellar spectroscopy, in which all three faces of the 60° prism were used as refracting surfaces [49, 50]. Here reflection of the collimated beam from the first prism face gave a white (undispersed) slit image in the field of view, which was seen superimposed on the spectrum. By rotating the prism assembly, this fiducial line could be seen at any desired wavelength, thus serving as a useful aid for line identifications or for measurements of line positions in spectra. Newall's arrangement was criticized by Frank Wadsworth[3] (1867–1936) [52], because

---

[3] Donald Osterbrock [51] made the following comments on Frank Wadsworth: 'Wadsworth dropped completely out of astronomy after 1904... He was at Clark University as an instrumental assistant to A. A. Michelson 1889–92, and then an assistant in charge of the astrophysical observatory under Langley 1892–94. Then he was assistant professor of physics (1894–96) and of astrophysics (1897–98) at Yerkes Observatory... He was never promoted to associate professor at Chicago, but wanted to be, and left a year after it was denied to him. He was director of Allegheny Observatory 1900–04 and a consulting engineer to John Brashear 1901–04... Wadsworth was a very self-confident person, who believed he knew all the answers about instrumental design and construction, and that no-one else was competent to discuss such issues with him.

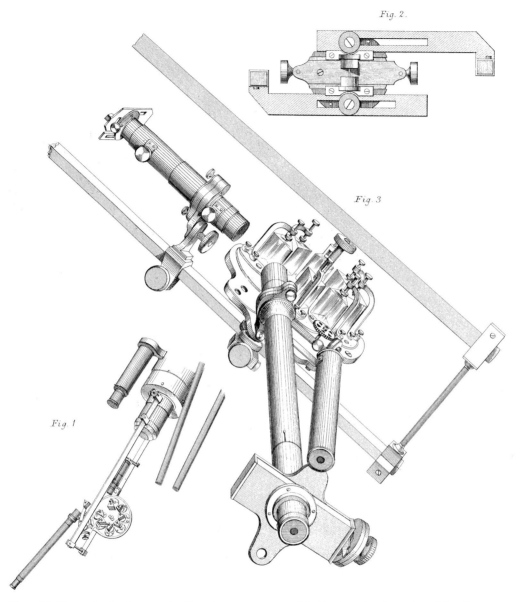

Figure 1.8. Norman Lockyer's seven-prism Browning spectroscope used for his pioneering observations of the solar chromosphere, 1869.

Michelson got tired of him and was very glad to pass him on to Hale, who was also evidently fed up with his arrogance and wrote a definitely luke-warm recommendation for a promotion, prompting Wadsworth's strong protests. There are signs that he was also 'on the outs' with Brashear very soon after getting there. Wadsworth wanted a lot of money for all kinds of new instruments; Brashear, who was chairman of the observatory committee [at Allegheny], knew there was no money and didn't want to raise any. Brashear had idolized Keeler, a very diplomatic person; Wadsworth was the opposite and there are almost no positive remarks from Brashear about him after his first year there [at Allegheny]... Wadsworth had never done any real astronomical research, and I suppose there was no place he could get a job in astronomy after Yerkes and Allegheny.' He was a consulting engineer for glass and metal companies after 1904.

Figure 1.9. Howard Grubb's automatic compound prism laboratory spectroscope as supplied to the Royal Dublin Society. This spectroscope appeared in Grubb's 1885 catalogue of astronomical instruments; a similar instrument also appeared in Grubb's catalogue of 1877.

the light did not necessarily traverse the prism at minimum deviation, which was usually assumed to be necessary for maximum spectroscope efficiency (see Section 2.3 for a discussion of the minimum deviation condition).

Wadsworth himself devised a spectroscope comprising a single prism of equilateral cross-section, which was traversed by the light rays six times (twice through each pair of faces) using a system of plane mirrors [52]. Such an arrangement, however, was wasteful of light and could only be justified in cases where the prism material (such as fluorite for ultraviolet spectroscopy) was scarce.

Wadsworth was also the advocate of what he called a fixed-arm spectroscope, meaning an instrument where the viewing telescope was fixed in its orientation relative to the collimator [53, 54]. The direct vision instrument (180° between collimator and telescope) is one special case of this type of spectroscope, as is also the Littrow instrument (with 0° between collimator and telescope). The advantage of a fixed-arm spectroscope for astronomy was its mechanical rigidity, which was always a factor for an instrument mounted on a moving telescope. In general the fixed-arm concept is ideal when the dispersion is low, thereby

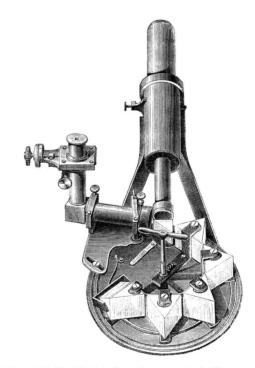

Figure 1.10. Secchi's high dispersion automatic double-pass spectroscope of about 1870. The first small prism reflects the light into the lower half of the prism train; it returns through the upper half into the viewing telescope on the left.

permitting the whole spectral region required to be seen in the eyepiece. However, Wadsworth designed fixed-arm instruments that were still able to scan in wavelength, by using a small rotatable mirror to reflect different spectral regions into the viewing telescope. He thus devised fixed-arm one-prism and two-prism instruments and also a Littrow spectroscope with two and a half prisms (the last half-prism being silvered) for astronomical use [54].

1.4.1   Spectroscope slits

A spectroscope slit is required to define the spectral resolution by admitting light from just a narrow angular range in the source. Albert Mousson (1805–90) in Zürich discussed the desirable properties of a slit mechanism, being one that prevents the delicate jaws from touching and maintains their parallelism for all jaw positions [55]. Mousson's slit mechanism, however, had one fixed jaw, the other movable, an arrangement that changed the mid-slit position for different slit widths.

More complex slits had two moving jaws, thereby maintaining the slit centre in a fixed location, which in turn left line positions unchanged for different resolutions. Such symmetrical slit mechanisms were described by Sigmund Merz [56], by H. Krüss [57], and by Frank Wadsworth [58, 59].

It is well known that the lines from a tall slit appear curved when viewed in a prism spectroscope (see for example [60, 61, 62]). The curvature is in the sense of the concave side of the lines being towards the shorter wavelength end of the spectrum. The effect arises from the light from the slit extremities not falling on the prism in the normal plane to the refracting surfaces. Such curved lines can make the measurement of line position more difficult.

In Dublin, Thomas Grubb (1800–78) devised a spectroscope in which the slit jaws were curved, so as to cancel their intrinsic curvature and hence render the lines straight [63]. Such a curved slit would be useful for high dispersion laboratory or solar telescopes where the source has large angular size (so as to illuminate a large slit height) and accurate line position measurements are required. On the other hand, Lockyer had a curved slit for solar spectroscopy for quite a different reason – in his case it was designed to match the curvature of the solar limb to enable observations of

the chromosheric spectrum in the absence of a solar eclipse [64].

Deckers, which are cover plates to reveal just a selected part of the slit height, became a standard tool for spectrum photography. They were usually adjustable, and mounted just ahead of the slit. They are described by Johannes Hartmann in 1900 [65]. Frequently a small reflecting prism in front of the slit was used to send light from a comparison lamp into the spectroscope or spectrograph. Such a device was employed by Kirchhoff and Bunsen [66] and by many others since. For photographic work, the deckers then allowed comparison and celestial spectra to be recorded side by side.

Frequently stellar spectroscopes employed a cylindrical lens either with or without a slit. Thus Donati in 1860 had such a lens, but no slit [16, 17], so as to widen the spectrum. Huggins, on the other hand, placed such a lens just ahead of the slit, so as to produce a line image on the slit jaws [19] – see Section 1.5.

1.4.2   Projected scales and automatic line recorders

The measurement of the positions of spectral lines for the purposes of element identification and wavelength determination was a major activity throughout the history of laboratory and astronomical spectroscopy. The provision of projected scales or at least of fiducial reference markers, seen in juxtaposition with the spectra, became common from the 1870s. Thus Browning projected an illuminated cross onto the spectrum of a direct-vision spectroscope, by reflecting rays from the side off the last prism face [67]. Henry Procter instead had an illuminated scale similarly reflected [68], while Alexander Herschel proposed a scale produced by a row of equally spaced small holes illuminated with sodium light, each hole producing a small reference mark on the spectrum being investigated [69]. Maurice de Thierry also, like Procter, projected a scale off the last prism surface from the side, but he also had a micrometer wire that could be set on any point in the scale so as to facilitate measurements of line position [70]. An instrument of this type is shown in Fig. 1.11.

Further ingenious devices were arranged for recording line positions in spectra. One was built by Howard Grubb for William Huggins, with a view to

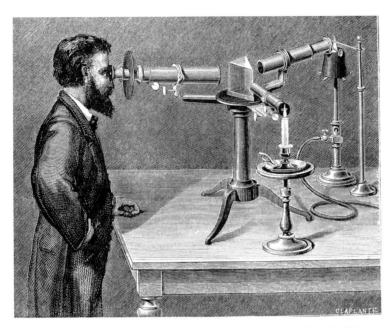

Figure 1.11. Early laboratory spectroscope with a projected scale, *circa* 1870. The light source in the foreground projects a scale into the viewing telescope after reflection off the prism face.

making a rapid recording of coronal line positions during the brief moments of a solar eclipse. The micrometer driving a moving pointer in the eyepiece was attached by a lever to a needle which could record a hole in a card at a place corresponding to the line's position in the spectrum [71]. Ten to twelve lines could be registered in only 15 seconds. In another device by Browning, the cross-wire micrometer was arranged also to drive a blackened glass plate onto which scratch marks could be made to record the line positions [72]. A similar device of mechanically inscribed positions was later used by Vogel [73].

1.5 THE DEVELOPMENT OF THE PRISM SPECTROGRAPH

The prism spectroscope was developed into a relatively refined instrument from the 1860s, by instrument makers such as Gustav Merz (1793–1876), John Browning and William Simms. Yet when it was used visually for stellar spectroscopy, only bright stars could readily be observed. Photography provided an integrating detector and therefore the promise of observing fainter stars. It also led to the ease of inspecting and measuring a glass plate rather than viewing a faint and flickering spectrum through an eyepiece.

Solar spectrum photography had been attempted almost immediately after the introduction of the photographic process in 1839. Sir John Herschel (1792–1871) in 1840 recorded solar continuum spectra on photographic paper, but did not have sufficient resolving power to see any dark lines, although he looked for them [74]. Successful solar spectra were, however, recorded soon afterwards by Edmond Becquerel (1820–91) in France in 1842 [75] and by John Draper (1811–82) in New York the following year [76]. The instrumentation used was rudimentary, simply comprising in Becquerel's case, a narrow slit, a flint glass prism, a camera lens and finally a daguerrotype plate mounted on a screen about one metre from the lens. Becquerel was able to record spectral lines from the Fraunhofer A band in the red all the way into the ultraviolet, and well beyond the H and K lines at the limit of visual vision. Further experiments in solar spectrum photography were made by Foucault and Fizeau in Paris in 1844 [77].

1.5.1 The spectrographs of Henry Draper and William Huggins

With the renewed interest in stellar spectra in the 1860s, the challenge was to apply the new photographic technique to the study of stellar spectra, so as to give a permanent stable record of the image and to explore the ultraviolet region for stars for the first time. As early as 1863 William Huggins, with his colleague William Miller, attempted to record the spectrum of Sirius using the wet collodion[4] plates then available [19, p. 428]. They did not find any lines. Details of how Huggins and Miller modified their spectroscope for photographic use are not given, and it is conceivable that their problem was simply that of focussing the spectrum on the plate, as the authors themselves suggested.

In the 1860s achromatic refractors were corrected for visual (green-yellow) wavelengths and hence gave poorer images in the blue and ultraviolet regions at which photographic plates were sensitive. The successful photography of the spectrum of Vega by Henry Draper (1837–82) in 1872 [78, 79] avoided this problem, as he made use of his 28-inch reflector, and this was equipped with a single quartz prism spectrograph that Draper had himself constructed.

He described his instrument as follows:

The especial spectroscope for stellar work that is now on the telescope is intended to satisfy the following conditions: 1st, to get the greatest practicable dispersion with the least width of spectrum that will permit the lines to be seen; 2nd, to use the entire beam of light collected by the 28-inch reflector or 12-inch achromatic without loss by diaphragms; 3rd, to permit the slit to be easily seen so that the star may be adjusted on it; 4th, to avoid flexure or other causes that might change the position of the spectrum on the sensitive plate in pointing the telescope first on one and then on another object; 5th, to admit of observing the spectrum on the sensitive plate at any time during an exposure without risk of shifting or disarrangement [79].

Although Draper's earliest experiments used the wet collodion plates, by 1876 he was using the new dry silver bromide gelatin[5] plates from Wratten and Wainwright in London. Unfortunately he gave few details concerning his instrument. In the last years of his life, Draper had a two-prism spectrograph acquired from John Browning, and he recorded spectra of the 1881 comet [80] and of the Orion nebula [81] with it.

On the other hand, William Huggins, who successfully photographed a stellar spectrum using dry gelatin plates in 1876, some four years after Draper, reported his work without delay [82]. Thus the first papers on the subject of stellar spectrum photography were both submitted by Draper and Huggins simultaneously in December of 1876.

Huggins went on to describe his instruments in some detail. His first successful observations were made on his 18-inch Grubb reflector, initially (1876–79) at the prime focus with a single prism spectrograph, but later at the Cassegrain focus. The first instrument had an Iceland spar 60° prism and quartz lenses, the optics being made by Adam Hilger [83] – see Fig. 6.1. It was mounted at the prime focus after removal of the secondary mirror, and the slit was observed through a small guide telescope that was mounted in the hole of the primary.

Huggins' second spectrograph was without ultraviolet-transmitting optics and was used on his 15-inch refractor. The dispersing element was either a prism (apex angle 37° and silvered on one face) or a Rowland plane reflection grating supplied by John Brashear [84]. This instrument was made by Troughton and Simms and is shown in Fig. 1.12. It was essentially a visual spectroscope which could, however, be adapted for photography when the eyepiece was replaced by a camera.

His final instrument was a two-prism Cassegrain spectrograph for the 18-inch reflector, with quartz lenses and 60° prisms of Iceland spar [85] – see Fig. 6.2. All these instruments were also described collectively in *An Atlas of Representative Stellar Spectra* [86].

The design of the Huggins two-prism ultraviolet Cassegrain spectrograph showed many advanced features, and it became a model for a generation of spectrographs at professional observatories. In particular, there were polished speculum jaws set slightly off-axis to reflect the periphery of the stellar image into a guiding eyepiece, and a slide mechanism to bring the optics for a comparison spectrum in front of the slit. The plate holder was tilted to compensate for chromatic aberration in the camera lens – a technique that

---

[4] Collodion is produced when guncotton (cellulose nitrate) is dissolved in a mixture of alcohol and ether.

[5] Gelatin is a protein material made from the bones and hides of cattle. As with collodion, it acts as the binder that supports the photosensitive silver salt.

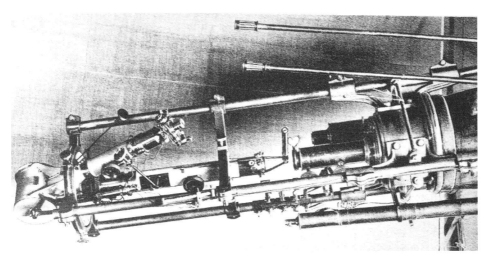

Figure 1.12. Huggins' composite spectroscope and spectrograph by Troughton and Simms, 1875. It was used on the 15-inch refractor at the Tulse Hill observatory. The dispersing element was either a single prism in double pass or a 4-inch Rowland grating. It is seen here in grating mode with an eyepiece.

was described in detail by Johannes Hartmann at Potsdam a few years later [87]. The whole instrument was in an aluminium box, so as to give a stiff, light construction. Spectrum widening was either by means of a cylindrical quartz lens or by trailing the star along the slit in right ascension, using eccentrically mounted spur gears which imparted a small periodic variation in the sidereal drive rate [88].

With this spectrograph Huggins produced some superb spectra of bright stars, which he published in his *Atlas of Representative Stellar Spectra* [86]. The wavelength range was from 330 nm (for the early-type stars of spectral type B) to about Hβ (486 nm).

1.5.2 Further prism spectrographs: Potsdam, Paris and Lick

The first spectrographs were developed by amateur astronomers with reflectors. But professional observatories soon entered the new field of stellar spectrum photography – in most cases using large visual refractors, which were then in vogue. Hermann Carl Vogel (1841–1907), the director of the Potsdam Astrophysical Observatory, undertook with Julius Scheiner (1858–1913) a major photographic radial-velocity programme on the 30-cm Schröder refractor from 1888 [89] (see also [90, 91]).

The Potsdam spectrograph had two Rutherfurd compound prisms. Collimator and camera were of equal focal length (408 mm) and guiding was from the starlight passing through the slit, reflected off the first prism face. A narrow hydrogen-filled Geissler tube placed in the telescope itself, 40 cm from the slit, provided the comparison source, although an iron spark comparison could also be used. The instrument is shown in Fig. 1.13 mounted on the 30-cm refractor at Potsdam.

The Potsdam instrument was followed by other prism spectrographs on the large refractors at Lick and Yerkes [92, 93]. The Lick instrument, known as the Mills spectrograph (after a Lick observatory benefactor) had three dense flint glass prisms. It was designed by W. W. Campbell (1862–1938) and was an instrument where temperature control and flexure reduction were major considerations. With a focal length of 724 mm, flexure could in principle cause spectral line shifts during an exposure on a moving telescope, which would blur the lines and make measurement of precise Doppler shifts impossible. The 180° total deviation of the beam in this spectrograph allowed the camera to be tied rigidly to the parallel and adjacent collimator to reduce flexure within the instrument to a minimum. However, flexure still prevented long exposures of plates for accurate radial-velocity measurements. Thus Vogel limited his exposures to one hour for this reason.

The temperature control of spectrographs was another important consideration, in particular because

Figure 1.13. Vogel's two-prism spectrograph on the 30-cm Schröder refractor. The instrument was used by Vogel and Scheiner for the pioneering photographic radial-velocity programme from 1889.

of the changing refractive index of the prism medium with temperature. The effect was important partly because of the variations in ambient temperature during a night, but also because of thermal effects from an observer's body when close to an unprotected instrument.

Henri Deslandres, working with the 1.2-m Paris Observatory reflector, recognized the importance of temperature stability as well as mechanical rigidity in his Cassegrain spectrograph [94]. He experimented with an internal heater, and with water circulation in copper pipes, in order to keep the temperature constant. He also preferred to use light flint glass prisms, which, although less dispersive than heavy flint glass, were also less susceptible to temperature variations.

Campbell at Lick also went to great trouble to overcome thermal effects in the three-prism Mills

spectrograph [95]. An internal heater, powered by a 10-volt battery, was designed to maintain temperature constancy inside the prism box as the night temperature fell. The whole instrument was surrounded by a heavy woollen blanket, which was inside a double-walled cedar-wood box lined with felt. The box was mounted on the telescope so as not to contribute to flexure of the optical components. In this way the prism temperature could be stabilized to about $\pm 0.05\,°C$ on timescales of an hour [95]. Figure 1.14 shows the Mills spectrograph on the Lick 36-inch refractor. According to Bowen, prism spectrographs can tolerate temperature changes no more than $0.5\,°C$ during an exposure, in contrast to grating spectrographs where even a $10\,°C$ change may be permissible [30, p. 46].

Similar conclusions were reached earlier by Hartmann at Potsdam, who studied the sensitivity of prism materials and of several Potsdam spectrographs to temperature variations [96]. Flint glass prisms typically gave an apparent $0.3\,Å/°C$ shift at $H\gamma$, corresponding to about $20\,\mathrm{km\,s^{-1}\,°C^{-1}}$. An elaborate dual heater with two thermostats in the prism box of the Potsdam spectrograph no. III (see Fig. 1.15) kept temperatures constant to better than $0.1\,°C$. The thermal time constant of the spectrograph was measured to be about 90 minutes.

Many of these principles used in the Mills spectrograph were adopted in the Bruce spectrograph at Yerkes. Here the even larger collimator focal length (958 mm) required especial care to avoid flexure in this large three-prism instrument [93]. A heater maintained the prisms' temperature drift to no more than $0.1\,°C$ over several hours. Figure 1.16 shows the Bruce spectrograph on a stand with the prism box removed. In Fig. 1.17 the spectrograph is seen on the Yerkes 40-inch telescope with the insulated casing in place.

In summary, the main problems that had to be tackled in prism spectrographs at this time were the avoidance of flexure, the need for temperature stability, the problem of chromatic aberration in visually corrected refractors, and the problem of how to guide on visual images, even though the blue-ultraviolet light was that used for spectrography. The question of converting a visual refractor, with a two-component achromatic visual objective, to one suitable for spectrum photography was considered by James Keeler (1857–1900) [97]. He inserted a small two-component lens inside the telescope's focus, so as to render the effective

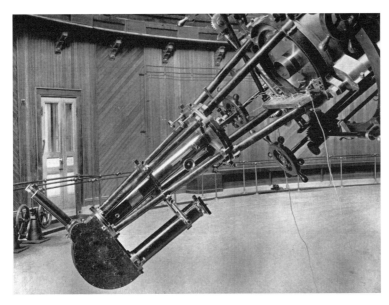

Figure 1.14. Campbell's original Mills spectrograph on the 36-inch refractor at Lick Observatory. The D-shaped prism box contains three dense flint glass prisms. Guiding is from reflection off the first prism face using the small viewing telescope at left. This spectrograph had first light in May 1895.

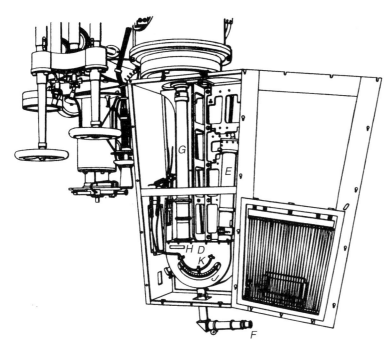

Figure 1.15. Potsdam three-prism spectrograph no. III. Here K and J are parts of a thermostat that controlled the current in the heater. One of the heater elements is seen on the inside of the door of the spectrograph enclosure, and comprised 20 m of silver wire. G is the collimator, D the prism box and E the camera.

Figure 1.16. The Bruce three-prism spectrograph at Yerkes on a stand showing the inside of the prism box. A heater comprising 40 feet of silver wire maintained the prism temperature to within $\pm 0.2\,^{\circ}$C over a night.

focal length nearly constant throughout the blue and near ultraviolet. Good light throughput could thereby be achieved over the entire wavelength range being recorded on the plate. Such a solution worked well for point-source spectrography, but of course was not suitable for wide-field direct photography.

Johannes Hartmann at Potsdam was an instrumentalist who considered various aspects of spectrograph design [98]. He pointed out that whereas both collimator and camera lenses should ideally be free of spherical aberration, chromatic aberration was also a problem for the collimator if a wide spectral range was to be recorded, but not nearly so much for the camera, where a plate tilt could, to a large extent, overcome this aberration. On the other hand, the collimator worked entirely on axis, whereas the camera had to receive ray pencils from the dispersing element over a wide angular field, so that control of off-axis aberrations then becomes important.

Hartmann was one of several authors who discussed prism efficiencies, and the light losses arising from both reflection and absorption for different prism glasses and angles in both planes of polarization.

Chromatic aberration was an important factor for collimators, yet paraboloid mirrors were seldom used in spectrographs in the nineteenth century, achromatic doublet lenses being the favoured option.

Frank Wadsworth, an optical engineer who was not an observational astronomer, and whose peers accused him of generally obfuscating many of the issues in spectroscope and spectrograph design,[6] did advocate mirrors for collimators to overcome the chromatic effects of lenses. However, stability of a telescope-mounted spectrograph was an even greater issue; a small deflection in the tilt of a mirror creates twice the angular displacement of the reflected rays, whereas a small lens tilt, to first order, causes no image movement. This advantage was clearly recognized by Keeler [99] and widely adopted.

Some of the important principles developed for prism spectrographs are seen in John Plaskett's single-prism instrument at the Dominion Observatory, Ottawa, in 1909. In Fig. 1.18 this spectrograph is seen on the 15-inch Brashear refractor. It featured a steel box supported through its centre of gravity by an external truss framework. Inside the box there are ribs to provide stiffness and reduce scattered light.

Multi-prism spectrographs gave high dispersion spectra with a wide angular range. These therefore required cameras able to accept a wide angular field

---

[6] See comments by James Keeler [99] and Arthur Schuster [100], as well as the footnote on Wadsworth in Section 1.4.

Figure 1.17. The Bruce three-prism spectrograph on the 40-inch refractor at Yerkes Observatory, showing the insulated aluminium enclosure. The instrument was used by Edwin Frost to measure stellar radial velocities from 1902.

with low spherical aberration. The focal plane needed to be flat so as to avoid bending glass plates. Hartmann discussed the requirements of camera lenses [87]. Early cameras, such as those used by Vogel at Potsdam or Campbell at Lick, could accept only $1\frac{1}{2}°$ or less total angular range, so the spectra were necessarily short.

The New Mills spectrograph at Lick, commissioned in 1903 for the great refractor, could accept a field of $2°\ 32'$ (see [101]) while the Bruce spectrograph at Yerkes [93] could accept $2°\ 50'$. In this last instrument the spectra were 30 mm in length. There were three-element cemented objectives to achieve the minimal spherical aberration and a wide field.

Vogel's new generation of spectrographs at Potsdam also had cameras able to accept a wide angular range. The so-called no. III spectrograph for the 80-cm refractor had a choice of two cameras by Steinheil and by Zeiss. The latter was a two-element cemented lens known as the 'Chromat' that gave high dispersion spectra over 100 mm in length from the b to the K lines; the focal length of this camera was 56 cm, and hence

the definition must have been good over a field angle of about 10° [87].

The New Mills spectrograph at Mt Hamilton was mounted on the telescope by attaching the spectrograph case to a large external truss frame in two places, at each end of the box, as seen in Fig. 1.19. This new type of mounting helped reduce flexure within the case, which contained all the optical components, and was therefore well-suited to radial-velocity work [101]. This type of mounting was copied widely in later spectrographs, including that at San Cristobal in Chile (see Fig. 1.20), and indeed on most of the Cassegrain prism spectrographs on the first reflectors in the first decades of the twentieth century.

1.6 THE DEVELOPMENT OF THE DIFFRACTION GRATING

The earliest references to diffraction grating phenomena were by David Rittenhouse (1732–96) in Pennsylvania [102] and by Thomas Young in England [3]. Rittenhouse commented on the multiple coloured images produced when sunlight was passed through

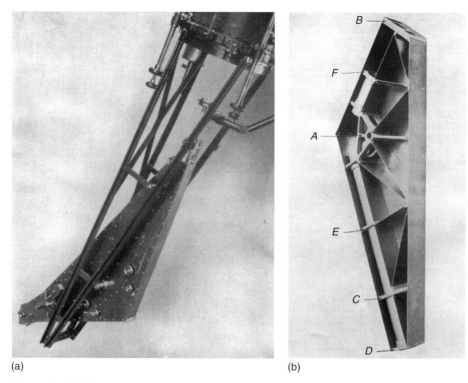

Figure 1.18. (a) Plaskett's single-prism spectrograph at the Dominion Observatory, Ottawa, in 1909 mounted on the 15-inch refractor. (b) The prism box was made of steel plate, 1.7 mm thick. The ribs provide stiffness and eliminate scattered light. Iron castings A to F support the optical components.

a screen comprising parallel hairs held between two fine screws, whereas Young, using a finely ruled glass micrometer scale, went further and measured the wavelengths of light in the solar spectrum and the diffraction angles of the first four orders – see Section 1.1.

Joseph Fraunhofer was easily the most prominent of the early experimenters with gratings. His first transmission gratings consisted of fine parallel wires between two screws. Initially there was only one wire per millimetre, but he soon produced wire gratings with up to 13 wires per millimetre, and with this device he was able to measure the deviations of several dark lines in the solar spectrum to obtain the wavelengths [4].

Fraunhofer went on to produce reflection gratings ruled with a diamond in a thin layer of gold leaf on glass. These achieved much finer groove spacings than was possible with wires – initially up to 20 gr/mm, but later in 1822, up to 130 gr/mm [5], and by 1823 with 300 gr/mm [6]. With such

finely ruled diffraction gratings, Fraunhofer was able to confirm the statement made by Young that the diffraction orders varied as $\sin \beta \propto n$ (rather than $\beta \propto n$).

One major problem encountered was that the rulings of the finest gratings could not be readily inspected using the microscopes then available; nor could inspection of the diamond tip provide a reliable guide to quality, with the result that trial and error prevailed to a large extent.

Fraunhofer ruled some gratings directly onto uncoated glass and he made the discovery that grooves sharp on one side but more diffuse on the other gave spectra in which the orders on one side were much more intense than the other. He ascribed this unequal intensity to the groove profile asymmetry, which was an early indication of a phenomenon that, a century later, would be exploited in the blazed grating, so important in astronomy.

When Fraunhofer died in 1826, progress in research in optics was greatly slowed. He had no

Figure 1.19. The New Mills spectrograph on the Lick refractor, commissioned in 1903. Note the external truss frame that supports the spectrograph in just two places, so as to minimize flexure.

contemporaries able to undertake painstaking optical experiments and with such refined instrumental skills to carry on his research in gratings. F. A. Norbert in Barth produced reflection gratings, generally ruled with a diamond on glass (though sometimes he used a glass base with a thin silvered layer for the grooves), and typically achieved gratings with 160 gr/mm [103, 104]. As with Fraunhofer's gratings, finer gratings were difficult to produce, given that microscopes of the day were unable to resolve the grooves. Norbert's finest gratings with up to 400 gr/mm were still unresolvable in the mid nineteenth century. However, this situation improved rapidly towards the end of the century, by which time gratings with 800 gr/mm were being

produced by Henry Rowland (1848–1901), and these were also just resolvable by microscopes of the day.

In New York, Lewis Rutherfurd began grating production in the 1870s. His gratings were all reflection gratings ruled on polished speculum metal (a very hard copper and tin alloy), typically with 680 gr/mm and 43 mm in width. Rutherfurd gratings were reputed to be superior to those of Norbert and their quality was confirmed by Charles Peirce (1839–1914) [105, 106]. They were among the first to be tested in astronomical spectroscopes, but like all nineteenth-century gratings, they were uncompetitive in comparison to prism instruments because of their low efficiency.

Figure 1.20. The two-prism spectrograph on the $36\frac{1}{4}$-inch reflector of the San Cristobal Observatory near Santiago, Chile. The telescope was installed in 1903 and operated by the D. O. Mills expedition from Lick until 1929. The external truss frame mounting supports the spectrograph through its centre of mass.

In Germany, Ferdinand Kurlbaum (1857–1927) made a careful study of Fraunhofer line wavelengths using speculum metal gratings by both Rutherfurd and Rowland [107]. This entailed careful measurement of the grating constants using a microscope. He obtained, for example, 680.2 gr/mm for the Rutherfurd grating at 20 °C. This grating, of size 43.4 mm, gave measurable spectra only in the first three orders.

In the 1870s Lord Rayleigh had proposed the photographic reproduction of gratings as a means for their mass production [108, 109], starting from either Norbert or Rutherfurd originals as masters. He had some success in copying a Norbert 6000 gr/in grating, as well as one by Rutherfurd with 17 280 gr/in, and he also considered drawing a much enlarged grating (one without periodic errors) and photographing this on a greatly reduced scale [110]. Such attempts in the end were not successful, because of lens aberrations in the reduction light train.

Lord Rayleigh later summarized his efforts in the photographic reproduction of gratings over nearly 25 years, when he reviewed progress in 1896 [111]. He

Figure 1.21. A ruling engine by Henry Rowland, circa 1882. The grating being ruled is on the right at e, riding on the grating carriage d.

reported some success in contact prints of 6000 gr/in gratings ruled on glass – but of course for rulings on speculum metal contact printing was not feasible.

Several major improvements in grating technology took place between the 1880s and the 1950s, and these eventually resulted in an efficient dispersing element that was clearly superior to the prism in nearly all astronomical applications. The first major advances came from Henry Rowland at the Johns Hopkins University in Baltimore; he was by far the greatest nineteenth-century pioneer in grating manufacture. Rowland perfected the screw of his ruling engine so as greatly to reduce the troublesome periodic ruling errors that resulted in ghosts [112]. He ruled by far the best reflection gratings then available, and also devised the concave grating. The latter was ruled on a spherical surface, thereby avoiding the use of refracting camera optics, and enabling spectra to be recorded in both ultraviolet and far-red to infrared regions [112]. Figure 1.21 shows a ruling engine built by Rowland.

Rowland emphasized the need to avoid periodic errors in his ruling engine so as to give spectra free of ghosts. He claimed that the lead screw that drove the diamond tool of his ruling engine was essentially free of periodic error and was able to maintain straight grooves to within 10^{-5} inches. Areas as large as $6\frac{1}{4} \times 4\frac{1}{4}$ inches could be ruled with 14 438 grooves per inch.

Joseph Ames (1864–1943), Rowland's colleague at Johns Hopkins, described some of the difficulties of grating production at this time:

A word should be said as to the difficulties of ruling gratings which may explain why so many orders for gratings remain unfilled. It takes months to make a perfect screw for the ruling engine, but a year may easily be spent in search of a suitable diamond point. The patience and skill required can be imagined. Most points make more than one 'furrow' at a time, thus giving a great deal of diffused light. Moreover, few diamond points rule with equal ease and accuracy up hill and down. This defect of unequal ruling is especially noticeable in small gratings, which should not be used for accurate work. Again, a grating never gives symmetrical spectra; and often one or two particular spectra take all the light. This is of course desirable, if these bright spectra are the ones to be used. Generally it is not so. It is not easy to tell when a good ruling point is found; for a 'scratchy' grating is often a good one; and a bright ruling always gives a 'scratchy' grating. When all goes well, it takes five days and nights to rule a 6 in. grating having 20,000 lines to the inch. Comparatively no difficulty is found in ruling 14,000 lines to the inch. It is much harder to rule a glass grating than a metallic one; for to all of the above difficulties is added the one of the diamond point continually breaking down. For this reason, Professor Rowland has ruled only three glass gratings. One of them has been lost, and the other two are kept in his own laboratory. These two were used by Dr. Bell in his determination of the absolute wave-length of the D-lines [113].

When Rowland died in 1901, his successor at the Johns Hopkins Laboratory was Joseph Ames, who acquired the three Rowland ruling engines which had been used to produce gratings of 14 438, 15 020 and 20 000 gr/in [114]. He restored the latter two of these, and in particular, the 15 020 gr/in machine produced better gratings in the early twentieth century than had been achieved by Rowland.

A major problem with grating production at this time was the appearance of ghosts due to periodic ruling errors of the grooves, which in turn resulted from errors in the lead screw driving the grating carriage in the ruling engine. According to A. A. Michelson, periodic errors of as little as 0.05 wavelengths in the groove location resulted in unacceptable ghost intensities. Michelson at the University of Chicago devised an interferometric technique of overcoming such errors, and consequently he was able to rule larger gratings than Rowland had achieved [115, 116]. In effect the perfect regularity of optical interference fringes imposed the same regularity in the grooves by means of a servo-control loop. Much later (after World War II) this became a standard technique, and as a result, lead-screw quality was no longer such a critical factor for grating manufacture. Michelson's two interferometric ruling engines were later acquired respectively by the Bausch and Lomb company and by MIT (where it was developed further by George Harrison and known as the MIT 'A' engine). A modern interferometrically controlled ruling engine was described by Harrison in 1955 [117], after the interferometric technique had been further perfected by Harrison from 1947 and in the 1950s.

Undoubtedly the greatest drawback to the use of the diffraction grating in astronomical spectroscopy, other than for solar work, was the low efficiency resulting from the intensity being distributed amongst several orders. The development of the blazed grating by Robert Wood (1868–1955) at Johns Hopkins University from 1910 was the key to overcoming this problem. Wood found that uniformly angled grooves with an asymmetric V-profile concentrated much of the light into a single order.

This property of the asymmetric V-shaped profile was not entirely new, from either experimental or theoretical points of view. Indeed, Henry Rowland had attempted to calculate the effect theoretically by summing an infinite array of point sources of electromagnetic radiation [118], while John Anderson (1876–1959) and C. M. Sparrow (also at Johns Hopkins), soon after the time of Wood's experiments, were able to estimate theoretically the relative intensities of light of different wavelengths in various orders, showing very clearly the strong action of the blaze [119].

Wood's initial experiments on blazed gratings were with coarsely ruled infrared gratings, which he termed echellette gratings, where the profile of the groove could be carefully controlled [120] – see also [121]. Together with John Anderson he went on to produce blazed reflection gratings for optical spectroscopy. This technology was continued some years later by Anderson and Clement Jacomini (1856–1940) at Mt Wilson Observatory, after the ruled grating section was established there in 1912 (see [122]). However, it was not until 1929 that the first blazed grating spectrograph was used for stellar spectroscopy, by Paul Merrill [123] – see Section 1.8.3.

The considerable time needed to rule diffraction gratings led to their high cost and limited availability. The technology of replicating gratings from a master eventually overcame this obstacle. After Lord Rayleigh's attempts at photographic reproduction, the British amateur Thomas Thorp experimented from 1898 with the deposition in solution of a celluloid layer onto a Rowland grating. When dry, this film-like layer could be peeled off and mounted on glass to give a transmission grating replica [124]. Non-uniform contraction of the celluloid layer when drying was a problem that was partly overcome in further experiments with a collodion film by Robert Wallace (1868–1945) at the Yerkes Observatory in 1905–6 [125, 126], and soon afterwards by John Anderson at Johns Hopkins University in 1910 [127]. Anderson found a technique of taking copies of collodion replicas using gums that dried to a hard layer on glass, and which could be coated with a thin platinum or nickel layer in a vacuum to give a reflection grating. However, the first step, that of the collodion replica, still suffered from the problems of uneven shrinkage.

Successful replicas that completely avoided wetting and the consequent problems of shrinkage in the films were developed in the 1930s. One such process was reported by Ernst Keil in California and was developed commercially, though in close collaboration with the staff at Mt Wilson [128]. Robert Wood meanwhile produced large ($4\frac{1}{2} \times 6\frac{3}{8}$ in) replicas at 15 000 gr/in from blazed masters [129]. Precise details of the process were not revealed, but Wood noted that moulded replicas using new plastics, such as Lucite, may be promising. By 1941 Wood reported the production of large replicated transmission gratings in a mosaic to form an objective grating for Schmidt telescopes [130]. The replicas comprised plastic films taken from a copper master and then mounted on glass.

Some of the problems of grating replication were discussed by Sir Thomas Merton (1888–1969) [131]. He proposed a technique of ruling a master grating as a helical path on a stainless steel cylinder, with replicas made by flattening out the thin pellicle taken from the cylinder and mounting it on a flat glass block. Gratings were produced in this way by L. A. Sayce at the National Physical Laboratory in the United Kingdom. They had coarse rulings and were suitable for infrared spectroscopy [132]. Finally in 1949 J. V. White and W. A. Fraser at the Perkin-Elmer Company took out a US patent for grating replication using epoxy resins. This marked the start of the modern era of the mass production of grating replicas. The process covered by the patent was described in some detail by R. F. Jarrell and G. W. Stroke [133]. In 1949 the Jarrell-Ash Company in Massachusetts acquired the right to use the process and develop it further.

The development of aluminized glass mirrors by John Strong in the early 1930s [134] using vacuum deposition techniques opened up a new era in grating production. Precise blazed gratings could be ruled much more readily in an aluminium layer (typically 1 to 1.5 micrometres thick) than in speculum, and such aluminized glass or Pyrex blazed gratings were used successfully in astronomy, initially at Mt Wilson [122] and Victoria [135]. Moreover, an aluminized grating surface gave significantly higher efficiency at the blaze wavelength than did speculum metal. Strong reported a 50 per cent intensity gain after aluminizing a ruled speculum grating [134], a result substantiated by the tests of C. P. Butler (1871–1952) and F. Stratton (1881–1960) at Cambridge [136]. The ultraviolet reflectivity of aluminium is also better than that of speculum, and this gain also appears in the improved ultraviolet efficiency of aluminized gratings.

The introduction of successful grating replica technology and, at about the same time, of vacuum deposition of aluminium layers on glass and other surfaces, naturally resulted in the aluminized replica becoming the standard type of grating of post-war 'mass' production. Gratings began to replace prisms for some astronomical spectroscopy from the 1930s, and their use in the form of aluminized blazed replicas was more or less universal by the 1950s. Details of the process are not available for commercial reasons, but the general outline is to take a flat optically polished substrate and coat it with an aluminium or gold layer in a vacuum. The master grating is ruled in this metallic layer with an interferometrically controlled diamond ruling engine. Sub-master replicas are obtained by coating the ruled master with a parting agent, aluminizing this coated surface and then placing an epoxy layer on top, which is then cemented to another substrate. When the epoxy is cured, the replica is removed from the master, taking with it the aluminized layer above the parting agent. Some information on this process is given in the Richardson Grating Laboratory's *Diffraction Grating Handbook* [137]. Replicas of sub-masters can then be produced, in each case the grooves being replicated in the epoxy layer.

In 1951 Harold and Horace Babcock (respectively 1882–1968 and 1912–2003) devised a grating figure of merit, which was a parameter that depended on grating size (sizes up to 8 inches were ruled at Mt Wilson) as well as performance, and took into account the resolving power (up to 5×10^5 was achievable for the best large gratings), luminous efficiency (60–70 per cent for good blazed gratings at the blaze wavelength), scattered light, ghost intensity and the presence of other defects (including white scattered light and other blemishes [122]). Their best gratings were on the Mt Wilson 'B' engine (built 1933) which was not interferometrically controlled, but nevertheless superbly engineered from the mechanical point of view. The air temperature during ruling was held to $24.0 \pm 0.005\,°C$.

In 1949 George Harrison (1898–1979) at MIT emphasized that the resolving power and efficiency of a spectrograph depended on having large gratings on which the collimated beam was incident at a relatively oblique angle. The number of grooves did not enter the equation [138]. This led Harrison to devise the échelle grating, which was a coarsely ruled grating with a large blaze angle. Between 1949 and 1974 Harrison constructed three ruling engines at MIT and successfully ruled échelle gratings with them, most notably with the MIT 'B' engine [139].

The development of so-called holographic gratings in the 1960s by several groups [140, 141, 142], based on the interference of monochromatic laser light, has not had a major impact on astronomy. Holographic gratings give essentially ghost-free spectra with very low scattered light. However, the groove profile is normally essentially sinusoidal, and the difficulty of controlling this to create a blazed grating, which is possible for ruled gratings, means that holographic gratings have not yet been able to compete with ruled ones in applications where efficiency is paramount. Nevertheless there have been promising experiments in generating more saw-tooth like groove profiles for holographic gratings. These can deliver very low scattered light in the first order and have high groove densities (≥ 1600 gr/mm) over large areas [143]. At the present time ion-etching techniques are delivering holographic blazed reflection grating with groove spacings of up to 3600 gr/mm. Their future application in astronomy seems promising.

All the gratings described hitherto have been surface relief gratings, in which the surface profile of the grating comprises grooves with a periodic structure. Recently a new type of grating has been developed, with no surface relief but a periodic modulation of the refractive index within a thin gelatin layer. The sinusoidal modulation of the refractive index is produced holographically in a dichromated gelatin layer, whose thickness is typically in the 4 to 20 micrometres range. This is the volume-phase holographic grating, which was first proposed for applications in astronomy by Sam Barden and his colleagues in 1998 [144, 145]. These so-called VPH gratings are usually operated in transmission, and can have high efficiencies, especially if the glass plates enclosing the gelatin layer have broad-band antireflection coatings. Refractive index modulation spatial frequencies can be as high as 6000 lines/mm. What is more, rotating the grating to give different angles of incidence produces a tunable grating in which the wavelength of peak efficiency can be varied.

VPH gratings are being fabricated by Kaiser Optical Systems in Michigan and by a new facility of the European Southern Observatory established at the Centre Spatial de Liège in Belgium [146]. Gratings up to 30 cm in size and with efficiencies of 70 to 90 per cent are possible. The first spectrographs using VPH grating technology were in design phase around the year 2000 or shortly thereafter [147, 148]. In addition VPH grisms are also being made, in which a VPH transmission grating is cemented to the face of a prism [149, 150].

1.7 GRATING MOUNTINGS FOR LABORATORY AND ASTRONOMICAL SPECTROGRAPHS

The mounting of a grating concerns the optical arrangement of how light from the slit is delivered to the grating and from there to the photographic plate or detector, the type and arrangement of the collimator and camera optical elements, and the mechanical movements of the grating or other components required for changing the wavelength region seen by the detector.

Early grating spectroscopes generally had light passing to the grating from a refracting collimator, and a small refracting telescope was used to view the spectrum. Both collimator and viewing telescope were independently rotatable about the grating box. Such instruments were developed from the traditional prism spectroscope, and indeed several grating spectroscopes of the late nineteenth century were developed to take either a prism or a grating as the dispersing element. Such was the case for Keeler's spectroscopes at Lick [151] and at Allegheny observatories [152], which were made by John Brashear (1840–1920) and equipped with plane Rowland gratings. Another composite grating and prism instrument of the same Brashear–Rowland pedigree was used by Huggins in London [84]. Both Keeler's and Huggins' instruments could be further adapted for spectrum photography.

Grating instruments for high resolution solar spectrography of the Littrow or autocollimating type were favoured by Hale at Mt Wilson. All three Mt Wilson solar telescopes (the horizontal Snow telescope, and the 60-foot and 150-foot tower telescopes) were equipped with such Littrow spectrographs [153, 154, 155]. In this arrangement a long focal length lens was placed immediately in front of the plane grating. For example, on the 60-foot tower telescope the slit was just above ground level, and the spectrograph was mounted vertically below ground in a pit. A Brashear visual objective of 9.1 m focal length (30 feet) was immediately in front of a Rowland 10.7-cm (4-inch) plane grating. Spectra were photographed in a plate-holder just to the side of the slit, and plates 43 cm long could be accommodated [154].

A very high resolving power laboratory spectro-graph of this Littrow type was described by Loomis and Kistiakowsky in 1932 [156].

Grating spectrographs using mirrors avoided chromatic aberrations and hence were better suited for spectrum photography. For plane gratings, one sim-ple arrangement was that described by Hermann Ebert (1861–1913) in Germany, which entailed mounting a reflection grating near the focal plane of a large concave spherical mirror [157]. A slit and photographic plate were immediately adjacent to each side of the grat-ing. The mirror served as both collimator and camera. Ebert obtained sharp spectra from this arrangement, but he was evidently unaware of the desirable can-cellation of the off-axis aberrations at each reflection from the mirror. Heinrich Kayser (1853–1940) did not favour the Ebert mounting, on the grounds that undis-persed rays from the slit could illuminate the plate after one reflection [44, p. 626]. In practice, baffling could easily overcome this drawback.

The Ebert mounting had been ignored for over half a century, when William Fastie (1911–2000) in the United States reinvented it in 1952 (at about the time of his appointment as professor of physics at Johns Hop-kins University). Fastie used the mounting in a small plane grating monochromator with an f/10 mirror. In this instrument the spectrum is scanned using a pho-tomultiplier as detector by rotating the grating [158]. Although there was some astigmatism in the optical system, the effects of this in degrading the resolv-ing power could be partly overcome by using curved slits for both the entrance and exit slits. A resolving power as high as 10^5 was achieved using a grating with 3-inch rulings and 30 480 gr/in. An infrared version of the Fastie instrument with an f/6 mirror was also developed [159].

A more popular mounting for a plane grating that has been widely used in astronomy is the Czerny–Turner mounting, which was first described in 1930 by M. Czerny and A. F. Turner [160]. It comprises two spherical concave mirrors, one acting as a col-limator, the other as a camera, and of equal focal length. The coma of one balances that of the other,

thus giving a good overall aberration correction. The Czerny–Turner mounting was used, for example, in the large McMath–Hulbert solar spectrograph at the University of Michigan in 1955 [161] – see Fig. 4.33. The instrument is mounted horizontally after divert-ing the beam from the McGregor tower telescope by a 45° plane mirror. Collimator and camera mirrors have apertures of 305 mm and focal lengths of 15.2 m. Either photographic or photoelectric recording of the solar spectrum is possible. The McMath–Hulbert spec-trograph is housed in a large vacuum tank, giving exceptional stability and a dust-free environment. The diffraction grating is 134×204 mm (600 gr/mm) and is rotated for photoelectric scanning. Another example of the Czerny–Turner mounting is at the McMath tele-scope and spectrograph at Kitt Peak, seen in Fig. 4.38.

1.7.1 Concave grating mountings

The original concave grating mounting was devised by Henry Rowland at the Johns Hopkins University in 1883 [162]. Here he showed that, for a spherical grat-ing of radius of curvature R_{gr}, a sharp spectral image was always located on a circle of diameter R_{gr}, pro-vided that the slit was also on this Rowland circle. The 'Rowland mounting' used for Rowland's classical study of the solar spectrum had a fixed slit and an adjustable assembly carrying both grating and plateholder, which had to be reset for different wavelength regions.

The theory and the practical method of using the instrument were described by Joseph Ames [163], Row-land's assistant at Johns Hopkins. Order separation was achieved by using appropriate filters, either glass plates or absorbing solutions.

The Rowland spectrograph was in practice quite bulky, given that a typical grating had a 21.5-foot (6.45-m) radius of curvature, which is therefore the diameter of the Rowland circle. This required it to be installed in a large room. The slit was in a fixed location, while the 6.45-m long arm carrying grating and plate had to be movable in a horizontal plane, as shown in Fig. 1.22. Gratings with 10 000, 14 438 and 20 000 gr/in were used by Rowland, many of them being ruled on 6-inch blanks.

Concave grating spectrographs suffer from astig-matism which broadens the spectrum perpendicular to the slit. A point on the slit gives a spectrum of width

$$z = (\sin^2 \beta + \sin \alpha \tan \alpha \cos \beta)l, \qquad (1.1)$$

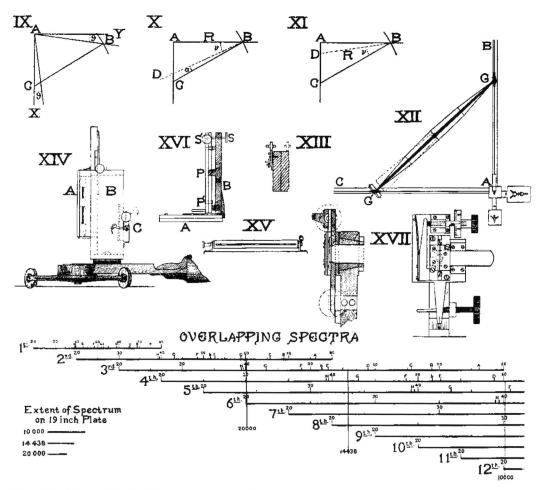

Figure 1.22. Diagram of Rowland's large solar spectrograph. In diagram XII the two arms AB and AC are wooden, 6.9 m in length and at right angles. The slit is at A and the light travels from there to the concave grating at G. The plateholder is at G′. G, A and G′ are on the Rowland circle. The arm GG′, of length 6.45 m, is always a diameter of the Rowland circle and can move to set different wavelengths on the plate, which is 48 cm long.

where l is the length of the rulings (see [164]). For the Rowland mounting, $\beta \approx 0°$, so $z \simeq \sin\alpha\tan\alpha.l$. The astigmatism increases greatly for larger angles of incidence, α, on the grating, and, as noted by Ames, spectrum widths of several centimetres are the result, especially in the higher orders where the angle of incidence is large. The astigmatism could be reduced over a certain range of wavelengths by using a cylindrical lens, as noted by Edison Pettit at Mt Wilson [165]. The astigmatism of concave gratings has prevented their use for much stellar work, given that the brightness of widened astigmatic spectra is much reduced. An exhaustive paper by H. G. Beutler in 1945 covered the theory of the concave grating and its aberrations [166].

Other mountings than the original Rowland mounting for concave gratings were soon devised, partly to overcome the difficulty of changing the wavelength region, and partly to reduce the effect of astigmatism. Captain William Abney's (1843–1920) mounting had a fixed grating and plate, but a movable slit that rotated on an arm about the Rowland circle [167]. Given that each movement of the slit entails a new direction for the incoming beam, this instrument has its limitations, even for solar work.

The mounting of Carl Runge (1856–1927) and Friedrich Paschen (1865–1947) (generally known as the Paschen–Runge mounting) was devised in Hannover in 1902 [168]. It has a fixed slit and fixed concave grating, but a moving plateholder on the Rowland circle. It therefore has a constant angle of incidence (often around $\alpha \approx 45°$, and a variable angle of diffraction, possibly from $\beta \simeq -40°$ to $+70°$. The Paschen–Runge mounting has been widely used, especially for photoelectric spectrometers, where the detector can readily be mounted on a movable radius arm.

In 1910 Albert Eagle at Imperial College, London, produced a much more compact mounting for the concave grating, which was based on the concept of the Littrow mounting for prism spectroscopes [169]. In the Eagle mounting, the angles α and β are about equal, and the light from the slit is fed to the spectrograph after reflection through 90° from a small prism. The slit and plateholder are close together on the Rowland circle, and the grating is along a chord to the other side of the circle; changing the wavelength setting entails moving the grating and plateholder closer or further apart, as well as rotating both to keep them tangential to the Rowland circle. The plateholder is tilted at an angle $i \simeq \alpha$ to the rays falling on it, giving a dispersion at the plate of

$$\frac{\mathrm{d}s}{\mathrm{d}\lambda} = \frac{R_{\mathrm{gr}} n}{d \cos i}. \tag{1.2}$$

Here R_{gr} is the grating radius of curvature, n is the order number, and d is the groove spacing. The spectrum is not exactly a 'normal spectrum' (i.e. constant dispersion) as the angle of incidence onto the plate varies slightly from one end of the plate to the other. Figure 1.23 shows a concave grating spectrograph with an Eagle mounting, as presented by Albert Eagle in 1910.

The Eagle mounting has the huge advantage of being a compact instrument that can be housed in a casing instead of a whole room, making it considerably more stable. It also has the property of much reduced astigmatism for higher order spectra. The astigmatism for the Eagle mounting is

$$z = \frac{n^2\lambda^2 L}{2d^2} = 2L\sin^2\alpha \tag{1.3}$$

(where L is the grating size), compared with

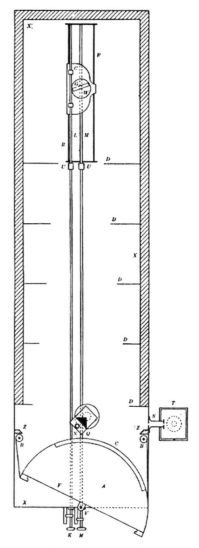

Figure 1.23. Eagle concave grating spectrograph, 1910. The light source enters the slit on the right, from where the small prism reflects the light to the concave grating at G. From there the beam returns down the spectrograph to the tiltable plateholder at F. Both plateholder and grating lie on the Rowland circle. A change of wavelength region involves tilting both plateholder and grating, and sliding the latter on its carriage mounted on rails that are a chord to the Rowland circle.

$$z = \frac{n^2\lambda^2 L}{d^2} \bigg/ \sqrt{1 - \frac{n^2\lambda^2}{d^2}} = L\tan\alpha\sin\alpha \tag{1.4}$$

for Rowland or Abney mounts. For a given order number and wavelength, the Eagle mounting astigmatism is always at least two times smaller. The gain is especially marked for higher orders, which are therefore much brighter in Eagle spectrographs. The Eagle mounting also accesses the longest wavelengths of any mounting, as is seen from the grating equation $n\lambda = d(\sin\alpha + \sin\beta)$, which becomes $\lambda \simeq \frac{2d}{n}\sin\alpha$ for the Eagle configuration.

Another mounting, known as the radius or Beutler mounting, was introduced at the University of Chicago for the vacuum ultraviolet laboratory spectrographs [170, p. 154]. Here the slit and plateholder are in fixed positions on the Rowland circle, while the grating moves on a radius arm on the Rowland circle. When the spectrograph is in a vacuum tank, the grating is one component that can be moved inside the vacuum without direct access from the user.

Two mountings that have been used for the concave grating do not use the Rowland circle. One devised very early was the Wadsworth mounting [171]. Here the grating received light from a collimated beam from a lens or concave mirror (a lens was used in Wadsworth's original configuration for this instrument, but Charles Fabry (1867–1945) and Henri Buisson (1873–1944) later introduced a concave mirror as a collimator to avoid chromatic aberration [172]). In the Wadsworth mounting, the eyepiece or plate is on the grating normal ($\beta = 0°$), while the slit and collimator illuminate the grating at an angle α which can be varied, as seen in Fig. 1.24. The advantage of this mounting is that astigmatism can be largely eliminated. Runge and Paschen also used this mounting, apparently devising it independently [173]. In the Wadsworth mounting the eyepiece or plateholder should move on a parabola, and at $\alpha \sim 0°$ the distance to the focus is $R_{gr}/2$. Thus in this position the linear dispersion is half that of Rowland circle mounts.

Finally, another mounting that also departs from the Rowland circle configuration was devised by M. Seya in Tokyo and analysed by T. Namioka at the University of Chicago [174, 175]. The Seya–Namioka mounting is a fixed arm spectrometer with the angle $(\alpha - \beta)$ between slit and detector directions being set at $70° 15'$ for a minimum of the astigmatic aberration over a wide range of wavelengths. The grating rotates but does not translate for wavelength scanning. The entrance slit and detector are at a distance of $0.818R_{gr}$ from the grating. The Seya–Namioka mounting is ideal for ultraviolet vacuum spectrometers, given the fixed

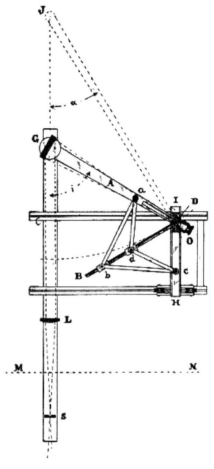

Figure 1.24. The Wadsworth mounting of a concave grating is shown in this diagram of a spectroscope from the original paper by Wadsworth in 1896. S is the slit, L the collimator lens, G the concave grating, and O is the eyepiece (or plateholder). The eyepiece is always normal to the grating and moves on a parabolic path as the angle of incidence i of the light on the grating is varied.

configuration of the optical elements, other than the grating rotation.

The Seya–Namioka spectrometer is one of the few concave grating mountings that have been used for stellar spectroscopy. An example is the photoelectric spectrometer designed by William Liller (b. 1927) [176], where a compact photoelectric scanning spectrometer was built for the f/5 Newtonian focus of the 1.55-m reflector at the Harvard College Observatory – see Fig. 1.25. The instrument scans between 320 and 500 nm with a resolution of about 1 nm.

Figure 1.25. The Seya–Namioka concave grating spectrometer at Harvard was built by William Liller in 1963, and used a 1P28 photomultiplier tube as detector. Rotation of the grating enabled a stellar spectrum to be scanned.

A much earlier attempt at photographic stellar spectroscopy with a concave Rowland grating in an Abney mounting was reported by Henry Crew (1859–1953) [177] using a Rowland grating on the 12- and 36-inch refractors at Lick. Little success was reported with this instrument, a failure that Donald Osterbrock (1924–2007) has attributed to Crew's lack of training and experience in astronomy, and his lack of support and funding from Lick Observatory [178].

George Ellery Hale had little more success at Yerkes. He attempted to record some stellar spectra with a Rowland concave grating in a Wadsworth mounting using the 24-inch horizontal solar telescope (a forerunner to the Snow telescope) in 1902, but these experiments were halted by a fire that destroyed the telescope in December of that year (see [178] for discussion and references). However, successful solar spectra were obtained with this instrument.

1.8 EARLY GRATING SPECTROSCOPES AND SPECTROGRAPHS

Gratings were rarely used for astronomical spectroscopy in the nineteenth and early twentieth centuries (prior to about 1920). The exception was for solar spectroscopy, where the abundance of light more than overcame the intrinsic inefficiency of early gratings,

with the result that several notable studies in solar spectroscopy used gratings.

1.8.1 Early solar grating spectroscopy

One of the most famous early studies of the solar spectrum using a transmission grating was undertaken by Anders Ångström (1814–74) in Uppsala in 1868 [181].[7] His drawing showed about 1000 lines with wavelengths calibrated from the grating equation for nine strong lines. He used two Norbert transmission gratings for this work.

Another notable paper in early solar grating spectroscopy was the determination of the solar rotation rate by Charles Young (1834–1908) at Dartmouth College in 1876, using the Doppler displacement of the solar spectra from the east and west limbs [182]. His spectroscope employed a Rutherfurd reflection grating (8640 gr/in) with a high angle of incidence. The grating used relatively high orders ($n = 6$ or 8) and a $45°$ prism for order separation. (This was the first use of a prism as a cross-disperser of high-order grating spectra, an arrangement that a century later became common in échelle spectrographs.) Young's paper was one of the early papers to demonstrate the applicability of the Doppler effect to light rays, by showing that the measured shift of spectral lines from the solar limbs was in concordance with the known solar rotation rate from sunspots.

By far the greatest exponent of gratings for solar spectroscopy in the nineteenth century was Henry Rowland at Johns Hopkins University. He recognized the benefits of the concave reflection grating for solar spectrum photography, especially in the ultraviolet, and he devised the elegant Rowland mount, which gave a normal (uniform dispersion) spectrum that was always in focus on the photographic plate for any wavelength setting [162]. The Rowland concave grating and its spectrographic mount were described further by Rowland's assistant, Joseph Ames, in 1892 [113]. This work by Rowland culminated in his celebrated table of solar spectrum wavelengths [183], which appeared

---

[7] But the work of Ångström was not the first after Fraunhofer to use a grating. See also [179] and [180] for early solar grating spectroscopy for wavelength determinations.

Figure 1.26. Keeler's spectroscope on the Lick 36-inch refractor, seen here in prism mode.

in eighteen papers in the first five volumes of the *Astrophysical Journal*.

After Rowland's death in 1901, further experiments in solar grating spectroscopy were carried out by George Ellery Hale, both at Yerkes Observatory [184] and, a few years later, from the newly established Mt Wilson Observatory [185]. Here he used plane gratings ruled by Ames' assistant in Baltimore, Lewis Jewell, and he successfully recorded the spectra of sunspots. But the exposures with the 24-inch Snow solar telescope (a fixed horizontal reflector) were inordinately long for stellar work, even on bright stars such as Arcturus [186].

Another notable solar grating spectrograph was the Porter spectrograph constructed by the Brashear Co. for Frank Schlesinger (1871–1943) at the Allegheny Observatory [187]. This instrument comprised a 10 × 15-cm Michelson grating (500 gr/mm) in a large vertical spectrograph to which sunlight was delivered from a coelostat. Schlesinger used this instrument to record the rotation rate of the Sun in 1911–12 as part of an international campaign.

1.8.2   Stellar grating spectroscopy before 1900

Early experiments at using gratings for stellar spectroscopy met with frequent failures. Both Angelo Secchi in Rome in 1877 (with a Rutherfurd grating) [188] and Hermann Carl Vogel in Potsdam (with several gratings from Wanschaff in Berlin) [189] found

that gratings were too inefficient for stellar work. Vogel made the interesting comment that gratings ruled on glass could then be silvered and these gave higher efficiency than those using speculum metal. Moreover they could be resilvered from time to time, once tarnished [189].

The only successful stellar grating spectroscope in the nineteenth century was that of James Keeler on the 36-inch Lick refractor [190]. His instrument was made by John Brashear and would operate visually, either with a Rowland grating (14 438 gr/inch) or with a choice of three prisms. In the grating mode of operation, in the third or fourth order, Keeler observed the spectra of gaseous nebulae and of several bright stars and planets, making wavelength and radial-velocity measurements during the years 1890–91. His success can be ascribed to his use of a large telescope and the restriction of his work to bright objects. The Keeler spectroscope is seen in Fig. 1.26 on the Lick refractor. Figure 1.27 is a diagram of the instrument in the grating mode.

After Keeler went to Allegheny Observatory in 1891 as director, he also had a similar spectroscope, which could operate in grating or prism mode, built for the Allegheny 13-inch refractor [152]. This is shown in Fig. 1.28 with a plane Rowland grating (14 438 gr/in, size 1.3 × 1.8 in).

Much less successful was Henry Crew, who at this time was attempting to photograph stellar spectra using a concave Rowland grating mounted in a

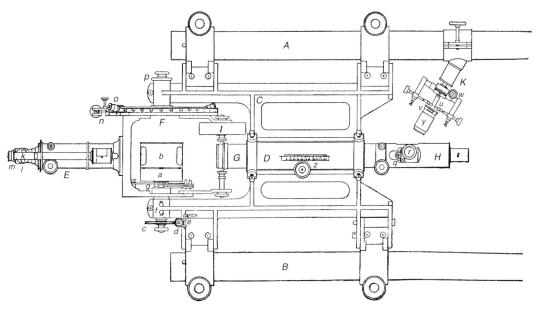

Figure 1.27. Diagram of Keeler's spectroscope at Lick Observatory. s:slit; G:collimator; a:grating table; K:comparison spectrum source; E:viewing telescope; A, B:brass supporting rods.

quickly assembled wooden spectrograph housing, also on the Lick refractor. The astigmatism introduced by the concave grating broadened the spectra normal to the dispersion, so rendering the spectral images too faint to record photographically. (See the discussion by D. E. Osterbrock on Crew's unsuccessful attempts at stellar spectrum photography [178].)

Further experiments with a concave grating were made at Johns Hopkins University by Charles Poor and Alfred Mitchell using a small 1×2-inch grating, which received the starlight directly and focussed the spectrum on a plate as a slitless objective grating spectrograph [191]. However, even Sirius required exposures of 40 minutes! This circumstance, nevertheless, appears to have given encouragement to Frank Wadsworth at Yerkes, who considered that larger concave reflection gratings and parallel wire objective gratings had a future role for stellar work [192]. Indeed, Mitchell continued these experiments with a larger Rowland concave grating ($2 \times 5\frac{3}{4}$ inches; 15 000 gr/in) at Johns Hopkins University and at Yerkes Observatory in 1898–9, and he obtained good spectra of 15 stars and of the Orion nebula. Sirius now took about five minutes to expose an excellent spectrum showing as many as 75 lines [193].

The simplicity of the apparatus, which was mounted on the side of a telescope but did not use the telescope's optics, the ultraviolet extent of the spectra, the normal dispersion and good definition (limited by the seeing) were cited as merits of the technique. In practice, the main disadvantage was that the light-gathering power was limited by the small grating size, since the grating itself served as the entrance pupil of the instrument. Dispersing light with a small grating after collecting it with a large telescope was the only practical way of reaching fainter stars.

1.8.3 Astronomical grating spectrographs and their development from 1900

In the first half of the twentieth century there were a number of milestones in the development of grating spectrographs. In the first decade of the century, gratings were used on relatively large telescopes for bright sources such as the Sun and Mars. Thus Walter Adams (1876–1956) used a 5-inch Rowland plane grating on the Mt Wilson Snow solar telescope (aperture 24 inches) to record a blue spectrum of Arcturus in 23 hours over five nights! The dispersion was 4.3 Å/mm [186]. Clearly such observations were

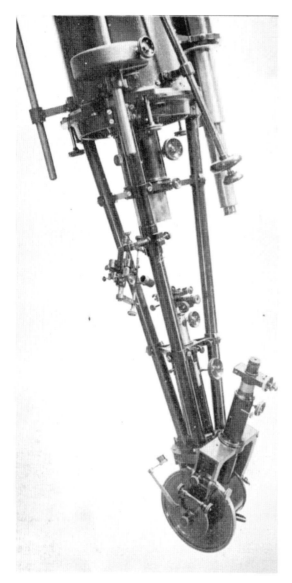

Figure 1.28. Keeler's Allegheny Observatory spectroscope on the 13-inch refractor, 1893, seen here with a plane Rowland grating.

unsuited to fainter objects. Indeed, the equipment had been developed by Hale for spectroscopy of sunspots [185] and continued an earlier programme initiated on the 40-inch refractor at Yerkes Observatory in 1902 [184].

This first phase of grating spectroscopy in the early twentieth century thus used unblazed gratings and relied on very bright objects, relatively large telescopes and sometimes very long exposure times to overcome both grating inefficiency and the inefficiency of the photographic emulsions then available.

At Lick Observatory, W. W. Campbell and Sebastian Albrecht used a large Michelson grating (98×68 mm) in second order on the 36-inch refractor to obtain high dispersion (11.2 Å/mm) spectra of Mars with a large grating spectrograph [194]. The exposures were long – typically around two hours.

What may be regarded as the second or intermediate stage in the development of grating spectrographs in the twentieth century came from the work of John Plaskett (1865–1941) during his time at the Dominion Observatory in Ottawa. He constructed a spectrograph there for the 15-inch Brashear refractor, shown in Fig. 1.29. The dispersing element could be either a plane grating or a silvered half-prism, thus enabling comparisons to be made [195, 196, 197]. The grating was plane ($2\frac{7}{8} \times 3\frac{3}{4}$ inches; 15 000 lines/in) and ruled by John Anderson at Johns Hopkins in 1911. This was about the time that Wood and Anderson were developing the blazed grating. This particular grating sent 50 per cent of the diffracted light into the first order. It is not clear whether it was an early example of a blazed grating as such, but was more probably selected for giving an unusually bright first order spectrum.

Several spectra of bright stars (including Procyon and Rigel) were obtained. Nevertheless Plaskett was still disappointed with the performance:

> . . . it may be said that although the spectra obtained from the grating are disappointingly weak . . . yet even under this handicap it can be used to advantage when the K line is required and if spectra of uniform intensity or of uniform dispersion are needed. It would also be useful in the red end where prismatic spectra are so unduly compressed. If a grating giving twice the intensity could be obtained it would be superior even to single-prism dispersion for most work [196].

Thus the prism spectra were far better exposed in the blue than those with the grating. Only in the ultraviolet did the high absorption in the prism give the grating an advantage. Plaskett exposed no spectra in the red beyond Hβ, but he anticipated that the grating might be advantageous here too, because of its higher dispersion.

The first truly successful stellar grating spectrograph was built at Mt Wilson Observatory for Paul

Figure 1.29. John Plaskett's grating spectrograph at the Dominion Observatory, Ottawa, 1914.

1922, and initially involved an unblazed concave grating. The key to his eventual success was a blazed 600 gr/mm plane grating, 4 inches square, ruled by Clement Jacomini at Mt Wilson, and which was able 'to concentrate the light in the first order on one side', thanks to a specially shaped diamond tool. The new instrument was a Cassegrain spectrograph constructed from cast aluminium alloy for lightness and stiffness. It was attached to the 100-inch telescope with box-section trusses. A commercial Bausch and Lomb telephoto lens served as collimator ($f_{coll} = 40$ inches), while three interchangeable camera lenses ($f_{cam} = 6, 10$ or 18 inches) gave a choice of dispersions (111, 66 or 34 Å/mm respectively). The new spectrograph was especially designed for red and far-red spectroscopy, taking advantage of the superior dispersion of a grating over a prism in this wavelength region. Figure 1.30 shows a diagram of this instrument.

Merrill presented initial spectra of Be stars around the Hα line. In addition, radial velocities were measured for a variety of late-type stars (to about $m_V = 7$), while Roscoe Sanford (1883–1958) obtained spectra of carbon stars as faint as ninth magnitude in two hours. A first survey of the spectral region 687–870 nm using near infrared sensitive emulsions was made with this instrument, which included the infrared calcium triplet.

With the 100-inch telescope Merrill and Sanford were able to record the spectra of relatively faint stars with the new grating spectrograph, sometimes after exposing for as long as six hours [198]. Merrill completed a classic survey of the near infrared spectra (700–900 nm) of bright stars of all spectral types with the plane-grating spectrograph in 1934, using spectra of both 66 and 34 Å/mm [199].

The fourth stage in the development of the grating spectrograph also originated at Mt Wilson, and was introduced by Walter Adams and Theodore Dunham (1897–1984) in 1935 [200]. This too involved a plane blazed grating, but from Robert Wood at Johns Hopkins (rather than from Jacomini at Mt Wilson), and was ruled on an aluminized layer on Pyrex at 14 400 gr/in. No doubt such gratings gave superior blaze efficiencies than those ruled on speculum, which is a much harder material. The Wood grating was used in first order in the red, or in second order for the blue-ultraviolet region. Adams and Dunham completely rebuilt the coudé spectrograph to accommodate

Merrill in 1928–9, from a design by E. C. Nicholls [123]. This can be designated as the third phase in the development of the astronomical grating spectrograph in the twentieth century. However, Merrill's experiments with grating spectroscopy extended back to

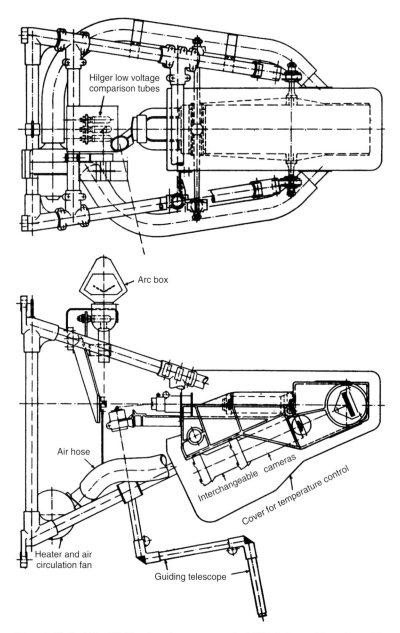

Figure 1.30. Paul Merrill's blazed grating spectrograph of 1929. It featured interchangeable cameras giving dispersions of 111, 66 or 34 Å/mm. This was the first grating spectrograph in routine use for stellar spectroscopy, and was designed to operate in the far-red region of stellar spectra. It was mounted at the Cassegrain focus of the 100-inch Hooker telescope at Mt Wilson Observatory.

the new blazed grating and a choice of three different Schmidt cameras ($f_{cam} = 32$, 73 or 114 inches). The longest focal length camera gave a dispersion of 2.9 Å/mm, but even then a blue spectrum of Mira at $m_{pg} = 6.0$ was recorded in $5\frac{1}{2}$ hours. On the other

hand, the shortest focal length camera could easily reach stars at $m_{pg} = 7$.

The development of the 100-inch coudé spectrograph at Mt Wilson and its conversion to a blazed-grating and Schmidt-camera spectrograph was a highly influential event in the history of astronomical spectrographs.[8] Other instruments, not only those for the coudé focus, adopted the high efficiency and broad wavelength coverage offered by the aluminized blazed grating and Schmidt camera combination. Among coudé spectrographs, those at McDonald in 1949, Palomar in 1952, Haute-Provence in 1959 and Lick in 1961 were notable.

In addition, Cassegrain instruments also followed this design concept of using a blazed grating and a Schmidt camera. An excellent example was built at Mt Wilson for the 60-inch telescope in 1956 (Fig. 1.31) [201]. It had three Babcock gratings giving different dispersions and blaze wavelengths, and also three different Schmidt cameras ($f_{cam} = 4$, 8 or 16 inches). With different settings, spectra from 330 nm to more than 1 μm could be recorded. An unusual feature was an inverted Cassegrain collimator, using two reflections to give a very compact arrangement of effective focal length 64 inches with unusually low flexure. In summary, this spectrograph allowed the spectra of faint stars to be recorded in any optical wavelength region – with the fastest camera, blue spectra at 80 Å/mm of twelfth magnitude stars being possible in two hours. It was largely free of flexure and required no temperature control. Apart from the Schmidt corrector plates, all optics were reflecting.

The design principles of reflection grating and Schmidt camera were also applied to the fast prime-focus nebular spectrograph at this time. Thus, on the 82-inch telescope at McDonald Observatory such a spectrograph was constructed for recording low dispersion (about 400 Å/mm) galaxy spectra for determining redshifts [202]. This instrument had a Pyrex paraboloid collimator mirror ($A_{coll} = 2$ in, f/4) and a fast f/0.65 UV-glass Schmidt camera ($A_{coll} = 2.5$ in). In 1951 a Bausch and Lomb replica grating was installed (15 000 gr/in, $\lambda_B = 650$ nm) – one of the first spectrographs to use a blazed replica.

The history of the Mt Wilson coudé and its development in the mid 1930s is told in detail by Dunham [203], who showed that the second order blazed grating could outperform flint glass prism dispersion over the whole visible spectral range from the near ultraviolet to

Figure 1.31. The compact Cassegrain grating spectrograph for the Mt Wilson 60-inch telescope, which was commissioned in 1956. The spectrograph had a choice of three fast Schmidt cameras, enabling faint stellar spectra to 330 nm to be recorded.

the far red, except for a small region in the blue where prisms had a slight advantage. This circumstance led to the eventual demise of the prism spectrograph. However, the fact that cheap grating replicas were not at that time (mid 1930s) readily available gave the prism a temporary reprieve until after World War II.

The introduction of the échelle grating by Harrison in 1949 [138] ushered in the fifth generation of twentieth-century grating spectrographs. The grating has been used in a variety of astronomical applications, including solar spectroscopy [204], stellar coudé spectroscopy [205, 206, 207, 208] and far ultraviolet spectroscopy from space [209, 210] – in each case the references cited mark the earliest introduction of the échelle in each astronomical application. However, the new grating's greatest impact in the two decades from about 1970 was in Cassegrain échelle spectrographs, such as that designed by Daniel Schroeder (b. 1933) at the Pine Bluff Observatory in Wisconsin [211]. Here a compact and relatively inexpensive spectrograph gave high dispersion (in the case of the Pine Bluff échelle, 2.5 Å/mm) and resolving power, comparable to the great coudé spectrographs, but possibly with somewhat better throughput.

All échelles require a cross-dispersion element (a low dispersion grating or prism being common) to separate orders, assuming that more than one order is to be simultaneously observed. The relative complexity of the resulting spectral format in the focal plane for spectral reductions is one price to pay. However, the advantage of this format for the new two-dimensional detectors of limited area (initially image intensifiers or television tubes; later solid-state detectors) resulted in their popularity. At least 14 Cassegrain échelle spectrographs were built between 1971 and 1986 for these reasons (see [212]). These all had high blaze angle échelles ($\theta_B = 63° 27' = \arctan 2$) operating in high orders, typically $n = 30$–75 for a 79 gr/mm instrument, and about twice these order numbers in the case of an échelle with 31.6 gr/mm (these two groove spacings were values produced by the Bausch and Lomb Company at that time).

A modification of the Cassegrain échelle, which confers some of the stability benefits of a coudé spectrograph, is the fibre-fed échelle spectrograph, developed by L. W. Ramsey (b. 1945) and S. C. Barden at Pennsylvania State University in the 1980s [213], and since copied elsewhere. This can be regarded as a sixth stage in the development of the astronomical grating spectrograph in the twentieth century.

The theory and development of the astronomical échelle spectrograph is covered in more detail in Sections 3.3 and 3.4.

1.9 PRISM SPECTROGRAPHS IN THE TWENTIETH CENTURY

In the early twentieth century a new generation of large reflecting telescopes was designed and built, as it was realized that the large refractors at Lick (36 in, 1887) and Yerkes (40 in, 1897) observatories were close to the limit of practical size. The Crossley 36-inch reflector (relocated to Lick in 1895) was an early forerunner of the new generation of reflectors. The Mt Wilson 60-inch reflector (completed 1908) was certainly the most famous example of the new breed in the first decade of the century. Both the 60-inch, and its successor, the 100-inch Hooker telescope, were notable for

providing three focal ratios corresponding to the Newtonian (or prime) focus, the Cassegrain and the coudé (respectively f/5, f/16 and f/30). Spectrographs used on these and other reflecting telescopes were, therefore, designed to operate at different focal ratios for different applications. An example of an early Cassegrain prism spectrograph is that on the Mt Wilson 60-inch telescope, shown in Fig. 1.32.

Since blazed gratings were not widely available, nor were they very efficient in the first half of the twentieth century, prism spectrographs remained in popularity throughout this period. The main features of their design that were important were as follows:

1. Different types of spectrograph were designed for the different foci of the large reflectors. In general these were high resolving power instruments for bright stars at the coudé focus, and low resolving power for faint or nebular objects at the prime or Newtonian focus.

2. Greater wavelength capability became an issue of importance. This in turn led to the search for
   (a) prisms whose transmission extended further into the ultraviolet,
   (b) cameras delivering a wide angular field with minimal spherical aberration and coma, and
   (c) cameras having a flat field (so as not to bend glass plates).

3. Larger telescopes require faster cameras of shorter focal length. The design of fast aberration-free cameras became a major issue. In particular, nebular spectroscopy requires a fast camera to allow the spectroscopy of faint extended objects with a wide slit.

4. It was realized that larger telescopes require larger dispersing elements, but that if these elements are prisms, then absorption in the glass gives a diminishing return which is no longer in proportion to the linear dimensions.

5. Radial-velocity observations continued to be a major application with reflectors. These required stiff flexure-free mechanical structures. However, spectrographs became more complex, sometimes with a choice of one, two or three prisms and three different cameras. Such multi-mode spectrographs became possible because of improved mechanical design. These spectrographs had earlier (at about

Figure 1.32. Three-prism Cassegrain spectrograph of Walter Adams on the Mt Wilson 60-inch telescope, 1912.

the turn of the century) generally not been considered practical for velocity work.

6. The exposure of the comparison spectrum was important for accurate radial-velocity work. The recording of two simultaneous comparison spectra on each side of a stellar spectrum facilitated the most accurate measurements.

Table 1.1 lists a selection of 16 of the best known prism spectrographs in use on reflectors in the first half of the twentieth century. For the first two decades these used mainly Jena O102 dense flint glass prisms. However, John Plaskett made a thorough comparative study of seven Jena glasses in 1914 and advocated at first O722, a light flint glass which, although less dispersive, gave four times the spectral intensity at the K line and five times at Balmer Hζ [214, 215]. Very soon thereafter he advocated another flint glass, O118, which gave more dispersion (about the same as O102) and slightly less ultraviolet absorption than O722. About 365 nm was the effective short wavelength limit for either of these glasses. As a result, the Cassegrain spectrograph on the 72-inch telescope at Victoria was the first to have O118 prisms [216], a choice adopted later elsewhere, for example at Berlin-Babelsberg [217].

Mt Wilson was the observatory where the first large coudé spectrographs were developed, on the 60-inch and 100-inch telescopes. It is interesting that early Cassegrain telescopes typically employed f/16 to f/18 focal ratios, thus limiting the collimated beam size to about 50 mm for an 800 mm collimator focal length (about the maximum if flexure was to be controlled). The coudé spectrograph had no such flexure limitation, and both the Mt Wilson coudé instruments employed large 152-mm high prisms and long focal length collimators in a Littrow arrangement, in which the beam was returned through the prism after reflection from a plane mirror. The same lens served as both collimator and camera. Such large instruments allowed high resolving powers to be reached, but absorption of light in the prism excluded the ultraviolet below about 400 nm.

On the other hand, for relatively low resolving power on faint or nebulous objects, the prime focus offered greater efficiency, because of only one reflection before the slit. For the prime focus of the Crossley reflector at Lick, Wright built an ultraviolet nebular spectrograph with quartz optics [218]. The concept was copied by Plaskett at Victoria, but using the less expensive ultraviolet crown glass [219]. Such

Table 1.1. *Table of prism spectrographs on reflecting telescopes, first half of the twentieth century*

Year	Observ.	Tel.	Focus	f_{coll} (mm)	Prisms	Camera(s)	P (Hγ) (Å/mm)	Notes	Ref.
1903	San Cristobal Santiago	$36\frac{1}{4}''$	Cass.	800	3 × 63° 28′ Jena O102	cemented triplet	10.3	—	[101]
1906	San Cristobal Santiago	$36\frac{1}{4}''$	Cass.	800	2 × 63° 28′	uncemented Hartmann triplet	20.3	—	[101]
1910	Allegheny	30″	Cass.	630	1 × 62° 45′	Brashear triplet			[221]
1910	Mt Wilson	60″	coudé	5500	1 × 63°	—	1.4	Littrow; prism in double pass	[224]
1911	Detroit	$37\frac{1}{2}''$	Cass.	686	1 × 63° 36′	Hastings–Brashear doublet	44.1 (450 nm)	—	[225]
1912	Mt Wilson	60″	Cass.	1020	(a) 3 × 63° 29′ Jena O102	Brashear uncemented triplet	5.2	$P = 15.7$ Å/mm with 1 prism	[223]
					(b) 2 × 63° 29′ Jena O102	Cooke astrographic triplet	18.0	2 prisms	
1916	Lick	36″	prime	279	2 × 60° quartz	quartz lens		UV nebular spec.	[218]
1919	DAO	72″	Cass.	1143	3 × 63° Jena O118	Hastings–Brashear cemented triplet	35	3 cameras available	[216]
1923	DAO	72″	prime	200	2 × 60° UV crown	single element lens	50 (360 nm)	UV spec.	[219]
1925	Mt Wilson	100″	coudé	(a) 4570	1 × 63.3° Jena O102	4-element Ross lens	2.9	Littrow; double pass	[200]
				(b) 2740	Jena O102	Ross lens	4.8	installed 1930	[203]
				(c) 2740	1 × 62.5° BK5	UV lens	7 (390 nm)	better UV range	

1930	Berlin–Babelsberg	1.25-m	Cass.	995	$1 \times 66.6°$ Jena O118	3 cameras	23 35 69		[217]
1930	Mt Wilson	100″	Cass.	600	$2 \times 60°$ light flint	Rayton f/0.6 7 elements	418	fast cam. for gal. z	[234]
1936	Potsdam	30-cm	Newtonian	150	$2 \times 30°$ half prisms Jena O102	Tessar	60	parabolic collimator mirror	[228]
1939	McDonald	82″	Cass.	1000	$2 \times 62° 11'$ UBK5, UZK5	500 mm Ross lens or Schmidt			[229]
1939	McDonald	82″	coudé	2130	$63\frac{1}{2}° + 32°$ Chance 623360 glass	4 or 2 element lenses	3.5	Littrow; prisms in double pass	[230]
1951	Radcliffe	74″	Cass.	1153	$2 \times 64°$		96, 50, 30 or 22	4 cameras	[235, 236]

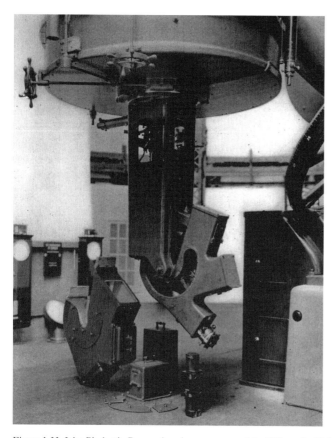

Figure 1.33. John Plaskett's Cassegrain prism spectrograph in 1919 on the 72-inch reflector at the Dominion Astrophysical Observatory in Victoria, Canada, had one, two or three prisms. Each configuration received the dispersed light into a different camera.

spectrographs, however, required extreme plate tilts (about 50°) to compensate for chromatic aberration in the camera, and field curvature was also a problem, limiting good focus to a wavelength coverage of no more than 50 nm in one exposure. Wright conducted a classic study of the ultraviolet spectra of planetary nebulae with the Lick instrument [220].

Most of the spectrographs of this era were however at the Cassegrain focus. Early observers found that the temperature sensitivity of this focus was a bothersome problem for spectroscopy. The Cassegrain reflector's focal plane is much more sensitive to temperature than that of the refractor, as a small change in mirror separation, arising from the contraction of an iron tube as the temperature falls, is greatly magnified by the optical arrangement. The early solution, adopted at San Cristobal and Allegheny, was to move the whole spectrograph backwards to maintain the star in focus on

the slit, which was no doubt a rather cumbersome process, but easier than adjusting an inaccessible secondary mirror [221, 222].

Cassegrain instruments certainly became more complex in this era than had been the case a few decades earlier on refracting telescopes. Thus the Cassegrain spectrograph on the Mt Wilson 60-inch telescope [223] could operate with one or two prisms and had a choice of two cameras, while the corresponding instrument at Victoria was designed for up to three prisms and had three different cameras [216]. This spectrograph, with its three cameras, is seen in Fig. 1.33. In 1946 the instrument was modified to become a Littrow spectrograph with the beam returning through the prisms in double pass after reflection from a plane mirror [226] – see Fig. 1.34. The collimator lens ($f_{coll} = 1.14$ m) then acted as a camera. This modified spectrograph could give high dispersions (up to 2 Å/mm at the K line),

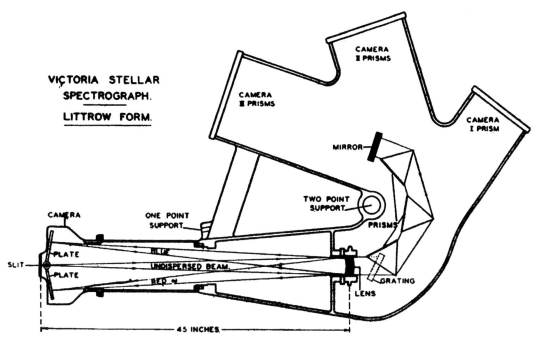

Figure 1.34. The Victoria spectrograph modified to become a high dispersion Littrow instrument. The photographic plates are mounted each side of the slit and the collimator lens also becomes the camera. This Littrow instrument was used at Victoria from 1946.

which was an important consideration at Victoria, as the 72-inch telescope had no coudé facility. The spectrograph could also use blazed Wood gratings as the dispersing element.

Camera design to achieve flat fields and accept a wide angular field, both essential for good wavelength coverage, was a major issue. Plaskett experimented with several new cameras using triplet lens designs by Moffitt, and also a Ross quadruplet [227, 231]. He had earlier tested a range of ten commercially available photographic lenses for spherical aberration, field flatness and angular width of field [232]. Compound lenses, such as the Zeiss 'Chromat' (as used by Hartmann) and the Brashear light crown, performed well among the longer focal length cameras for higher dispersion with three prisms. For faster lenses the Zeiss Tessar and Ross Special Homocentric were the top performers. The last named operated at f/5.6, gave good definition over 8° with a focal length of 254 mm. The angular field was four times that commonly achieved by cameras in the late nineteenth century.

For nebular spectroscopy, in particular for measuring the redshifts of ever fainter galaxies, faster cameras

were imperative so as to allow wider slits, yet they had to minimize coma and spherical aberration. William Rayton at the Bausch and Lomb Optical Company designed an f/0.6 camera based on a much enlarged microscope objective [233]. This was incorporated into the Cassegrain nebular spectrograph on the Mt Wilson 100-inch telescope used by Hubble and Humason for their pioneering galaxy work [234]. The dispersion was only 418 Å/mm at Hγ with this two-prism spectrograph. The ratio f_{coll}/f_{cam} was 19, allowing for wide slits to be used, the slit image being greatly demagnified on the photographic plate.

One of the last large Cassegrain prism spectrographs was commissioned in 1951 at the 74-inch Radcliffe Observatory reflector in South Africa. The design, however, went back to the 1930s, but construction was delayed by the war [235, 236]. This two-prism instrument had four cameras for dispersions between 22 and 96 Å/mm and a collimator focal length of over one metre (1153 mm), possibly the largest of any Cassegrain spectrograph yet built. It was designed to undertake radial-velocity measurements of the much neglected southern stars.

REFERENCES

[1] Newton, I., *Phil. Trans. R. Soc.* **6** (no. 80), 3075 (1672)

[2] Newton, I., *Optical Lectures, University of Cambridge, 1669*, London: F. Fayram (1728)

[3] Young, T., *Phil. Trans. R. Soc.* **92**, 12 (1802)

[4] Fraunhofer, J., *Denkschr. der königl. Acad. zu München für 1821 und 1822*, **8**, 1 (1822)

[5] Fraunhofer, J., *Astron. Nachrichten* **1**, 295 (1822)

[6] Fraunhofer, J., *Gilberts Ann. der Phys.* **74**, 337 (1823)

[7] Fraunhofer, J., *Denkschr. der Münch. Akad. der Wiss.* **5**, 193 (1817); also in *Gilberts Ann. der Phys.* **56**, 264 (1817)

[8] Fraunhofer, J., *Edin. Phil. J.* **9**, 296 (1823)

[9] Fraunhofer, J., *Edin. Phil. J.* **10**, 26 (1824)

[10] Wollaston, W., *Phil. Trans. R. Soc.* **92**, 365 (1802)

[11] Arago, F., *Comptes Rendus de l'Acad. Sci., Paris* **8**, 710 (1839)

[12] Simms, W., *Mem. R. astron. Soc.* **11**, 165 (1840)

[13] Swan, W., *Phil. Mag.* **11**(4), 448 (1856)

[14] Kirchhoff, G., *Abhandl. der Berliner Akad. Part I*, p. 63 (1861); p. 227 (1862)

[15] Kirchhoff, G., *Abhandl. der Berliner Akad. Part II*, p. 225 (1863)

[16] Donati, G. B., *Nuovo Cim.* **15**, 292 (1862)

[17] Donati, G. B., *Mon. Not. R. astron. Soc.* **23**, 100 (1863)

[18] von Steinheil, C. A., *Astron. Nachrichten* **59**, 253 (1963)

[19] Huggins, W. and Miller, W. A., *Phil. Trans. R. Soc.* **154**, 413 (1864)

[20] Huggins, W. and Miller, W. A., *Phil. Trans. R. Soc.* **154**, 437 (1864)

[21] Janssen, J., *Comptes Rendus de l'Acad. Sci., Paris* **55**, 576 (1862)

[22] Secchi, A., *Astron. Nachrichten* **59**, 193 (1863)

[23] Secchi, A., *Comptes Rendus de l'Acad. Sci., Paris* **63**, 364 (1866)

[24] Secchi, A., *Comptes Rendus de l'Acad. Sci., Paris* **64**, 738 (1867)

[25] Secchi, A., *Les Etoiles*, Paris: Librairie Germer Baillière et Co. (1879). See p. 74.

[26] Secchi, A., *Comptes Rendus de l'Acad. Sci., Paris* **69**, 1053 (1869)

[27] Rutherfurd, L. M., *Amer. J. Sci. Arts* **35**, 71 (1863)

[28] Rutherfurd, L. M., *Amer. J. Sci. Arts* **35**, 407 (1863)

[29] Rutherfurd, L. M., *Amer. J. Sci. Arts* **39**(2), 129 (1865)

[30] Bowen, I. S., in *Astronomical Techniques*, ed. W. A. Hiltner, Chapter 2: Spectrographs, p. 34. University of Chicago Press (1962)

[31] Airy, G. B., *Mon. Not. R. astron. Soc.* **23**, 188 (1863)

[32] Rutherfurd, L. M., *Amer. J. Sci. Arts* **36**, 154 (1863)

[33] Kirchhoff, G. and Bunsen, R., *Poggendorfs Ann. der Phys.* **110**, 160 (1860)

[34] Kirchhoff, G. and Bunsen, R., *Phil. Mag.* **20**(4), 89 (1861) and **22**(4), 329 (1861)

[35] Fizeau, H. and Foucault, L., *Comptes Rendus de l'Acad. Sci., Paris* **26**, 680 (1848)

[36] Huggins, W., *Phil. Trans. R. Soc.* **154**(II), 139 (1864)

[37] Browning, J., *Mon. Not. R. astron. Soc.* **30**, 198 (1870)

[38] Proctor, R. A., *Mon. Not. R. astron. Soc.* **30**, 215 (1870)

[39] Browning, J., *Mon. Not. R. astron. Soc.* **30**, 214 (1870)

[40] Browning, J., *Mon. Not. R. astron. Soc.* **32**, 213 (1872)

[41] Browning, J., *Mon. Not. R. astron. Soc.* **33**, 410 (1873)

[42] Lockyer, J. N., *Proc. R. Soc.* **17**, 131 (1869)

[43] Lockyer, J. N., *Phil. Trans. R. Soc.* **159**, 425 (1869)

[44] Kayser, H., *Handbuch der Spectroscopie*, vol. 1, Leipzig: Verlag S. Hirzel (1900)

[45] von Littrow, O., *Sitzungsber. der Wiener Akad. der Wiss.* **47**(II), 26 (1863)

[46] Grubb, H., *Mon. Not. R. astron. Soc.* **31**, 36 (1870)

[47] Cornu, A., *J. de Phys., Paris* **2**(2), 53 (1883)

[48] Thollon, L., *Comptes Rendus de l'Acad. Sci., Paris* **86**, 329 (1878)

[49] Newall, H. F., *Proc. Cambridge Phil. Soc.* **8**, 138 (1894)

[50] Newall, H. F., *Astron. & Astrophys.* **13**, 309 (1894)

[51] Osterbrock, D. E., private communication to the author, 17 October (1996)

[52] Wadsworth, F. L. O., *Astrophys. J.* **2**, 264 (1895)

[53] Wadsworth, F. L. O., *Astron. & Astrophys.* **13**, 835 (1894); also in *Phil. Mag.* **38**(5), 337 (1894)

[54] Wadsworth, F. L. O., *Astrophys. J.* **1**, 232 (1895)

[55] Mousson, A., *Arch. sci. phys. (Genève)* **10**(2), 221 (1861)

[56] Merz, S., *Phil. Mag.* **41**(4), 129 (1871)

[57] Krüss, H., *Rep. für phys. Tech.* **18**, 217 (1882)

[58] Wadsworth, F. L. O., *Amer. J. Sci. Arts* **48**(3), 19 (1894)

[59] Wadsworth, F. L. O., *Astron. & Astrophys.* **13**, 527 (1894)

[60] Bravais, A., *J. école polytech.* **18**(30), 77 (1845)

[61] Bravais, A., *J. école polytech.* **18**(31), 1 (1847)

[62] Ditscheiner, L., *Poggendorfs Ann. der Phys.* **129**, 336 (1866)

[63] Grubb, T., *Proc. R. Soc.* **22**, 308 (1874); also in *Phil. Mag.* **48**(4), 532 (1874)

[64] Lockyer, J. N. and Seabroke, G. M., *Proc. R. Soc.* **21**, 105 (1873)

[65] Hartmann, J., *Zeitschr. für Instr. Kunde* **20**, 47 (1900)

[66] Kirchhoff, G. and Bunsen, R., *Zeitschr. für analyt. Chemie* **1**, 139 (1862)

[67] Browning, J., *Mon. Not. R. astron. Soc.* **30**, 71 (1870)

[68] Procter, H., *Nature* **6**, 473 (1872)

[69] Herschel, A., *Nature* **18**, 300 (1878)

[70] de Thierry, M., *Comptes Rendus de l'Acad. Sci., Paris* **101**, 811 (1885)

[71] Huggins, W., *Proc. R. Soc.* **19**, 317 (1871)

[72] Browning, J., *Mon. Not. R. astron. Soc.* **33**, 411 (1873)

[73] Vogel, H. C., *Zeitschr. für Instr. Kunde* **1**, 391 (1881)

[74] Herschel, J., *Phil. Trans. R. Soc.* **130**, 1 (1840)

[75] Becquerel, E., *Biblio. univers. de Genève* **40**, 341 (1842)

[76] Draper, J. W., *Phil. Mag.* (3) **22**, 360 (1843)

[77] Foucault, L. and Fizeau, H., *Comptes Rendus de l'Acad. Sci., Paris* **23**, 679 (1846)

[78] Draper, H., *Amer. J. Sci. Arts* (3) **13**, 95 (1877)

[79] Draper, H., *Amer. J. Sci. Arts* (3) **18**, 419 (1879)

[80] Draper, H., *Amer. J. Sci.* **22**, 134 (1881)

[81] Draper, H., *Amer. J. Sci.* **23**, 339 (1882)

[82] Huggins, W., *Proc. R. Soc.* **25**, 445 (1877)

[83] Huggins, W., *Phil. Trans. R. Soc.* **171**, 669 (1880)

[84] Huggins, W., *Astron. & Astrophys.* **12**, 615 (1893)

[85] Huggins, W., *Astrophys. J.* **1**, 359 (1895)

[86] Huggins, W. and Huggins, M., *Publ. Sir Wm. Huggins Observ.* **1**, 1 (1899). London: W. Wesley & Son, see Chapter IV

[87] Hartmann, J., *Zeitschr. für Instr. Kunde* **24**, 257 (1904)

[88] Huggins, W., *Astrophys. J.* **5**, 8 (1896)

[89] Vogel, H. C., *Publik. der astrophysik. Observ. Potsdam* **7**, 1 (1892)

[90] Vogel, H. C., *Mon. Not. R. astron. Soc.* **52**, 87 (1891)

[91] Vogel, H. C., *Mon. Not. R. astron. Soc.* **52**, 541 (1892)

[92] Campbell, W. W., *Astrophys. J.* **8**, 123 (1898)

[93] Frost, E. B., *Astrophys. J.* **15**, 1 (1902)

[94] Deslandres, H. A., *Bull. Astron.* **15**, 49 (1898)

[95] Campbell, W. W., *Astrophys. J.* **11**, 259 (1900)

[96] Hartmann, J., *Astrophys. J.* **15**, 172 (1902)

[97] Keeler, J. E., *Astrophys. J.* **1**, 101 (1895)

[98] Hartmann, J., *Astrophys. J.* **11**, 400 (1900)

[99] Keeler, J. E., *Astrophys. J.* **1**, 248 (1895)

[100] Schuster, A., *Astrophys. J.* **21**, 197 (1905)

[101] Campbell, W. W. and Moore, J. H., *Publ. Lick Observ.* **16**, 1 (1928)

[102] Rittenhouse, D., *Trans. Amer. Phil. Soc.* (II), 201 (1786)

[103] Norbert, F. A., *Poggendorfs Ann. der Phys.* **67**, 173 (1846)

[104] Norbert, F. A., *Poggendorfs Ann. der Phys.* **85**, 80 and 83 (1852)

[105] Peirce, C. S., *Amer. J. Math.* **2**, 330 (1879)

[106] Peirce, C. S., *Nature* **24**, 262 (1881)

[107] Kurlbaum, F., *Wiedemanns Ann. der Phys.* **33**, 159 (1888)

[108] Rayleigh, Lord, *Proc. R. Soc.* **20**, 414 (1872)

[109] Rayleigh, Lord, *Phil. Mag.* **47**(4), **81**, 193 (1874)

[110] Rayleigh, Lord, *Phil. Mag.* **44**(4), 392 (1872)

[111] Rayleigh, Lord, *Nature* **54**, 332 (1896)

[112] Rowland, H. A., *Phil. Mag.* **13**(5), 469 (1882); also in *Nature* **26**, 211 (1882)

[113] Ames, J. S., *Astron. & Astrophys.* **11**, 28 (1892)

[114] Ames, J. S., *Astrophys. J.* **23**, 348 (1906)

[115] Michelson, A. A., *Proc. Amer. Phil. Soc.* **54**(217), 137 (1915)

[116] anon., *Nature* **96**, 154 (1915)

[117] Harrison, G. R. and Stroke, G. W., *J. Opt. Soc. Amer.* **45**, 112 (1955)

[118] Rowland, H. A., *Astron. & Astrophys.* **12**, 129 (1893); also in *Phil. Mag.* (5) **35**, 397 (1893)

[119] Anderson, J. A. and Sparrow, C. M., *Astrophys. J.* **33**, 338 (1911)

[120] Wood, R. W., *Phil. Mag.* **20**, 770 (1910)

[121] Trowbridge, A. and Wood, R. W., *Phil. Mag.* **20**, 886 (1910)

[122] Babcock, H. D. and Babcock, H. W., *J. Opt. Soc. Amer.* **41**, 776 (1951)

[123] Merrill, P. W., *Astrophys. J.* **74**, 188 (1931)

[124] Thorp, T., *Pop. Astron.* **14**, 93 (1906); also in *Brit. J. Photog.* (Dec 1905)

[125] Wallace, R. J., *Astrophys. J.* **22**, 123 (1905); see also anonymous review in *Nature* **73**, 21 (1905)

[126] Wallace, R. J., *Astrophys. J.* **23**, 96 (1906)

[127] Anderson, J. A., *Astrophys. J.* **31**, 171 (1910)

[128] Keil, E., *Pop. Astron.* **41**, 378 (1933)

[129] Wood, R. W., *Nature* **140**, 723 (1937)

[130] Wood, R. W., *Publ. Amer. Astron. Soc.* **10**, 218 (1941)

[131] Merton, T., *Proc. R. Soc.* **201A**, 187 (1950)

[132] Sayce, L. A., *Endeavour* **12**, 210 (1953)

[133] Jarrell, R. F. and Stroke, G. W., *Appl. Optics* **3**, 1251 (1964)

[134] Strong, J., *Publ. astron. Soc. Pacific* **46**, 18 (1934)

[135] Beals, C. S. and McKellar, A. M., *J. R. astron. Soc. Canada* **32**, 369 (1938)

[136] Butler, C. P. and Stratton, F. J. M., *Nature* **134**, 810 (1934)

[137] Palmer, C., *Diffraction Grating Handbook*, Rochester, NY: Richardson Grating Lab., 4th edn. (2000)

[138] Harrison, G. R., *J. Opt. Soc. Amer.* **39**, 522 (1949)

[139] Harrison, G. R. and Thompson, S. W., *J. Opt. Soc. Amer.* **60**, 591 (1970)

[140] Burch, J. M., *Research* **13**, 2 (1960)

[141] Rudolph, D. and Schmahl, G., *Umschau Wiss.* **67**, 225 (1967)

[142] Labeyrie, A. and Flamand, J., *Opt. Commun.* **1**, 5 (1969)

[143] Schmahl, G. and Rudolph, D., in *Auxiliary Instrumentation for Large Telescopes*, ed. S. Laustsen and A. Reiz, Geneva:ESO/CERN, p. 209 (1972)

[144] Barden, S. C., Arns, J. A. and Colburn, W. S., *Proc. Soc. Photo-instrumentation Engineers (SPIE)* **3355**, 866 (1998)

[145] Barden, S. C., Arns, J. A., Colburn, W. S. and Williams, J. B., *Publ. astron. Soc. Pacific* **112**, 809 (2000)

[146] Blanche, P.-A., Habraken, S. L., Lemaire, P. C., Jamer, C. A. J., *Proc. Soc. Photo-instrumentation Engineers (SPIE)* **4842**, 31 (2003)

[147] Robertson, J. G., Taylor, K., Baldry, I. K., Gillingham, P. R. and Barden, S. C., *Proc. Soc. Photo-instrumentation Engineers (SPIE)* **4008**, 194 (2000)

[148] Bernstein, G. M., Athey, A. E., Bernstein, R. *et al.*, *Proc. Soc. Photo-instrumentation Engineers (SPIE)* **4485**, 453 (2002)

[149] Oka, K., Klaus, W., Fujino, M. *et al.*, *Proc. Soc. Photo-instrumentation Engineers (SPIE)* **4829**, 569 (2003)

[150] Hill, G. J., Wolf, M. J., Tufts, J. R. and Smith, E. C., *Proc. Soc. Photo-instrumentation Engineers (SPIE)* **4842**, 1 (2003)

[151] Keeler, J. E., *Astron. & Astrophys.* **11**, 140 (1892)

[152] Keeler, J. E., *Astron. & Astrophys.* **12**, 40 (1893)

[153] Hale, G. E., *Astrophys. J.* **21**, 151 (1905)

[154] Hale, G. E., *Astrophys. J.* **27**, 204 (1908)

[155] Hale, G. E., *Publ. astron. Soc. Pacific* **24**, 223 (1912)

[156] Loomis, F. A. and Kistiakowsky, G. B., *Rev. Sci. Inst.* **3**, 201 (1932)

[157] Ebert, H., *Wiedemanns Ann. der Phys.* **38**, 489 (1889)

[158] Fastie, W. G., *J. Opt. Soc. Amer.* **42**, 641 (1952)

[159] Fastie, W. G., *J. Opt. Soc. Amer.* **42**, 647 (1952)

[160] Czerny, M. and Turner, A. F., *Zeitschr. für Physik* **61**, 792 (1930)

[161] McMath, R. R., *Astrophys. J.* **123**, 1 (1956)

[162] Rowland, H. A., *Amer. J. Sci.* (3) **26**, 87 (1883); *Phil. Mag.* (5) **16**, 197 (1883)

[163] Ames, J. S., *Phil. Mag.* (5) 27, 369 (1889); also in *Astron. & Astrophys.* **11**, 28 (1892)

[164] Runge, C. R. and Mannkopff, R., *Zeitschr. für Physik* **45**, 13 (1927)

[165] Pettit, E., *Publ. astron. Soc. Pacific* **43**, 75 (1931)

[166] Beutler, H. G., *J. Opt. Soc. Amer.* **35**, 311 (1945)

[167] Abney, W. de W., *Phil. Trans. R. Soc.* **177** (II), 457 (1886)

[168] Runge, C. R. and Paschen, F., *Abhandl. Akad. Wiss. Berlin*, Anhang 1 (1902)

[169] Eagle, A., *Astrophys. J.* **31**, 120 (1910)

[170] Sawyer, R. A., *Experimental Spectroscopy*, New York: Prentice Hall (1944)

[171] Wadsworth, F. L. O., *Astrophys. J.* **3**, 47 (1896)

[172] Fabry, C. and Buisson, H., *J. de Phys.* (4) **9**, 940 (1910)

[173] Runge, C. R. and Paschen, F., *Ann. der Phys.* **61**, 641 (1897)

[174] Seya, M., *Sci. of Light (Tokyo)* **2**, 8 (1952)

[175] Namioka, T., *J. Opt. Soc. Amer.* **49**, 951 (1959)

[176] Liller, W., *Appl. Optics* **2**, 187 (1963)

[177] Crew, H., *Astron. & Astrophys.* **12**, 156 (1893)

[178] Osterbrock, D. E., *J. Hist. Astron.* **17**, 119 (1986)

[179] Müller, J., *Poggendorfs Ann. der Phys.* **118**, 641 (1863)

[180] Mascart, E., *Comptes Rendus de l'Acad. Sci., Paris* **56**, 138 (1863)

[181] Ångström, A. J., *Recherches sur le spectre normal du soleil*, Upsala: W. Schultz (1868)

[182] Young, C. A., *Amer. J. Sci. Arts* (3) **12**, 321 (1876)

[183] Rowland, H. A., *Astrophys. J.* **1**, 29, 131, 222, 295 and 377 (1895)

[184] Hale, G. E., *Astrophys. J.* **16**, 211 (1902)

[185] Hale, G. E., *Astrophys. J.* **23**, 11 (1906)

[186] Adams, W. S., *Astrophys. J.* **24**, 69 (1906)

[187] Schlesinger, F., *Publ. Allegheny Observ.* 3(13), 99 (1914)

[188] Secchi, A., *l'Astronomie in Roma*, p. 27 (1877)

[189] Vogel, H. C., *Zeitschr. für Instr. Kunde* **1**, 47 (1881)

[190] Keeler, J. E., *Publ. Lick Observ.* **3**, 161 (1894)

[191] Poor, C. L. and Mitchell, S. A., *Astrophys. J.* **7**, 157 (1898)

[192] Wadsworth, F. L. O., *Astrophys. J.* **7**, 198 (1898)

[193] Mitchell, S. A., *Astrophys. J.* **10**, 29 (1899)

[194] Campbell, W. W. and Albrecht, S., *Lick Observ. Bull.* **6**, 11 (1910)

[195] Plaskett, J. S., *J. R. astron. Soc. Canada* **6**, 290 (1912)

[196] Plaskett, J. S., *Astrophys. J.* **37**, 373 (1913)

[197] Plaskett, J. S., *Publ. Dominion Observ. Ottawa* 1(7), 171 (1914)

[198] Adams, W. S., *Mt Wilson Observ. Ann. Rep. 1930–31.* Carnegie Yearbook **30**, 171 (1932)

[199] Merrill, P. W., *Astrophys. J.* **79**, 183 (1934)

[200] Adams, W. S., *Astrophys. J.* **93**, 11 (1941)

[201] Wilson, O. C., *Publ. astron. Soc. Pacific* **68**, 346 (1956)

[202] Page, T. L., *Astrophys. J.* **116**, 63 (1956)

[203] Dunham, T., *Vistas Astron.* **2**, 1223 (1956)

[204] Pierce, A. K., McMath, R. R. and Mohler, O., *Astron. J.* **56**, 137 (1951)

[205] Kopylov, I. M. and Steshenko, N. V., *Izvestia Crimean Astrophys. Observ.* **33**, 308 (1965)

[206] Fujita, Y., *Vistas Astron.* **7**, 71 (1965)

[207] Liller, W., *Appl. Optics* **9**, 2332 (1970)

[208] Butcher, H. R., *Publ. Astron. Soc. Australia* **2**, 21 (1971)

[209] Tousey, R., Purcell, J. D. and Garrett, D. L., *Appl. Optics* **6**, 365 (1967)

[210] Boland, B. C., Burton, W. M., Jones, B. B. and Reay, N. K., *Intl. Astron. Union Symp.* **41**, 254 (1971)

[211] Schroeder, D. J., *Appl. Optics* **6**, 1976 (1967)

[212] Hearnshaw, J. B., *Intl. Astron. Union Symp.* **118**, 371 (1986)

[213] Barden, S. C., Ramsey, L. W., Huenemoerder, D. P. and Buzasi, D., *Bull. Amer. Astron. Soc.* **19**, 1099 (1987)

[214] Plaskett, J. S., *Astrophys. J.* **40**, 127 (1914)

[215] Plaskett, J. S., *Publ. Dominion Observ. Ottawa* **1**, 171 (1914)

[216] Plaskett, J. S., *Astrophys. J.* **49**, 209 (1919)

[217] Guthnick, P., *Sitzungsber. Preuss. Akad. der Wiss.* **1**, 3 (1930)

[218] Wright, W. H., *Lick Observ. Bull.* **9**, 52 (1917)

[219] Plaskett, J. S., *Pop. Astron.* **31**, 20 (1923)

[220] Wright, W. H., *Publ. Lick Observ.* **13**, 191 (1918)

[221] Schlesinger, F., *Publ. Allegheny Observ.* **2**, 1 (1910)

[222] Schlesinger, F., *Publ. Allegheny Observ.* **2**, 197 (1913)

[223] Adams, W. S., *Astrophys. J.* **35**, 163 (1912)

[224] Adams, W. S., *Astrophys. J.* **33**, 64 (1911)

[225] Curtiss, R. H., *Publ. Observ. Univ. Michigan* **1**, 37 (1912)

[226] Beals, C. S., Petrie, R. M. and McKellar, A., *J. R. astron. Soc. Canada* **40**, 349 (1946)

[227] Plaskett, J. S., *Pop. Astron.* **31**, 659 (1923)

[228] Herrmann, W. and Brück, H., *Zeitschr. für Instr. Kunde* **56**, 459 (1936)

[229] Moffitt, G. W., *Contrib. McDonald Observ.* **1**, 74 (1943)

[230] van Biesbroeck, G., *Contrib. McDonald Observ.* **1**, 103 (1943)

[231] Plaskett, J. S., *Astrophys. J.* **59**, 65 (1924)

[232] Plaskett, J. S., *Astrophys. J.* **29**, 290 (1909)

[233] Rayton, W. B., *Astrophys. J.* **72**, 59 (1930)

[234] Humason, M. L., *Astrophys. J.* **71**, 351 (1930)

[235] Thackeray, A. D., *Mon. Not. astron. Soc. S. Africa* **10**, 29 (1951)

[236] Jackson, J., *Nature* **167**, 169 (1951)

2 · The theory of spectroscopes and spectrographs

2.1 GENERAL PROPERTIES OF A SPECTROGRAPH

A spectrograph is an instrument that receives light from a source, disperses the light according to its wavelength into a spectrum, and focusses the spectrum onto a detector, which records the spectral image. In the astronomical case the source might be a star or galaxy, and the light will first be collected by a telescope. Many telescopes produce an image of the source on a spectrograph slit (although slitless objective prism instruments are also possible). After the slit, a collimator renders the rays almost parallel, and a dispersing element, usually a grating or a prism, sends photons of different wavelengths into different directions. A camera then records a continuous succession of monochromatic slit images, each displaced in the dispersion direction according to its wavelength. This array of slit images constitutes the spectrum.

The simplest possible slit spectrograph (Fig. 2.1) therefore comprises a slit, a collimator (either a mirror or a lens system), a dispersing element (typically a grating or a prism) and a camera (again a mirror or lens system) and finally a detector (perhaps a charge-coupled device or CCD, perhaps a photographic plate, but in early instruments it was the human eye in conjunction with an eyepiece).

2.2 THE SPECTROGRAPH FIGURE OF MERIT

The figure of merit of a given spectrograph is a parameter indicating the rate at which that instrument is able to acquire spectral information. Two critical factors that make up the figure of merit are the throughput and resolving power.

The throughput in practice always depends on wavelength. It can be defined as the fraction of the source photons in the focal plane of the telescope that are recorded by the detector. Not all photons delivered by the telescope are recorded in the spectrograph, as some are wasted through unwanted reflection, absorption, scattering or vignetting. Therefore throughput is a measure of spectrograph efficiency in not wasting the light it receives.

The concepts of throughput and resolving power as the most important parameters of a spectrometer were emphasized by the French physicist Pierre Jacquinot (1910–2002) in 1954 [1]. He wrote:

> Every monochromator or spectrometer can be characterized by two fundamental quantities: 1. The resolving power $R = \lambda/\delta\lambda$ ... 2. The luminosity, that is to say the value of the energy or of the flux collected by the receiver ... [1].

The resolving power is a measure of the spectral detail recorded in the spectrum. Lord Rayleigh in 1879 [2] considered the concepts of wavelength resolution and resolving power. They are terms frequently interchanged, but whose meanings are quite distinct. The wavelength resolution is the smallest separation, $\delta\lambda$, of two 'monochromatic' spectral lines that can just be resolved as two by a given spectrograph. Even with an infinitesimally narrow slit, Rayleigh showed that diffraction effects in the dispersing element inevitably result in finite wavelength resolution.

Rayleigh considered the intensity distribution, $I(\theta)$, of the diffraction pattern from a rectangular aperture, A, which is the well-known sinc^2 function:

$$I(\theta) \propto \left[\frac{\sin\left(\frac{\pi a \sin\theta}{\lambda}\right)}{\frac{\pi a \sin\theta}{\lambda}} \right]^2. \qquad (2.1)$$

The first minima each side of the principal maximum occur at angles $\theta \simeq \sin\theta = \lambda/A$. Rayleigh's criterion for marginal resolution is that the principal maximum of the line at wavelength λ shall just coincide with the first minimum of the line at $\lambda + \delta\lambda$, if the lines are just

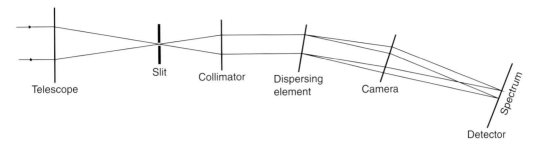

Figure 2.1. Schematic diagram of an astronomical slit spectrograph showing the various optical elements.

resolved. Rayleigh's criterion is an arbitrary one, but leads to a simple expression for resolving power. Even though it gives a line separation slightly greater than the true resolution limit, his expression is generally adopted.

The resolving power is a dimensionless quantity, $\lambda/\delta\lambda$, where $\delta\lambda$ is the wavelength resolution. The angular dispersion of the dispersing element is $\frac{d\theta}{d\lambda}$, where θ is the angle of deviation of a prism or angle of diffraction of a grating. Hence the angular separation of two monochromatic lines separated at the resolution limit is $\delta\theta = \frac{d\theta}{d\lambda}\delta\lambda$ and hence, for a dispersing element of aperture A,

$$R = A\frac{d\theta}{d\lambda}. \qquad (2.2)$$

Lord Rayleigh [2] derived this simple expression, at first explicitly for gratings, but then also for prism spectroscopes. He concluded: 'a double line cannot be fairly resolved unless its components subtend an angle exceeding that subtended by the wavelength of light at a distance equal to the horizontal aperture.'

Rayleigh's expression for the resolving power is of great generality, as it makes no assumptions concerning the nature of the dispersing element, nor concerning the design parameters of the spectrograph other than A and $\frac{d\theta}{d\lambda}$. It represents the diffraction-limited resolving power.

In practice, the resolving power will normally, for astronomical spectrographs, be slit-limited. It is also possible, but undesirable, for R to be detector-limited. Slit-limited resolving power is normal in astronomical work, because a wider slit admits more light for a given object, and hence increases throughput. The image in the camera focal plane of a uniformly illuminated slit will be the convolution of a square box function with the sinc^2 diffraction pattern (see Fig. 2.2). For wide

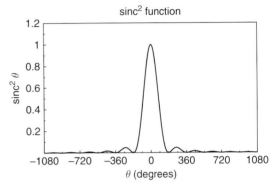

Figure 2.2. Plot of the sinc^2 function that plays a key role in diffraction theory and the resolving power of spectrographs.

slits, the width of this will be essentially that of the box, which in turn is determined by the slit width, w.

The figure of merit can be defined as the product of throughput, T, and resolving power, R, or $M = RT$. Needless to say, other performance parameters (such as wavelength coverage) may also be of importance. Suppose one is observing a star whose image produced by the telescope is spread over a finite diameter of characteristic angular size $\theta_\star$ by the effects of atmospheric seeing. The slit width in angular measure is $\theta_s = w/f_{tel}$, where f_{tel} is the telescope's effective focal length ($f_{tel} = A_{tel}F_{tel}$, where A_{tel} is the telescope's aperture and F_{tel} is the focal ratio of the rays converging to the focal plane). If $\theta_s < \theta_\star$, which is normally the case, then the throughput T is approximately proportional to θ_s. In this case, $R\theta_s$ can be taken as a relative figure of merit. Such an expression for M was given by Bingham (1940–2005) [3].

Another possible factor influencing spectrograph figure of merit is the total wavelength coverage, $\Delta\lambda$, of one exposure. Bingham [3] has also defined a second figure of merit, which is essentially $\Delta\lambda RT$, which is proportional to $N\theta_s$, where N is the number of wavelength resolution elements in the recorded range ($N = \Delta\lambda/\delta\lambda$). Whether this is a better figure of merit than just RT depends on the application; for some purposes, such as stellar abundance analysis, large $\Delta\lambda$ may be desirable, so that the lines of many chemical elements can be simultaneously recorded. At other times only one line may be of interest, and large wavelength coverage becomes unnecessary.

2.3 RESOLVING POWER AND THROUGHPUT OF PRISM INSTRUMENTS

Prisms are rarely used as the primary dispersing elements in modern spectrographs, because large prisms have high absorption losses and all prisms give a non-linear dispersion. The angular dispersion at longer red wavelengths is generally small unless large absorption losses are also tolerated, so the figure of merit is low, at least in the red. Historically, however, prism spectrographs played a major rôle in the development of astrophysics in the nineteenth and first third of the twentieth centuries. The elementary theory of their use as dispersing elements is therefore given here.

2.3.1 Dispersive properties of prisms and the diffraction-limited resolving power

For the prism in Fig. 2.3, α is the apex angle and t is the base length. A ray is shown with angles of incidence or refraction of i_1, i_2, r_1 and r_2, all relative to their local normals. The angle of deviation is θ given by

$$\theta = (i_1 - r_1) + (i_2 - r_2). \tag{2.3}$$

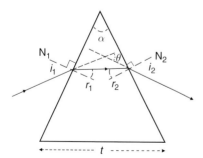

Figure 2.3. Diagram of a prism of apex angle α and base t, showing various angles a light ray makes with the face normals.

In addition

$$\alpha = (r_1 + r_2) \tag{2.4}$$

so

$$\theta + \alpha = (i_1 + i_2) \tag{2.5}$$

and Snell's law of refraction gives

$$\sin i_1 = n \sin r_1 \tag{2.6}$$

$$\sin i_2 = n \sin r_2 \tag{2.7}$$

for a prism of refractive index n used in air.

The deviation θ is a minimum when when $i_1 = i_2 = i$ and $r_1 = r_2 = r$, in which case

$$i = \frac{1}{2}(\alpha + \theta) \tag{2.8}$$

and $r = \alpha/2$, giving

$$n = \frac{\sin i}{\sin r} = \frac{\sin \frac{1}{2}(\alpha + \theta)}{\sin \frac{1}{2}\alpha}. \tag{2.9}$$

Thus minimum deviation occurs whenever a ray traverses the prism symmetrically, parallel to the base. The proof of this follows in two steps. In general it can be shown that

$$n = \frac{\sin \frac{1}{2}(\alpha + \theta)}{\sin \frac{1}{2}\alpha} \frac{\cos \frac{1}{2}(i_1 - i_2)}{\cos \frac{1}{2}(r_1 - r_2)}. \tag{2.10}$$

Next it is seen that the second factor is always less than unity if $i_1 \neq i_2$, because $|i_1 - i_2| > |r_1 - r_2|$, regardless of whether $i_1 > i_2$ or $i_1 < i_2$. Hence

$$n < \frac{\sin \frac{1}{2}(\alpha + \theta)}{\sin \frac{1}{2}\alpha} \tag{2.11}$$

if the ray is not at minimum deviation. That requires that $\theta > \theta_{min}$ if $i_1 \neq i_2$.

Incidentally, the concept of minimum deviation was known to Newton at the time of his famous experiments on the dispersion of white sunlight [4], and Joseph Fraunhofer used the relationship between refractive index and the minimum deviation to determine refractive indices of different glasses for spectral lines of different wavelength [5].

The angular dispersion of a prism is

$$\frac{d\theta}{d\lambda} = \frac{d\theta}{dn} \frac{dn}{d\lambda} \tag{2.12}$$

and at minimum deviation this becomes

$$\frac{d\theta}{d\lambda} = \frac{2\sin\frac{\alpha}{2}}{\sqrt{(1 - n^2\sin^2\frac{\alpha}{2})}}\frac{dn}{d\lambda}. \qquad (2.13)$$

The second factor, $\frac{dn}{d\lambda}$, is the dispersion of the glass, and is purely a property of the material. Since at minimum deviation

$$\frac{d\theta}{dn} = \frac{2\sin r}{\cos i}, \qquad (2.14)$$

an alternative expression for the angular dispersion at minimum deviation is

$$\frac{d\theta}{d\lambda} = \frac{2}{n}\tan i\frac{dn}{d\lambda} \qquad (2.15)$$

where the angle of incidence is wavelength dependent to achieve the minimum deviation condition.

Yet another formula uses the geometrical relationship between the beam diameter, A, filling the prism and the prism base dimension, t, which is

$$t = 2A\sin\frac{\alpha}{2}\sec i, \qquad (2.16)$$

from which one at once obtains

$$\frac{d\theta}{d\lambda} = \frac{t}{A}\frac{dn}{d\lambda} \qquad (2.17)$$

at minimum deviation. This immediately gives a useful expression for the diffraction-limited resolving power, using Equation 2.2:

$$R_{\text{diff}} = A\frac{d\theta}{d\lambda} = t\frac{dn}{d\lambda}, \qquad (2.18)$$

a result that was derived by Lord Rayleigh in 1879 [2]. This result is valid for any arbitrary angle of incidence, provided t is defined as the length of those rays in the glass that travel the nearest to the base.

In the case of a prism train, the total angular dispersion is simply the sum of the dispersions of the individual prisms:

$$\frac{d\theta}{d\lambda} = \sum_i \frac{d\theta_i}{d\lambda} = \sum_i \frac{d\theta_i}{dn_i}\frac{dn_i}{d\lambda} \qquad (2.19)$$

and for N prisms of the same glass and all used at minimum deviation

$$\frac{d\theta}{d\lambda} = \frac{1}{A}\frac{dn}{d\lambda}\sum_i t_i, \qquad (2.20)$$

while the diffraction-limited resolving power is

$$R_{\text{diff}} = \left(\sum t_i\right)\frac{dn}{d\lambda}. \qquad (2.21)$$

This therefore depends only on the glass and the sum of the base dimensions.

The conceptual simplicity of the dispersion of a prism belies the relative mathematical complexity of the analysis when the prism is not used in the symmetrical configuration of minimum deviation. The angular dispersion $\frac{d\theta}{d\lambda}$ also goes through a minimum as the angle of incidence i_1 is varied, a fact known to John Herschel [6]. The phenomenon was studied by A. Mousson [7] and an approximate solution for the position of minimum dispersion was derived by L. Thollon in 1879 [8]. Provided the apex angle α is not too large, minimum dispersion occurs when $r_1 \simeq n^2 r_2$, that is, for a larger angle of incidence i_1 than the position of minimum deviation. On the other hand, as i_1 increases further, $\frac{d\theta}{d\lambda}$ increases without limit, becoming infinite when the rays emerge at grazing angle ($i_2 = 90°$).

Since rays of different wavelength all have the same angle of incidence (i_1), it follows, from differentiating Snell's law for the first prism surface, that

$$0 = n\cos r_1\, dr_1 + \sin r_1\, dn \qquad (2.22)$$

and at the second prism surface

$$\cos i_2\, di_2 = \sin r_2\, dn + n\cos r_2\, dr_2. \qquad (2.23)$$

But $\alpha = (r_1 + r_2)$ so $dr_1 + dr_2 = 0$. Hence

$$\frac{d\theta}{d\lambda} = \frac{di_2}{dn}\frac{dn}{d\lambda} = \frac{d\theta}{dn}\frac{dn}{d\lambda} = \frac{\sin\alpha}{\cos r_1\cos i_2}\frac{dn}{d\lambda}. \qquad (2.24)$$

This is the general expression for prismatic angular dispersion. The denominator, $\cos r_1\cos i_2$, is a maximum when $r_1 \simeq n^2 r_2$, as shown by Thollon [8], giving the dispersion minimum.

However, angular dispersion of a prism spectrograph is not in itself a factor in the figure of merit, but the resolving power. Thus the diffraction-limited resolving power is $R_{\text{diff}} = A_1\frac{d\theta}{d\lambda}$, where A_1 is the refracted beam dimension in the prism's principal plane, as shown in Fig. 2.4. The incident beam, in the general case of departure from minimum deviation, has a dimension $A_0 \neq A_1$. Geometry gives

$$A_1 = \left(\frac{\cos r_1\cos i_2}{\cos i_1\cos r_2}\right)A_0 \qquad (2.25)$$

and hence the general diffraction-limited resolving power is

$$R_{\text{diff}} = A_0\left(\frac{\sin\alpha}{\cos i_1\cos r_2}\right)\frac{dn}{d\lambda}. \qquad (2.26)$$

Table 2.1. *Deviations, dispersions and resolving powers for a prism with $\alpha = 50°$, $n = 1.6$*

Angle of incidence $i_1(°)$	Angle of emergence $i_2(°)$	Deviation θ (°)	Dispersion $\frac{d\theta}{dn}$ (°)	Relative R for fixed A_0 $R \propto \frac{\sin \alpha}{\cos i_1 \cos r_2}$	Rel. R for given prism $R \propto \frac{1}{\cos r_1}$	Comments
90	18.30	58.30	59.24	∞	1.281	max. R
69.96	22.85	42.81	58.84	2.304	1.235	min. $\frac{d\theta}{dn}$
42.55	42.55	35.09	65.72	1.147	1.103	min. θ
22.85	69.96	42.81	132.00	1.027	1.031	min. R (fixed A_0)
18.30	90	58.30	∞	1.034	1.020	max. $\frac{d\theta}{dn}$

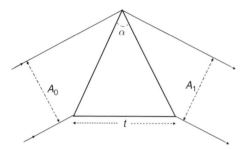

Figure 2.4. Diagram of a prism of apex angle α and base t, showing the beam sizes A_0 and A_1.

This has a different denominator from the angular dispersion equation (2.24). If one assumes that A_0 is constant and considers the dependence of the resolving power on the angle of incidence, then as i_1 increases, R_{diff} has a minimum value when ($\cos i_1 \cos r_2$) is a maximum, or when $r_2 \simeq n^2 r_1$. This minimum of resolving power occurs when i_1 is less than the value for minimum deviation, as Table 2.1 shows.

The maximum of resolving power is, from the table, at grazing incidence, because as i_1 increases, a larger prism is required for a fixed beam size, A_0, and the prism's dimensions and aperture, A_1, eventually become infinite as i_1 tends to $90°$. The reflection loss from the first surface also becomes 100 per cent at grazing incidence, so in fact no light enters the prism!

A more meaningful situation is to take a prism of fixed dimensions and to consider the diffraction-limited resolving power as a function of the angle of incidence.

$$R_{\text{diff}} = A_1 \left(\frac{\sin \alpha}{\cos r_1 \cos i_2} \right) \frac{dn}{d\lambda} = \frac{l \sin \alpha}{\cos r_1} \frac{dn}{d\lambda} \quad (2.27)$$

where l is the dimension of the second prism face which is filled with light. This results in

$$R_{\text{diff}} \simeq \left(\frac{t \cos \frac{\alpha}{2}}{\cos r_1} \right) \frac{dn}{d\lambda} \quad (2.28)$$

where the approximation arises because l is taken to be the dimension of the second prism face, even if not filled with light. This still reaches a maximum when r_1 is a maximum, which is for grazing incidence; but now R_{diff} is seen to vary only slowly with the angle of incidence. The maximum value of $1/\cos r_1$ is $(1 - \frac{1}{n^2})^{-\frac{1}{2}}$, which is 1.28 for a refractive index of $n = 1.6$, so the diffraction-limited resolving power for a given prism only increases by 28 per cent in going from normal to grazing incidence. On the other hand, the reflection losses increase dramatically over the same range, and the lower throughput makes use of high angles of incidence inefficient. Figure 2.5 shows how the deviation, dispersion and resolving power vary with the angle of incidence for a typical prism.

2.3.2 Slit-limited resolving power of prism spectrographs

Before discussing throughput further, it is necessary to evaluate the slit-limited resolving power for a spectrograph, because this is generally the mode in which an astronomical spectrograph is operated.

If the slit is of width w, then at the prism it subtends an angle $\delta i_1 = w/f_{\text{coll}}$, where f_{coll} is the collimator focal length. The small angle δi_1 can be thought of as the angular spread of nearly collimated rays arriving at the prism. By differentiating the Snell's law equations, $\sin i_1 = n \sin r_1$ and $\sin i_2 = n \sin r_2$, one obtains

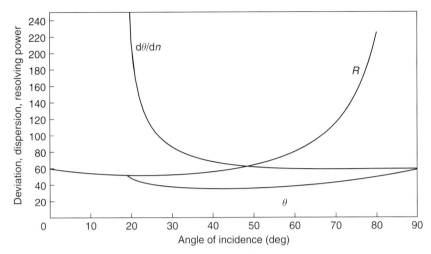

Figure 2.5. Plot of the variation of deviation θ, dispersion $\frac{d\theta}{dn}$, and resolving power R as a function of angle of incidence i_1 for a prism with apex angle $50°$ and refractive index $n = 1.6$.

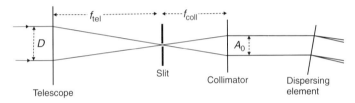

Figure 2.6. Schematic diagram of a telescope and spectrograph defining the parameters D, f_{tel}, f_{coll} and the beam size A_0.

$$\delta i_2 = -\left(\frac{\cos i_1 \cos r_2}{\cos r_1 \cos i_2}\right) \delta i_1 \qquad (2.29)$$

where the factor in brackets is known as the anamorphic magnification. For prisms at minimum deviation, this factor is unity. For the prism described in Table 2.1, the anamorphic magnification is 0.446 at the minimum of dispersion, and 2.243 at the minimum of resolving power at fixed aperture A_0 (ignoring the minus sign).

After passing through the prism, the monochromatic rays will have a spread in angular values δi_2, which degrades the wavelength resolution to a value

$$\delta\lambda = \delta i_2 \left/ \frac{d\theta}{d\lambda}\right. . \qquad (2.30)$$

Using Equation 2.24 for $\frac{d\theta}{d\lambda}$ results in

$$\delta\lambda = \left(\frac{\cos i_1 \cos r_2}{\sin \alpha}\right) \frac{w}{f_{coll}} \left/ \frac{dn}{d\lambda}\right. \qquad (2.31)$$

or

$$R_{slit} = \lambda/\delta\lambda = \left(\frac{\sin \alpha}{\cos i_1 \cos r_2}\right) \frac{\lambda f_{coll}}{w} \frac{dn}{d\lambda} = R_{diff} \frac{\lambda f_{coll}}{A_0 w}. \qquad (2.32)$$

Hence the slit-limited resolving power is a minimum for the same angle of incidence that minimizes the diffraction-limited resolving power, R_{diff}, provided A_0 is held constant. In general $R_{slit} \ll R_{diff}$ because $w \gg \lambda f_{coll}/A_0$, where A_0 is the aperture of the collimator. This can be seen because $w = \theta_s f_{tel}$, where θ_s is the angular size of the slit projected back onto the sky, and f_{tel} is the telescope's effective focal length. The diagram in Fig. 2.6 shows, by similar triangles, that

$$\frac{A_0}{D} = \frac{f_{coll}}{f_{tel}}. \qquad (2.33)$$

Therefore

$$R_{slit} = R_{diff} \frac{\lambda}{\theta_s D}. \qquad (2.34)$$

The factor λ/D is just the angular size of the central diffraction disk, or Airy disk, produced by the telescope of aperture D, which is designated here as θ_A (more accurately, $\theta_A = 1.2\lambda/D$ for a circular aperture), so

$$R_{\text{slit}} = R_{\text{diff}}\frac{\theta_A}{\theta_s}. \qquad (2.35)$$

For ground-based spectrographs atmospheric seeing leads to angular stellar image sizes of order of an arc second, so $\theta_\star \simeq 10^{-5}$ rad and for good throughput $\theta_s \simeq \theta_\star$, but (except for very small telescopes $D \leq 1$ cm) $\theta_s \gg \theta_A$. So throughput requirements dictate that spectrographs cannot efficiently operate in the diffraction-limited régime.

Equation 2.34 relating R_{slit} and R_{diff} is always valid, and since $R_{\text{diff}} = t\frac{dn}{d\lambda}$, it follows that

$$R_{\text{slit}} \simeq t\frac{dn}{d\lambda}\left(\frac{\lambda}{\theta_s D}\right). \qquad (2.36)$$

This is a useful relationship because it shows that for a prism of given dimensions, the slit-limited resolving power is essentially independent of the angle of incidence. In practice

$$R_{\text{slit}} \simeq t\left(\frac{\cos\frac{\alpha}{2}}{\cos r_1}\right)\frac{dn}{d\lambda}\left(\frac{\lambda}{\theta_s D}\right) \qquad (2.37)$$

is a better approximation if $r_1 \neq \alpha/2$.

It is noted in passing that three diffraction effects all come into play for very narrow slits at about the same slit width $w \simeq \lambda f_{\text{tel}}/D$, as was also pointed out by Bingham [3]. These are:

1. The slit width is about the diameter of the diffraction-produced Airy disk.
2. The dispersing element gives diffraction-limited resolving power, instead of slit-limited.
3. Diffraction effects at the slit first become important, causing some light to be diffracted beyond the geometrical cone of rays from the telescope, and hence to miss the collimator.

The expressions derived for the angular dispersion of a prism and for the resolving power of a prism spectrograph all contain the factor $\frac{dn}{d\lambda}$ for the dispersion of the glass.

2.3.3   Properties of glasses used in prisms

The refractive index of any glass depends on the glass type and on the wavelength. The French mathematician and physicist Augustin Cauchy (1789–1857), in his *Memoir on the Dispersion of Light* found an approximate formula for the refractive index as a function of wavelength:

$$n = A + B/\lambda^2 + C/\lambda^4 + \cdots \qquad (2.38)$$
$$\simeq A + B/\lambda^2, \qquad (2.39)$$

where A, B and C are constants for a given glass [9].

The Cauchy formula is thus in accordance with the observation that all glasses show strongly rising refractive index and dispersion ($\frac{dn}{d\lambda}$) at shorter wavelengths. The dispersion is given by

$$\frac{dn}{d\lambda} \simeq \frac{-2B}{\lambda^3}. \qquad (2.40)$$

The diffraction-limited resolving power at minimum deviation (Equation 2.18) therefore becomes

$$R_{\text{diff}} = \frac{2Bt}{\lambda^3} \qquad (2.41)$$

while the slit-limited expression is

$$R_{\text{slit}} = \frac{2Bt}{\lambda^2\theta_s D}, \qquad (2.42)$$

while in the general case of any angle of incidence

$$R_{\text{diff}} = A_0\left(\frac{\sin\alpha}{\cos i_1 \cos r_2}\right)\frac{2B}{\lambda^3} \qquad (2.43)$$

and

$$R_{\text{slit}} = \frac{2Bt}{\lambda^2\theta_s D}\left(\frac{\cos\frac{\alpha}{2}}{\cos r_1}\right). \qquad (2.44)$$

Following the practice of Fraunhofer, glass refractive indices are usually measured at wavelengths of bright emission lines of easily produced laboratory spectra, in particular at the F(Hβ) line at 486.1 nm, at the C(Hα) line – 656.3 nm, and at the d(HeI) line – 587.6 nm.

The average dispersion is just the quantity ($n_F - n_C$), and the relative dispersion is ($n_F - n_C$)/($n_d - 1$). The Abbe number is defined as

$$\nu_d = \frac{n_d - 1}{n_F - n_C} \qquad (2.45)$$

and is the reciprocal of the relative dispersion. It is a crude measure of the dispersive ability of a particular glass, and is thus related to the Cauchy constant, B. A high Abbe number represents a low dispersion, which is a characteristic of crown glasses ($\nu_d > 50$). The denser lead-oxide-containing flint glasses have generally higher refractive indices, smaller Abbe numbers ($\nu_d < 50$) and higher dispersion.

2.3.4 Prism spectrograph throughput

The second important parameter (after resolving power) in the performance of a prism spectrograph is throughput. There are several reasons why photons can be lost at the dispersing element, thereby reducing the throughput. The main ones are that light is absorbed in traversing the glass, and also that light is reflected instead of refracted from both prism surfaces, on entering and again on leaving the glass. In addition, the beam from the collimator may overfill the dispersing element, causing light loss beyond its periphery. As will be discussed later (see Section 2.3.5) there may be an advantage in the figure of merit in so doing.

Equations 2.41 and 2.42 show that the resolving power depends only on the base dimension t for a given type of glass. Therefore all prisms with the same base but different apex angles α give the same resolving power. This applies both to slit- and diffraction-limited resolving power, because they are proportional to each other (see Section 2.3.2). They also give the same absorption, because the average light path through all such prisms is the same (being $t/2$ if the beam fills the prism aperture). The fractional light loss by absorption is then

$$\frac{I}{I_0} = e^{-ks} \qquad (2.46)$$

or

$$\left\langle \frac{I}{I_0} \right\rangle = e^{-kt/2} \qquad (2.47)$$

where k is the volume absorption coefficient of the glass.

Moreover, Equation 2.21 shows that the resolving power depends on the sum of the base dimensions for a multiple-prism train. The choice of the number of prisms will not depend on the absorption, because, for a given resolving power, one prism of large apex angle or several of smaller angle all result in the same absorptive losses.

If a prism is not used at minimum deviation, then Equation 2.44 still shows that the resolving power is proportional to the size of the prism, a statement of general validity, for both diffraction- and slit-limited

cases. High resolving power thus necessarily leads to a high absorptive loss. This is an important reason why prisms are not competitive with diffraction gratings for high resolving power spectroscopy, especially at shorter wavelengths, where many glasses have increasingly large absorption coefficients.

Frank Wadsworth in 1903 showed that the resolving power of a prism instrument could be affected by absorption in the glass. If the absorption is high, then the intensity of the beam near the base is diminished, to the extent that the effective beam size is also less, and hence the full aperture of the prism (or prisms) is no longer being utilized [10]. He gave as an example the three-prism Bruce spectrograph (prism base 5.1 cm; glass absorption coefficient $k_{390\,nm} = 0.37$) at Yerkes Observatory, where at 390 nm the resolving power was reduced to half the theoretical value predicted by Equation 2.41. Of course, the figure of merit ($M = RT$) will be reduced by more than this factor (by approximately four times), showing the severe effect of absorption on spectrograph performance, especially at shorter wavelengths in multiple flint glass prisms.

The reflection loss from both surfaces of a prism is another factor contributing to a decrease in throughput. The fraction of light reflected in the two orthogonal polarization planes from a plane interface between air and glass was given by Augustin Fresnel (1788–1827) in 1821, based on the boundary conditions for electromagnetic waves. For a plane wave polarized at right angles to the plane of incidence, the fraction reflected is $X = \sin^2(i - r)/\sin^2(i + r)$, while for a plane of polarization in the plane of incidence, this fraction is $Y = \tan^2(i - r)/\tan^2(i + r)$. Both expressions reduce to $X = Y = (n - 1/n + 1)^2$ for normal incidence, while Y becomes zero for $(i + r) = 90°$, leaving the reflected ray 100 per cent polarized for unpolarized incident light (Brewster's law).

Under conditions of minimum deviation, the symmetry of the light path ensures equal reflection losses at both surfaces, and the intensity fraction of initially unpolarized light that is transmitted is just

$$T_1 = \frac{1}{2}[(1 - X)^2 + (1 - Y)^2]. \qquad (2.48)$$

For a train of N identical prisms, all at minimum deviation, this becomes

$$T_N = \frac{1}{2}[(1 - X)^{2N} + (1 - Y)^{2N}]. \qquad (2.49)$$

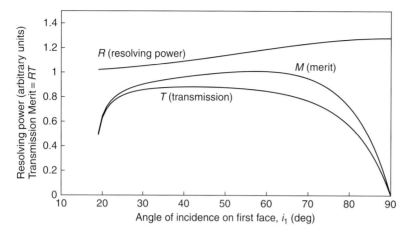

Figure 2.7. Plot of the resolving power, transmission and figure of merit of a prism spectrograph as a function of the angle of incidence on the first prism face of a prism with apex angle $50°$ and refractive index $n = 1.6$.

Given that, if the apex angle is

$$\alpha = \tan^{-1}\left(\frac{1}{n}\right), \tag{2.50}$$

then i will be at the Brewster angle, which gives no reflection loss at all for the rays whose polarization plane is in the plane of incidence, it then follows that unpolarized light will have at least 50 per cent of its intensity transmitted by a train of prisms with this apex angle, irrespective of their number. These principles were noted by Edward Pickering in 1868, in the days before he became the director of the Harvard College Observatory, in an early study of deviations, dispersions and reflection losses involving prism trains of various apex angles [11]. He noted that if long prism trains of ten prisms were used, then more dispersion and lower light loss resulted from using $\alpha = 64°$ prisms with the Brewster angle of incidence, than those with $\alpha = 45°$.

Reflection losses from single prisms of equal base but different apex angle and refractive index were also studied by Frank Wadsworth in 1895 [12]. In all cases the prisms were used at minimum deviation. The reflection losses for unpolarized light always increase at larger angles of incidence, dramatically so for $\alpha > 60°$ and $n = 1.8$, or $\alpha > 72°$ for $n = 1.5$, thus making such large apex angle prisms inefficient.

2.3.5 Figure of merit for prism spectrographs

The figure of merit for a spectrograph is $M \propto RT$, and the question of maximizing this can now be considered.

Figure 2.7 shows how T, R and M vary for a prism with apex angle $\alpha = 50°$ and refractive index $n = 1.6$, for different angles of incidence. Only reflection losses are taken into account. Absorption losses can be substantial for large prisms, but these do not vary rapidly with angle of incidence.

The transmission was calculated using the general equation

$$T = \frac{1}{2}(1 - X_1)^N(1 - X_2)^N + \frac{1}{2}(1 - Y_1)^N(1 - Y_2)^N \tag{2.51}$$

for a train of N identical prisms, where the subscripts on the reflection coefficients refer to the first and second prism faces.

For a single prism, the results show that the figure of merit is a maximum close to but not at the position of minimum deviation. For this particular prism, $i_1 \simeq 58°$ at the maximum of RT, but $i_1 = 42.55°$ for the minimum deviation. But operating at the minimum deviation implies only a 2.6 per cent decrease in the figure of merit.

If two identical prisms are used, each with the same angle of incidence, then the resolving power is doubled, but the transmission is now a lower and more sharply peaked function of the angle of incidence. The result is to shift the maximum of the figure of merit closer to the symmetrical position of minimum deviation. For two prisms, this occurs at $i_1 = 50°$ and for a longer train the angle would decrease further towards the limiting value of $42.55°$.

In general, it can be concluded that prisms always operate close to maximum efficiency when at minimum deviation. Small gains can be made for a one-prism spectrograph by increasing the angle of incidence, but for multiple-prism instruments, such gains become increasingly smaller, making minimum deviation operation effectively the optimum and in practice the solution of choice for spectrograph designers.

The principles of the efficient design of prism spectrographs were discussed in a lucid summary by G. A. Boutry in his textbook *Instrumental Optics* [13]. In particular, Boutry has combined the diffraction-limited resolving power R_{diff} with the slit-limited value R_{slit} to give the total resolving power R as

$$\frac{1}{R} = \frac{1}{R_{\text{diff}}} + \frac{1}{R_{\text{slit}}} \qquad (2.52)$$

or

$$R = R_{\text{diff}} \left(\frac{\lambda}{\lambda + \theta_s D} \right), \qquad (2.53)$$

a result first derived by Arthur Schuster (1851–1934) [14]. The value of R is then slightly less than the smaller of R_{diff} and R_{slit}. Instruments operating with $R_{\text{diff}} = R_{\text{slit}}$, and hence $R = R_{\text{diff}}/2$, have a high brightness level in a monochromatic emission-line spectrum as well as a high resolving power. This condition is therefore ideal for visual spectroscopes, where a high contrast of the emission lines is important.

2.4 RESOLVING POWER AND THROUGHPUT OF DIFFRACTION GRATING SPECTROGRAPHS

In this section, the general properties of the diffraction grating spectrograph will be developed. Paul Merrill's Cassegrain grating spectrograph of 1929 [15] and the coudé grating spectrograph of Adams and Dunham from 1936 [16], both on the Mt Wilson 100-inch telescope, were the first really successful grating instruments. Since the mid twentieth century, blazed diffraction gratings have been the preferred dispersing elements in astronomical spectrographs. They outperform prisms at longer red wavelengths because of higher angular dispersion, and they outperform prisms in the ultraviolet because they do not suffer from the high absorption losses of large prisms at short wavelengths.

2.4.1 The diffraction-grating equation and angular dispersion

Fraunhofer's diffraction-grating equation is

$$n\lambda = d(\sin\alpha + \sin\beta). \qquad (2.54)$$

Here n is an integer, the order number, d is the spacing of the parallel grooves or rulings in the plane of the grating, α is the angle of incidence and β the angle of diffraction, both measured relative to the grating normal. When operating a grating in collimated light, α is constant for all wavelengths, whereas β is wavelength dependent. The sign convention is that α and β are both positive if they are on the same side of the grating normal (N). In Fig. 2.8, a reflection grating is shown for positive values of both α and β. Incidence of the light in the normal plane, which is a plane perpendicular to both the grating's surface and to its grooves, is implicit in the diffraction-grating equation. If incidence occurs at an angle γ to the normal plane, then the equation becomes

$$n\lambda = d \cos\gamma(\sin\alpha + \sin\beta). \qquad (2.55)$$

Such a mode of operation with $\gamma \neq 0$ is employed in some échelle spectrographs with large angles of incidence and diffraction, for reasons to be discussed later, but for most grating spectrographs, the grating is illuminated in the normal plane, and hence the simple form of the equation with $\cos\gamma = 1$ suffices.

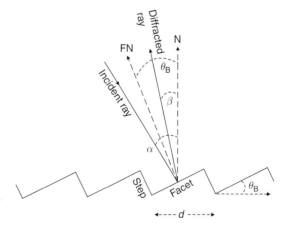

Figure 2.8. Diagram of a conventional diffraction grating showing the grating normal (N), facet normal (FN) and profile of the blazed grooves.

The blazed diffraction grating was proposed by Robert Wood in 1910 [17] and then exploited by John Anderson, both at Johns Hopkins University in Baltimore. It has become the standard type of reflection grating universally employed in modern instruments. The blaze refers to the uniformly angled facets, tilted at the blaze angle θ_B to the grating normal, which direct much of the light into a single order.

For blazed gratings, it is a common practice to use them in what is known as a quasi-Littrow mode. The Littrow spectroscope was originally a prism instrument with the light passing through the prism twice after being reflected internally from a silvered face, and returning back through the collimator lens along its original path [18]. In modern spectrographs, the Littrow arrangement has come to signify any instrument in which the dispersed beams from the dispersing element return close to the path of the incident light ($\alpha \simeq \beta$). A quasi-Littrow arrangement is one in which the Littrow condition is approximately satisfied.

Grating spectrographs function more efficiently in a quasi-Littrow mode than far from the Littrow condition, especially échelle grating spectrographs. For a blazed grating one may write

$$\alpha = \theta_B + \theta \qquad (2.56)$$

and

$$\beta = \theta_B - \theta. \qquad (2.57)$$

The second equation is only applicable in an order centre where the intensity is a maximum. Here θ_B is the blaze angle and θ is referred to as the Littrow angle, as it measures the small departure from the Littrow condition.

Using this notation results in the diffraction-grating equation having the form

$$n\lambda = 2d \cos \gamma \sin \theta_B \cos \theta \qquad (2.58)$$

and for the quasi-Littrow case with also incidence in the normal plane (θ small and γ zero), this becomes

$$n\lambda \simeq 2d \sin \theta_B. \qquad (2.59)$$

The diffraction-grating equation can be differentiated to find the angular dispersion. Thus

$$n = d \cos \beta \left(\frac{d\beta}{d\lambda} \right) \qquad (2.60)$$

or

$$\left(\frac{d\beta}{d\lambda} \right) = \frac{n}{d \cos \beta} \qquad (2.61)$$

$$= \frac{\sin \alpha + \sin \beta}{\lambda \cos \beta} \qquad (2.62)$$

$$\simeq \frac{2 \tan \beta}{\lambda} \simeq \frac{2 \tan \theta_B}{\lambda} \qquad (2.63)$$

in the quasi-Littrow mode of operation.

It is important to note that the angular dispersion at a given wavelength depends only on the angle of diffraction β and not at all on the groove spacing d, a point emphasized by George Harrison in 1949 [19]. A grating with a finer groove spacing may operate in a lower order for a given wavelength and value of β; or it may operate with a larger angle of β, and in this latter case, the angular dispersion will be higher.

2.4.2 Fourier analysis of diffraction gratings

The most elegant analysis of the properties of diffraction gratings involves Fourier transform theory. Expositions are given, for example, by M. C. Hutley [20, pp. 42–51] and by D. F. Gray (b. 1938) [21, pp. 44–53]. The analysis here follows that given by Gray for a transmission grating onto which a plane wave

$$F(x, t) = F_0(t) \exp[2\pi i(x/\lambda) \sin \alpha] \qquad (2.64)$$

is incident. The x-axis is in the plane of the grating perpendicular to the rulings, while α is the angle of incidence. The grating is defined by $G(x) = 1$ in the transparent slits, but $G(x) = 0$ for the opaque spaces elsewhere.

Secondary point sources in the plane of the grating can be considered to reradiate the incident energy. The electric field from one of them in the direction of propagation β is

$$F(x, t)G(x) \exp[2\pi i(x/\lambda) \sin \beta]dx, \qquad (2.65)$$

while the whole grating contributes to the resultant wave in the direction β with electric field amplitude

$$g(\beta) = \int_{\text{whole grating}} F(x, t)G(x) \exp[2\pi i(x/\lambda) \sin \beta]dx \qquad (2.66)$$

$$= F_0 \int G(x) \exp[2\pi i(x/\lambda)(\sin \alpha + \sin \beta)]dx. \qquad (2.67)$$

Writing $\sigma = (\sin\alpha + \sin\beta)/\lambda$ gives

$$g(\sigma) = F_0 \int_{-\infty}^{\infty} G(x) \exp[2\pi i x\sigma]\,dx, \qquad (2.68)$$

which shows that the intensity $|g(\sigma)^2|$ in the direction specified by the parameter σ is just proportional to the Fourier transform of the grating function $G(x)$, provided $G(x)$ is defined over an infinite range by

$$G(x) = (B_S(x) \star \text{Ш}(x))B_L(x). \qquad (2.69)$$

where $\star$ represents a convolution. $B_S(x)$ and $B_L(x)$ are box functions, which are unity within widths of respectively t and L, but zero elsewhere; here t is the width of the transparent slot ($t < d$), and L is the total width of the whole grating of N slots ($L = Nd$). The function $\text{Ш}(x)$ is an infinite array of equally spaced Dirac delta functions $\delta(x)$, sometimes known as a *shah* function (after the Russian letter Ш, which resembles delta functions) or as a Dirac comb function. Clearly

$$\text{Ш}(x) = \sum_{-\infty}^{\infty} \delta(x - nd). \qquad (2.70)$$

Using the convolution theorem in Fourier theory shows that

$$g(\sigma) = (b_S(\sigma).\,\text{Ш}(\sigma)) \star b_L(\sigma), \qquad (2.71)$$

where $b_S(\sigma)$ is the Fourier transform of $B_S(x)$ given by

$$b_S(\sigma) = \frac{t\sin\pi t\sigma}{\pi t\sigma} = t\,\text{sinc}(\pi t\sigma) \qquad (2.72)$$

and similarly

$$b_L(\sigma) = \frac{L\sin\pi L\sigma}{\pi L\sigma} = L\,\text{sinc}(\pi L\sigma) \qquad (2.73)$$

is the transform of $B_L(x)$. Also, the transform of $\text{Ш}(x)$ is another array of delta functions, namely

$$\text{Ш}(\sigma) = \sum_{-\infty}^{\infty} \delta(\sigma - n/d). \qquad (2.74)$$

Thus

$$g(\sigma) = \left(\text{Ш}(\sigma) \star \frac{L\sin\pi L\sigma}{\pi L\sigma}\right)\left(\frac{t\sin\pi t\sigma}{\pi t\sigma}\right) \qquad (2.75)$$

$$= \sum_{-\infty}^{\infty}\left(\frac{L\sin\pi L(\sigma - n/d)}{\pi L(\sigma - n/d)}\right)\left(\frac{t\sin\pi t\sigma}{\pi t\sigma}\right) \qquad (2.76)$$

from which the intensity ($\propto |g(\sigma)|^2$) in the direction specified by the parameter σ can be found. Note that

$b_L(\sigma)$ is an exceedingly narrow function, because L is large, and it governs the diffraction-limited resolving power of the grating. For an infinitesimally narrow slit, $|b_L(\sigma)|^2$ is related to the point spread function or instrumental profile. On the other hand, $b_S(\sigma)$ is a broad function governed by the diffraction pattern of a single grating slit or groove. The square of its modulus, $|b_S(\sigma)|^2$, is related to the intensity distribution over one entire diffraction order from the grating.

The last equation shows that, apart from the overall efficiency function $b_S(\sigma)$ over a whole order, the intensity distribution for perfectly collimated monochromatic radiation from the grating is a series of functions $b_L(\sigma)$ repeating at equal intervals of $1/d$ in the variable σ, or whenever $(\sigma - \frac{n}{d}) = 0$. This expression at once results in

$$n\lambda = d(\sin\alpha + \sin\beta), \qquad (2.77)$$

which is Fraunhofer's diffraction-grating equation.

2.4.3 Diffraction-limited resolving power of a grating

Rayleigh's equation ($R = A\frac{d\theta}{d\lambda}$) (see Section 2.2) for the diffraction-limited resolving power of a spectrograph applies for any type of dispersing element. In general, larger dispersing elements give a higher resolving power. The aperture presented by a grating for the light leaving the grating at an angle of diffraction β is just $L\cos\beta$, so Rayleigh's equation becomes

$$R_{\text{diff}} = L\cos\beta\left(\frac{n}{d\cos\beta}\right) \qquad (2.78)$$

where the angular dispersion $\frac{d\beta}{d\lambda}$ from Equation 2.61 has been used. Thus

$$R_{\text{diff}} = L\left(\frac{n}{d}\right) = nN \qquad (2.79)$$

as found by Lord Rayleigh in 1879 [2].

If the alternative expression for angular dispersion of Equation 2.63 ($\frac{d\beta}{d\lambda} \simeq \frac{2\tan\beta}{\lambda}$) is used, then

$$R_{\text{diff}} = L\cos\beta\left(\frac{2\tan\beta}{\lambda}\right) \qquad (2.80)$$

$$= \frac{2L\sin\beta}{\lambda} \simeq \frac{2L\sin\theta_B}{\lambda} \qquad (2.81)$$

if the spectrograph operates in the quasi-Littrow mode ($\beta \simeq \alpha \simeq \theta_B$, the blaze angle). Otherwise the exact solution is

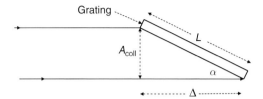

Figure 2.9. Diagram of a grating defining its dimension L perpendicular to the grooves, the beam size A for full illumination of the grating aperture and the path difference between marginal rays Δ.

$$R_{\text{diff}} = \frac{2L \sin \theta_B \cos \theta}{\lambda}. \qquad (2.82)$$

The equation for diffraction-limited resolving power is a consequence of Heisenberg's uncertainty principle, namely $\Delta x . \Delta p \simeq h$. The uncertainty in the position of a photon diffracted by a reflection grating (see Fig. 2.9) is $\Delta x = 2\Delta = 2L \sin \theta_B$, while the uncertainty in momentum is $\Delta p = \Delta(h/\lambda) = (h/\lambda^2)\Delta\lambda$. Hence

$$\Delta x . \Delta p = \frac{2hL \sin \theta_B}{\lambda^2} \Delta\lambda \simeq h, \qquad (2.83)$$

which becomes

$$\frac{\lambda}{\Delta\lambda} \simeq \frac{2L \sin \theta_B}{\lambda}, \qquad (2.84)$$

which is essentially Equation 2.81 for R_{diff}.

2.4.4 Slit-limited resolving power of grating spectrographs

Figures 2.1 and 2.6 show a schematic and idealized astronomical grating spectrograph with light being delivered to the spectrograph slit from a telescope of aperture D.

The whole optical system is shown with single refracting (lens) elements for simplicity, for telescope, collimator and camera. In practice, these are often reflecting elements and may comprise multiple surfaces. The slit width in the focal plane of the telescope is w and the monochromatic image of the slit in the focal plane of the camera has width w'. A_{coll} and A_{cam} are the beam diameters respectively before and after the grating, whose dimension perpendicular to the grooves is L.

The width of the slit is assumed to dominate the wavelength resolution and hence the resolving power. As for the case of a prism spectrograph, the slit will

dominate the resolving power provided the angular spread of monochromatic photons leaving the grating, which is $\delta\beta$ and due to the finite slit, is much greater than the angular spread λ/A_{cam} due to diffraction. The finite slit width w imparts an angular spread $\delta\alpha$ in monochromatic photons arriving at the grating, where

$$\delta\alpha = w/f_{\text{coll}}. \qquad (2.85)$$

Differentiating the Fraunhofer diffraction-grating equation at constant wavelength λ shows that

$$\cos \alpha \, \delta\alpha + \cos \beta \, \delta\beta = 0 \qquad (2.86)$$

or

$$|\delta\beta| = \frac{\cos \alpha}{\cos \beta} \delta\alpha = \frac{\cos \alpha}{\cos \beta} \frac{w}{f_{\text{coll}}}. \qquad (2.87)$$

The factor $\left(\frac{\cos \alpha}{\cos \beta}\right)$ is the anamorphic magnification, and the equation above (2.87) can be compared with the corresponding prism equation, 2.29.

Thus, for slit-limited operation

$$\frac{\cos \alpha}{\cos \beta} \frac{w}{f_{\text{coll}}} \gg \frac{\lambda}{A_{\text{cam}}}. \qquad (2.88)$$

Now $A_{\text{cam}}/\cos \beta = A_{\text{coll}}/\cos \alpha$, so the slit-limited condition becomes

$$w \gg \lambda f_{\text{coll}}/A_{\text{coll}}, \qquad (2.89)$$

which is identical to the result for the prism instrument. For the reason discussed in Section 2.3.2, this inequality is generally satisfied in astronomical work.

The wavelength resolution in this slit-limited case is

$$\delta\lambda = \left[1 \Big/ \frac{d\beta}{d\lambda} \right] \delta\beta \qquad (2.90)$$

$$= \frac{d \cos \beta}{n} \left(\frac{\cos \alpha}{\cos \beta} \right) \frac{w}{f_{\text{coll}}} \qquad (2.91)$$

$$= \frac{\lambda}{(\sin \alpha + \sin \beta)} \frac{w \cos \alpha}{f_{\text{coll}}}. \qquad (2.92)$$

Thus

$$R_{\text{slit}} = \frac{\lambda}{\delta\lambda} = \frac{(\sin \alpha + \sin \beta) f_{\text{coll}}}{w \cos \alpha} \qquad (2.93)$$

$$= \frac{2 f_{\text{coll}}}{w} \frac{\tan \theta_B}{(1 - \tan \theta_B \tan \theta)} \qquad (2.94)$$

at the central position in an order (where $\beta = \theta_B - \theta$).

In the case of quasi-Littrow operation ($\alpha \simeq \beta$), this expression can be approximated to

$$R_{\text{slit}} = \frac{2f_{\text{coll}}}{w} \tan \theta_{\text{B}} \qquad (2.95)$$

as θ is, in this case, small. This expression for the slit-limited resolving power, R_{slit}, can also be written in terms of the collimator beam size, A_{coll}, by noting (from similar triangles in Fig. 2.6) that $f_{\text{coll}}/f_{\text{tel}} = A_{\text{coll}}/D$. Hence

$$R_{\text{slit}} = \frac{2A_{\text{coll}}f_{\text{tel}}}{wD} \frac{\tan \theta_{\text{B}}}{(1 - \tan \theta_{\text{B}} \tan \theta)} \qquad (2.96)$$

$$= \frac{2A_{\text{coll}}}{\theta_s D} \frac{\tan \theta_{\text{B}}}{(1 - \tan \theta_{\text{B}} \tan \theta)} \qquad (2.97)$$

$$\simeq \frac{2A_{\text{coll}} \tan \theta_{\text{B}}}{\theta_s D} \qquad (2.98)$$

in the quasi-Littrow case, as $w = \theta_s f_{\text{tel}}$, where θ_s is the angular size of the slit in radians projected back onto the sky.

A third and possibly the most useful form of the same expression is

$$R_{\text{slit}} = \frac{2L \sin \theta_{\text{B}} \cos \theta}{\theta_s D}, \qquad (2.99)$$

which becomes

$$R_{\text{slit}} = \frac{2L \sin \theta_{\text{B}}}{\theta_s D} \qquad (2.100)$$

in the quasi-Littrow mode. Here L is the size of the illuminated beam in the plane of the grating perpendicular to the grooves. This relationship was derived by Richard Bingham in his 1979 article reviewing the design of grating spectrographs [3]. Similar expressions have been given by Robert Tull (b. 1929) [22] and by Daniel Schroeder and Fred Chaffee (b. 1941) [23].

All three expressions for the resolving power are fundamental to the design philosophy of astronomical spectrographs. Equation 2.95 shows that high resolving power comes from long collimator focal lengths, from narrow slits and from gratings with high blaze angles. A larger collimator focal length of course implies a larger beam size, A_{coll}, which may consequently overfill the grating. The equation is still valid in this case, and indeed the light loss from overfilling the grating may be more than compensated by the increased throughput from a wider slit – a point first made by Robert Tull in 1972 [22] and used, for example, by Francisco Diego and David Walker in the design of the coudé échelle

spectrograph on the Anglo-Australian telescope [24]. Equation 2.95 also shows that R_{slit} goes as $\tan \theta_{\text{B}}$. Of course, for a given value of collimator focal length, f_{coll}, a larger blaze angle grating requires a correspondingly larger dimension L, or else the collimator beam will overfill the échelle. For a given size of grating, larger $\tan \theta_{\text{B}}$ will result in rapidly diminishing throughput.

It is noted that Equation 2.94 shows that high resolving power requires long collimator focal lengths, which was the principal reason for building large coudé spectrographs. A Cassegrain instrument with a single concave parabolic collimator is generally limited in its collimator focal length to about a metre, otherwise flexure problems become insurmountable. On the other hand, a coudé spectrograph is a stationary instrument and has no such limit to its dimensions.

Equation 2.98 shows that large beam sizes, A_{coll}, also favour high resolving power. Once again, this is the beam size from the collimator, regardless of whether this overfills or underfills the grating.

Finally, Equation 2.100, which is probably the most useful form of the expression for slit-limited resolving power, shows that R_{slit} is proportional to the size L of the grating normal to the grooves. This equation is only valid if $A_{\text{coll}} = L \cos \alpha$, or the grating is not overfilled. In the case of overfilling, then the equation can still be used provided L is understood to represent the size of that grating which would have filled the collimator beam.

This equation clearly shows that for very high resolving power, either one requires very large gratings, or gratings with large blaze angles, that is échelle gratings. Large conventional (low blaze angle) gratings can only be mounted in coudé spectrographs, whereas échelles may be used in smaller Cassegrain instruments and still deliver high resolving power.

Echelle spectrographs with $\theta_{\text{B}} = \arctan 2$ have $\sin \theta_{\text{B}}$ equal to 0.89, so huge gains can clearly not be made by yet further increases in blaze angle. There may, however, be other advantages in having gratings with still larger blaze angle (see Section 3.3). It is noted here that the resolving power of a spectrograph is proportional to the path difference, Δ, between marginal rays striking the grating (see Fig. 2.9). All three equations, 2.94, 2.96 and 2.99, are essentially statements of this fact.

Both Equations 2.96 and 2.99 show an inverse dependence on telescope aperture D, which is simply because larger telescopes (for a given collimator beam size) have longer focal lengths and hence a smaller focal plane scale, ρ, given by ρ (arc s/mm) $= 206\,265/f_{tel}$ (mm). A telescope of twice the size may collect four times as much light, but in passing this light through a spectrograph slit, the light is used only half as efficiently, giving an increase in the figure of merit of only a factor of two (i.e. $M \propto D$).

If diffraction-limited resolving power is not completely negligible, which may be true for small telescopes at long wavelengths, then the overall resolving power is given by

$$\frac{1}{R} = \frac{1}{R_{slit}} + \frac{1}{R_{diff}} \tag{2.101}$$

$$= 1 \Big/ \left(\frac{2L \sin\theta_B \cos\theta}{\theta_s D} \right) + 1 \Big/ \left(\frac{2L \sin\theta_B \cos\theta}{\lambda} \right). \tag{2.102}$$

Therefore

$$R = \frac{2L \sin\theta_B \cos\theta}{\theta_s D + \lambda}. \tag{2.103}$$

Note that $R_{slit} = R_{diff}(\frac{\lambda}{\theta_s D}) = R_{diff}(\theta_A/\theta_s)$ where θ_A is the angular size of the Airy disk. These relations are identical to Equations 2.52 and 2.53 derived for prism spectrographs. Thus these equations, $R_{slit} = R_{diff}(\theta_A/\theta_s)$ and $(\frac{1}{R} = \frac{1}{R_{slit}} + \frac{1}{R_{diff}})$, do not depend on the nature of the dispersing element.

The above equation for the overall resolving power R can also be written as

$$R = R_{diff} \left(\frac{\lambda}{\lambda + \theta_s D} \right). \tag{2.104}$$

This result, which is valid for both grating and prism spectrographs, is sometimes referred to as Schuster's equation, after Arthur Schuster, who presented it in his *Encyclopedia Britannica* article on spectroscopy [14]; see also [25].

2.4.5   Comparison of the figure of merit of prism and grating spectrographs

Pierre Jacquinot's analysis in 1954 [1] compared the performance of prism and grating spectrometers. For a prism instrument the slit-limited resolving power is

$$R_{slit} = t \frac{dn}{d\lambda} \left(\frac{\lambda}{\theta_s D} \right) \tag{2.105}$$

while for a Littrow grating instrument it is

$$R_{slit} = \frac{2L \sin\theta_B}{\theta_s D}. \tag{2.106}$$

The ratio of resolving powers of prism to grating for the same slit is thus

$$\frac{R_{prism}}{R_{grating}} = \frac{\lambda t \frac{dn}{d\lambda}}{2L \sin\theta_B}. \tag{2.107}$$

Jacquinot considered dispersing elements of the same size ($t \simeq L$) and showed that for all common glasses, wavelengths and a reasonable blaze angle, that blazed gratings outperformed prisms. Absorption and reflection losses in prisms and grating efficiency were not taken into account. The advantage was shown to be a factor of 50–100 in the near infrared, and 5–10 times in the near ultraviolet and in the infrared. Jacquinot's paper was published at a time when low-cost blazed gratings were becoming available. This availability, and Jacquinot's theoretical demonstration of the superiority of such gratings over prisms contributed to the decline in the use of prism spectrometers in the second half of the twentieth century.

2.5   CAMERAS AND DETECTORS: SOME SIMPLE REQUIREMENTS

So far, the resolving power of a spectrograph has been analysed without reference to either the camera or detector system. It has therefore been treated as a property intrinsic to the wavelengths and angular distribution of the photons as they leave the dispersing element.

The function of the camera is to convert an angular distribution of photons with wavelength into a linear distribution on the detector, which in turn records the spectral image, ideally without further degradation of the resolving power.

The details of camera optics and detectors are not considered in this section. For the present purpose, we consider a camera to have a focal length f_{cam}, and that it accepts a focal ratio of $F_{cam} = f_{cam}/A_{cam}$ for any monochromatic beam, where A_{cam} is the aperture or beam size of the white pupil at any wavelength as the light leaves the dispersing element. The camera is taken to be a paraxial (that is, aberration-free) system, that produces an image of width w' of the slit, which has width w.

The detector in the camera's focal plane has an array of discrete contiguous pixels whose dimension in the dispersion direction is Δs_{pix} and which number N_{pix} in this dimension. The overall detector length is therefore $\Delta S_{\text{det}} = \Delta s_{\text{pix}} N_{\text{pix}}$. In the case of photographic detectors, then Δs_{pix} should be interpreted as the typical grain size in the emulsion, which is comparable to the limiting photographic spatial resolution.

The immediate problem is to find the appropriate camera focal length and focal ratio for a given instrument, as well as a suitable overall dimension for the detector and for the pixel size.

2.5.1 Reciprocal dispersion and free spectral range

The angular dispersion is $\frac{d\beta}{d\lambda}$, which equals $\frac{n}{d\cos\beta} \simeq \frac{2\tan\beta}{\lambda}$ for a grating instrument. This gives a reciprocal dispersion on the detector of $P = \frac{d\lambda}{dx} = 1/(f_{\text{cam}}\frac{d\beta}{d\lambda})$, which for a grating spectrograph is $\frac{d\cos\beta}{nf_{\text{cam}}}$, where x is the distance along the spectrum in the dispersion direction on the detector. Commonly used units for P are Å/mm or nm/mm. This parameter is sometimes referred to as the plate factor, a reference to the use of photographic plates to record spectra.

The free spectral range is defined as the change in wavelength at a given angle of diffraction β in going from one order n of a grating to the adjacent order of the grating, $n \pm 1$. Since the grating equation is $n\lambda =$ constant, clearly $n\delta\lambda + \lambda\delta n = 0$ and for $\delta n = 1$ the result is (ignoring the minus sign)

$$\Delta\lambda_{\text{FSR}} = \frac{\lambda}{n} \tag{2.108}$$

for the free spectral range. A reflection grating has grooves of spacing d which, in an idealized model, have facets of width $f = d\cos\theta_B$ and steps of height $s = d\sin\theta_B$ as shown in Fig. 2.8.

Hence

$$\Delta\lambda_{\text{FSR}} \simeq \frac{\lambda^2}{2d\sin\theta_B} = \frac{\lambda^2}{2s}. \tag{2.109}$$

As shown in Section 2.4.2, the intensity distribution over an order for a transmission grating goes roughly as $|b_s|^2 = \left(\frac{t\sin\pi t\sigma}{\pi t\sigma}\right)^2$, a sinc-squared function (see Fig. 2.2). For a reflection grating, the slot width t of a transmission grating can be replaced by the full groove width d. The full-width at half maximum of a

sinc-squared function is when $\sigma = (\sin\alpha + \sin\beta)/\lambda = 1/d$ or

$$\Delta\beta_{\text{FWHM}} = \frac{\lambda}{d\cos\beta}. \tag{2.110}$$

Writing $\beta = (\theta_B - \theta)$ and assuming θ to be small gives an angular spread of an order to be $\Delta\beta \simeq \frac{\lambda}{f\cos\theta}$. Thus the projected width of a groove ($d\cos\beta$) or of a facet ($f\cos\theta$) controls the intensity distribution over a whole order.

The wavelength range within one order is clearly

$$\Delta\lambda_{\text{FWHM}} = \Delta\beta_{\text{FWHM}} \bigg/ \left(\frac{d\beta}{d\lambda}\right) \tag{2.111}$$

$$= \frac{\lambda}{d\cos\beta}\frac{d\cos\beta}{n} = \frac{\lambda}{n} \tag{2.112}$$

$$= \Delta\lambda_{\text{FSR}}. \tag{2.113}$$

Therefore the free spectral range also equals the useful wavelength range within a spectral order. For échelle spectrographs this simple relationship breaks down, because the effective projected facet width is less than $f\cos\theta$ as a result of the phenomenon of shadowing – in this case $\Delta\lambda_{\text{FWHM}}$ can exceed $\Delta\lambda_{\text{FSR}}$ with undesirable consequences (see Section 3.3.2).

An observer may wish to record a spectrum over a wavelength range of $\Delta\lambda_{\text{FWHM}}$. In the first or any low order, this may be so large that it is impractical to record all in one exposure. For high order échelle spectrographs, the order's angular width and wavelength range are much less, and recording a whole order may be feasible. But it is not useful to go beyond the half-intensity limits of the range $\Delta\lambda_{\text{FWHM}}$, as the adjacent orders will always be more intense at such wavelengths. (An exception might be if spectral orders are co-added, thereby recovering the weak extremities of an order.)

The physical length of an order produced by the camera is

$$\Delta S_{\text{order}} = f_{\text{cam}}\Delta\beta_{\text{FWHM}} \tag{2.114}$$

$$= f_{\text{cam}}\frac{\lambda}{d\cos\beta} \tag{2.115}$$

$$= \frac{\lambda}{nP} = \Delta\lambda_{\text{FWHM}}/P. \tag{2.116}$$

On the other hand, the actual wavelength range recorded will be $\Delta\lambda_{\text{order}} = P\Delta S_{\text{det}}$ if $\Delta S_{\text{det}} < \Delta S_{\text{order}}$, that is, the detector is smaller than the order length.

It is emphasized that the intensity distribution over an order, or the efficiency function of a grating, is only very crudely given by a sinc-squared function. The actual curves are complex functions of the polarization, the blaze angle and angle of incidence and of the groove profile. Theoretical results have been presented over a range of blaze angles by E. G. Loewen *et al.* [26].

2.5.2 Detector resolution and the Nyquist sampling theorem

The Nyquist sampling theorem states that, in order to extract all the information from a discretely sampled function, the spatial sampling frequency must be at least twice the highest spatial frequency present in the Fourier transform of the function. The theorem is often attributed to Harry Nyquist (1889–1976), a Swedish–American engineer who worked for many years at the Bell Telephone Laboratories. His famous paper of 1928, 'Certain topics in telegraph transmission theory' [27], concluded that the maximum pulsed data rate on a telegraph line was just twice the bandwidth (defined as the highest frequency signal that the line could carry). This is not a complete statement of the sampling theorem, which is more correctly attributed to the American engineer Claude Shannon (1916–2001), in 1949 [28], though others also have a claim (see [29]). Shannon considered the sampling frequency necessary to convey all the information to reproduce a function that is discretely sampled. In the context of stellar spectroscopy, this corresponds to the spatial frequency of pixels that is required in the detector.

For stellar spectra recorded with slit-limited resolving power, the monochromatic slit image of width w' corresponds to a wavelength resolution $\delta\lambda$ given by

$$\delta\lambda = w'/P = \lambda/R_{\text{slit}}. \tag{2.117}$$

In the unlikely situation of diffraction-limited resolving power

$$\delta\lambda = \lambda/R_{\text{diff}} = \frac{\lambda^2}{2L\sin\beta}. \tag{2.118}$$

To satisfy the Nyquist theorem, the spatial density of pixels must be at least $2/(P\delta\lambda)$ pixels mm^{-1} or each pixel must be of size $\Delta s_{\text{pix}} \leq w'/2$. If Δs_{pix} is larger than this limit, then the spectrograph becomes detector limited in its resolving power and high

spatial frequencies (i.e. features narrow in wavelength), if present, are lost. If such narrow features are not in the spectrum, then a wider slit would deliver higher throughput without loss of intrinsic spectral information.

Although two or three pixels per resolution element are acceptable, much smaller pixels will contribute to more noise in the spectrum. Most astronomical spectroscopy with CCD detectors has been readout-noise limited rather than photon-noise limited (though with modern CCDs recording high signal-to-noise data this is often now no longer the case). In the circumstance of readout noise dominating, the increased detector noise from more pixels in effect results in a lower detector quantum efficiency (DQE), which is defined in terms of signal-to-noise ratios (S/N) for each pixel by

$$\text{DQE} = \frac{(\text{S/N})^2_{\text{recorded signal}}}{(\text{S/N})^2_{\text{incident photons}}}, \tag{2.119}$$

and hence a higher exposure is required to achieve a given result of (S/N) in the recorded spectrum.

The conclusion is that departures from the Nyquist condition in either direction imply a lowering of overall spectrograph efficiency below the optimum. If it is assumed that the Nyquist condition is satisfied with $w' = 2\Delta s_{\text{pix}}$ for a grating spectrograph, where

$$\frac{w'}{w} = \frac{f_{\text{cam}}}{f_{\text{coll}}}\left(\frac{\cos\alpha}{\cos\beta}\right) \tag{2.120}$$

and

$$w = \theta_s f_{\text{tel}}, \tag{2.121}$$

then the required camera focal length is given by

$$f_{\text{cam}} = \frac{2\Delta s_{\text{pix}}}{\theta_s f_{\text{tel}}}\left(\frac{\cos\beta}{\cos\alpha}\right)f_{\text{coll}} \tag{2.122}$$

$$= \frac{2\Delta s_{\text{pix}} L\cos\beta}{\theta_s D}, \tag{2.123}$$

where L is the size of the grating perpendicular to the grooves, θ_s is the angular size of the slit and D is the telescope aperture.

The last equation has a number of important consequences. These are that:

1. the camera focal length must match the pixel size;
2. larger telescopes require faster spectrograph cameras (smaller f_{cam}) to satisfy the Nyquist theorem;

3. large gratings require longer focal length cameras;
4. échelle gratings (large β) use shorter focal length cameras than conventional gratings (other parameters being the same);
5. if spectra of different slit size (and hence resolving power) are recorded, they should ideally have different camera focal lengths;
6. a nebular spectrograph requires a wide slit and therefore a fast camera.

The result for f_{cam} can be combined with Equation 2.100 for the slit-limited resolving power to give

$$f_{cam} = \Delta s_{pix} R \cot\theta_B (1 + \tan\theta \tan\theta_B) \simeq \Delta s_{pix} R \cot\theta_B \qquad (2.124)$$

(as θ is small), which at once shows that longer focal length cameras are needed for high resolving powers from narrow slits.

Finally the focal ratio of the camera is defined by

$$F_{cam} = f_{cam}/A_{cam} = f_{cam}/L\cos\beta \qquad (2.125)$$

for monochromatic light, and is given by

$$F_{cam} = \frac{2\Delta s_{pix}}{\theta_s D} \qquad (2.126)$$

if the Nyquist theorem is satisfied. The camera focal ratio does not depend on any properties of the grating, nor on the focal ratio of the telescope and collimator, but only on the angular slit width, the pixel size and the telescope aperture.

Clearly the camera focal length is required to satisfy both the Nyquist theorem as well as to deliver the required wavelength range to the detector. It may not be possible simultaneously to meet both conditions. Normally wavelength range is sacrificed if the detector is unable to record all that is required in one order, as detector-limited resolving power is never justified.

Finally it is noted that the plate factor $\frac{d\lambda}{dx}$ can be written as

$$P = \frac{d\lambda}{dx} = \frac{\lambda}{2\Delta s_{pix} R_{slit}} \qquad (2.127)$$

provided that the Nyquist theorem is satisfied. This expression emphasizes that in this case high resolving power should also accompany a low plate factor, commonly referred to as 'high dispersion'. Because of the ambiguity of this term, and because ultimately resolving power and not dispersion is a measure of

spectrograph performance, the use of the terminology 'high dispersion' is not recommended.

2.5.3   Number of detector pixels required

The ideal number of detector pixels required to record a spectral order in a grating spectrograph depends on the number of resolution elements in a free spectral range. This can be analysed from the point of view of angular measure. The angular resolution is the angle subtended by the slit after the grating, $\delta\beta = (\frac{\cos\alpha}{\cos\beta})\frac{w}{f_{coll}}$, while the angular width of an order (FWHM), which is here assumed to equal one free spectral range, is $\Delta\beta_{FWHM} = (\frac{\lambda}{d\cos\beta})$. The number of resolution elements in a free spectral range is therefore

$$\frac{\Delta\beta}{\delta\beta} = \frac{\lambda}{d\cos\beta}\left(\frac{\cos\beta}{\cos\alpha}\right)\frac{f_{coll}}{w} = \frac{\lambda L}{d\theta_s D}, \qquad (2.128)$$

where we have used $w = \theta_s f_{tel}$ and $L\cos\alpha/D = f_{coll}/f_{tel}$. This can be written as

$$\frac{\Delta\beta}{\delta\beta} = \frac{\theta_A N}{\theta_s} \qquad (2.129)$$

where $\theta_A \simeq \lambda/D$ is the Airy disk angular size and $N = L/d$ is the number of grooves in the grating.

The number of pixels required to record all the information in an order is just twice this, since the Nyquist theorem demands just two pixels per resolution element. Hence

$$N_{pix} = 2\left(\frac{\theta_A}{\theta_s}\right)N. \qquad (2.130)$$

This simple result is independent of order number and many other spectrograph parameters. Simply stated, larger gratings with more grooves require larger detectors with more pixels. This requirement was readily met in the days of large coudé spectrographs with photographic recording, but not so easily with CCD detectors. Values of $\frac{\theta_A}{\theta_s}$ for stellar point sources might fall in the range $0.02 \leq \frac{\theta_A}{\theta_s} \leq 1$, with values of a few tenths being common.

2.6   GRATING EFFICIENCY, SHADOWING AND THE QUASI-LITTROW ANGLE

The relationship for (w'/w) (Equation 2.120) contains an anamorphic magnification factor, $(\frac{\cos\alpha}{\cos\beta})$, which arises from the change in shape of the beam (i.e. in the eccentricity of the beam's elliptical cross-section) after being diffracted at an angle β. For many spectrographs, $\alpha \simeq \beta$; if α and β are also fairly small, the anamorphic magnification will be close to unity, as pointed out by Ira Bowen [30]. But these conditions are not always met. Some spectrographs operate with a substantial angle ($30°$ to $40°$) between incident and diffracted beams, and échelle spectrographs have large values of both α and β, making the factor $(\frac{\cos \alpha}{\cos \beta})$ no longer close to one.

Two questions are considered here: Is it better to have $\alpha > \beta$ or $\beta > \alpha$? And what are the effects of departing significantly from the Littrow condition of $\alpha \approx \beta$? These questions have also been discussed extensively in the paper by Schroeder and Hilliard [31].

A grating can be operated in a given order in two different positions, by interchanging the angles α and β. If one has a given spectrograph (fixed f_{cam} and f_{coll}) and $\alpha > \beta$, and then changes the grating orientation with a $180°$ rotation about the facet normal (FN), then the anamorphic magnification will go from less than unity to greater than unity. The slit image w' will therefore increase and no longer span two pixels, or, to restore the Nyquist condition, a narrower slit admitting less light would have to be employed. The benefits of operating with $\alpha > \beta$ for a given spectrograph (i.e. fixed f_{cam} and f_{coll}) have been emphasized by François Schweizer [32].

If, on the other hand, f_{cam} and f_{coll} can at the outset be regarded as free parameters, then Equation 2.99 shows that the resolving power R is an even function of the angle θ, and hence R does not depend on the sign of θ. If $\theta < 0$ ($\alpha < \beta$), then the collimator focal length and the collimated beam size will be larger than otherwise, and this will both increase the size of the spectrograph as well as the expense for no apparent gain. However, the effect is probably negligible for all but échelle spectrographs.

Bingham has also discussed the effects of interchanging α and β on spectrograph performance [3]. Except for échelle spectrographs, there is in fact little loss in performance whether $\alpha > \beta$ or $\alpha < \beta$, provided that f_{cam} and f_{coll} are treated as free parameters. The effects of groove shadowing may result in slightly lower peak efficiency for the $\alpha < \beta$ configuration, due to some light being lost after it impinges on the groove steps, and also because the effective facet width is less, which results in a broader angular distribution of the

light over an order, with $\Delta\lambda_{FWHM} > \Delta\lambda_{FSR}$. These effects therefore favour the $\alpha > \beta$ mode of operation.

Bingham however makes the point that the expression for the resolving power (Equation 2.99) can be written as

$$R = \frac{2L(\sin \alpha + \sin \beta)}{\theta_s D}. \qquad (2.131)$$

This shows a symmetry in α and β, so the product $R\theta_s$, which determines the figure of merit, is unaffected by interchanging the angles α and β. Also Loewen et al. [26] show that efficiency is not greatly sensitive to the sign of $(\alpha - \beta)$.

Departures from the Littrow condition may not be serious for conventional gratings at low blaze angle. Firstly, it is noted that the $R\theta_s$ product in Equation 2.99 goes as $\cos\theta$. Even if $(\alpha - \beta) = 40°$, then $\cos\theta = 0.94$, implying just a small reduction in the figure of merit. The blaze wavelength for the centre of an order decreases only slightly at larger values of θ as the equation $n\lambda = 2d \sin\theta_B \cos\theta$ shows. The angular dispersion, $\frac{d\beta}{d\lambda}$, and hence also the reciprocal dispersion (or plate factor, P) vary slightly with θ, as shown by the expression

$$\lambda \frac{d\beta}{d\lambda} = \frac{2 \tan \theta_B}{1 + \tan \theta \tan \theta_B}, \qquad (2.132)$$

which is valid for $\beta = \theta_B - \theta$ at the blaze wavelength in the centre of an order. For small values of θ_B, this declines only slightly with the angle θ.

Charles G. Wynne (1911–99) and S. P. Worswick [33] have shown that large departures from the Littrow condition can lead to light loss in some circumstances. They take the case of the Cassegrain spectrograph on the 4.2-m William Herschel telescope where $\alpha = 75°$, $\beta = 40°$ (hence $\theta = 17.5°$), for which a circular collimator beam of 15 cm diameter is dilated to an ellipse of nearly 45 cm major axis as it leaves the grating. This beam dilation requires the camera to operate at a much higher numerical aperture (f/0.9 instead of f/2.7) if all the light is to be accepted, which would be impossibly unwieldy for a Cassegrain spectrograph.

The arguments that show a moderate advantage in adopting $\alpha > \beta$ and $(\alpha - \beta)$ being small for conventional gratings become much more compelling for high blaze angle échelle gratings, as shown by Schroeder and Hilliard [31]. This is a consequence of groove shadowing, which can be analysed approximately using an

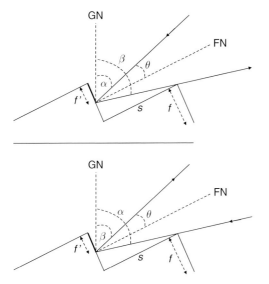

Figure 2.10. Profile of grooves of an échelle grating showing the effect of groove shadowing for $\alpha < \beta$ (top) and $\alpha > \beta$ (bottom). The effective facet width in both cases is f'. GN is the grating normal and FN is the facet normal.

idealized rectangular groove profile and simple geometry [34]. Figure 2.10 shows the situations for $\alpha < \beta$ and for $\alpha > \beta$ in the order centre, where β lies in the range $\theta_B \pm \theta$. The facets have width $f = d \cos \theta_B$, while the steps are of size $s = d \sin \theta_B$.

In the first diagram for $\alpha < \beta$ light is lost which strikes the step after the facet. This is referred to as case B by Schroeder and Hilliard. The light in the diffracted beam is that which comes from a reduced facet width $f' = f(1 - \tan |\theta| \tan \theta_B)$, while the fraction of the light falling on the grating which is lost is

$$\phi = \left(1 - \frac{\cos \beta}{\cos \alpha}\right) \simeq 2 \sin |\theta| \tan \theta_B \qquad (2.133)$$

(from purely geometrical arguments, when $\alpha < \beta$). This parameter would be 35 per cent for $\theta = 5°$ and $\theta_B = 63° \, 26'$ (R2 échelle), but 70 per cent loss for a $76°$ blaze angle (R4 échelle).

On the other hand, when $\alpha > \beta$ for an échelle (this is Schroeder and Hilliard's case A), then $\phi = 0$, as no light is lost at the groove steps. But the useful facet width is still reduced by the same factor of $(1 - \tan \theta \tan \theta_B)$. The reduction in effective facet width in both cases causes an undesirable spreading in the angular distribution of the light intensity within an order, which is approximately a sinc-squared function of full

width $\Delta\beta(\text{FWHM}) = \lambda/f' \cos \theta \simeq \lambda/d \cos \beta$ (see Section 3.3). These circumstances mean that $\alpha > \beta$ is nearly always adopted for échelle spectrographs, as here there is no light loss.

The increase in the angular width of the orders in an échelle spectrograph transfers light from the order centres near the blaze wavelengths to the edges of an order, well beyond the free spectral range. In effect, light of a given wavelength, which in the strict Littrow condition ($\theta = 0°$) only goes at most to two adjacent orders, is now (for $\theta > 0$) divided up amongst three or more orders. The far order wings increase in intensity, but these regions are probably not observed, as they lie beyond the limits of the detector. The intensities in the order centres are accordingly diminished. This effect has been discussed by several authors, notably by Daniel Schroeder [35], and by Schroeder and Hilliard [31]. Schroeder [35] also noted that the ratio $\Delta\lambda_{\text{FWHM}}/\Delta\lambda_{\text{FSR}}$ is equal to

$$\frac{\cos \beta}{\cos \alpha} = \frac{(1 + \tan \theta \tan \theta_B)}{(1 - \tan \theta \tan \theta_B)}, \qquad (2.134)$$

which not only explains why $\alpha > \beta$ is necessary for échelles, but also why θ must be small.

Walker and Diego have computed the sinc-squared functions for various values of $\theta \geq 0°$ for the 31.6 gr/mm $\tan \theta_B$ coudé échelle on the Anglo-Australian telescope [36], where $\theta = 6°$ has been adopted. Figure 2.11 shows these functions for increasing departures from the Littrow condition. Further discussion of échelle grating efficiency as a function of the angles α and θ is to be found in Section 3.3.2.

One further comment applies explicitly to Cassegrain échelle spectrographs. In a Cassegrain instrument, the collimator focal length is limited to about one metre (or less in some cases) in order to control flexure. This might be the limiting factor in determining the resolving power, rather than the overall grating size. In this case, Equation 2.94 for resolving power should be used, rather than Equation 2.99. Equation 2.94 for *fixed* collimator focal length is much more sensitive to the quasi-Littrow angle θ than Equation 2.99. What is more, R is much lower for échelle spectrographs of given f_{coll} if $\alpha < \beta$ (i.e. if θ were negative). In effect, negative θ makes the expression $(1 - \tan \theta \tan \theta_B)$ in the denominator greater than unity, so lowering the resolving power significantly

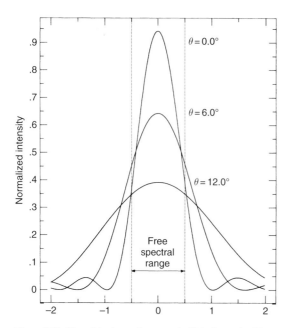

Figure 2.11. Plot of the increasing spread of échelle grating blaze functions for increasing values of the angle θ.

(by 30 per cent for $\theta = 5°$ and $\theta_B = 63° \, 26'$) for increasing departures from the Littrow condition. These comments only apply to Cassegrain échelle spectrographs, because only here is the resolving power limited by the collimator focal length rather than by the largest échelle grating available.

2.7   GRATINGS NOT ILLUMINATED IN THE NORMAL PLANE

The general grating equation is

$$n\lambda = d \cos \gamma (\sin \alpha + \sin \beta), \qquad (2.135)$$

which contains a factor of $\cos \gamma$. Here γ is the angle between the incident rays on the grating and the normal to the grating, the angle being measured in a plane perpendicular to the normal plane. Note that out-of-plane illumination with $\theta = 0°$ is referred to as case C by Schroeder and Hilliard [31].

The first effect of out-of-plane illumination is that the effective groove spacing d of the grating is reduced by the $\cos \gamma$ projection factor. The angular dispersion then becomes

$$\frac{d\beta}{d\lambda} = \frac{n}{d \cos \gamma \cos \beta}, \qquad (2.136)$$

which therefore increases very slightly for small non-zero values of γ from the in-plane situation, when evaluated at a given value of the diffraction angle, β. However, $\frac{d\beta}{d\lambda}$ is essentially unchanged from $\frac{2 \tan \beta}{\lambda}$ (Equation 2.63) when evaluated at a given wavelength, as β is insensitive to γ.

Using out-of-plane illumination can be useful in order to preserve a rigorous Littrow configuration ($\theta = 0°$), for the incident and diffracted beams now diverge by an angle of 2γ, thereby allowing the collimator and camera optics to be separated. Such a technique is only useful for échelle spectrographs, where large departures from a Littrow condition are objectionable. However, it is also in échelle spectrographs that out-of-plane illumination causes a significant line tilt, which may be an impediment in the subsequent reductions.

Differentiating the diffraction grating equation results in

$$\frac{d\beta}{d\gamma} = \left(\frac{\sin \alpha + \sin \beta}{\cos \beta} \right) \tan \gamma \qquad (2.137)$$

$$\simeq 2 \tan \theta_B \tan \gamma. \qquad (2.138)$$

If the illuminated height of the slit is h, then the slit image height will be $h' = h f_{cam}/f_{coll}$. The height h subtends an angle $\delta \gamma = h/f_{coll}$ at the collimator, which results in a range of diffraction angles

$$\delta\beta = 2 \frac{h}{f_{coll}} \tan \theta_B \tan \gamma, \qquad (2.139)$$

which corresponds to a deviation along the dispersion direction of

$$\delta x = 2 \frac{h}{f_{coll}} f_{cam} \tan \theta_B \tan \gamma \qquad (2.140)$$

$$= 2h' \tan \theta_B \tan \gamma. \qquad (2.141)$$

The result is sloping lines in the spectrum, sloping by an angle ϕ given by

$$\tan \phi = 2 \tan \theta_B \tan \gamma \qquad (2.142)$$

or

$$\phi \simeq 2\gamma \tan \theta_B \qquad (2.143)$$

(see Chaffee and Schroeder [23]). Echelle spectrographs in any case have sloping lines as a consequence of the need for cross-dispersion (see Section 3.3). By arranging for the out-of-plane angle to be on the appropriate side of the normal plane, these two effects can be made to cancel, giving perpendicular lines over at least some orders (Hearnshaw [37]). In addition, the spectrograph slit can be rotated, to compensate for any residual line tilt. For a slit rotated through an angle ψ, the effective slit width becomes $W \sec \psi$, so the resolving power is slightly degraded for no gain in throughput. These effects are discussed by Schroeder and Hilliard [31], who find that case A ($\theta = 6°$) still has a marginally higher figure of merit (resolving power times throughput) than case C ($\gamma = 6°$) with a rotated slit.

2.8 THE GRISM AS A DISPERSING ELEMENT

The previous sections discuss the application of prisms and gratings as dispersing elements in astronomical spectrographs. Since the 1980s, the grism has become an increasingly popular tool for astronomical spectroscopy. A good example of its use is in the European Southern Observatory's Faint Object Spectrograph and Camera – EFOSC (see Section 8.2).

A grism comprises a transmission diffraction grating attached to one face of a prism. Figure 2.12 shows the optical arrangement. Often the transmission grating is replicated in a thin resin layer from a ruled master, and this thin layer is then attached with an optical cement to the prism glass. Alternatively, a grism can be produced by ruling directly onto the prism, thereby avoiding the use of a resin layer, but incurring the additional task of ruling each grism individually.

The concept of a grism was first proposed by Ira Bowen and A. H. Vaughan in 1973 [38]. They described

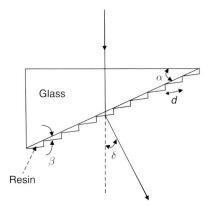

Figure 2.12. Optical diagram of a simple grism. Here α is the apex angle of the prism, β is the angle of the grating facets to the adjacent prism face, δ is the deviation of rays of wavelength λ and d is the groove spacing.

the device as a non-objective grating. Their device was mounted in the converging beam of a telescope, and the prism ensured that the effect of off-axis aberrations was greatly reduced.

The theory of the grism has been presented by, for example, Jacques Beckers (b. 1934) and Ian Gatley [39]. For a prism of apex angle α and refractive index n_G and with a grating of groove spacing d, and normal illumination on the first prism face,

$$n\lambda = (n_G - 1)d \sin \alpha \qquad (2.144)$$

is the condition on the wavelength λ for zero deviation, $\delta = 0$. The dispersion of a grism is given by

$$\frac{d\delta}{d\lambda} = \frac{(n_G - 1)\tan \alpha}{\lambda}. \qquad (2.145)$$

Normally the grism is operated in first order, $n = 1$.

The advantage of the grism over a transmission or reflection grating is that it disperses the light in a straight-through optical path. In effect, the zero deviation direction is also the blaze direction. The function of the prism within the grism arrangement is to refract the first order spectrum to approximately zero deviation, rather than as a dispersing element in its own right. This property means that an imaging camera can be quickly converted to a low resolution multi-object spectrograph, simply by inserting a grism into the optical train near the camera's entrance pupil, as is done, for example, in EFOSC.

In recent years there has been much interest in using grisms as infrared dispersing elements. This is because materials such as silicon and germanium are transparent in the near infrared and have very high refractive indices of respectively $n = 3.4$ for silicon and $n = 4.0$ for germanium, thereby giving a satisfactory dispersion [40, 41, 42].

Yet higher dispersion has been made possible with the échelle grism, as discussed by Michel Nevière [43]. Silicon échelle grisms allow for medium to high resolving power infrared spectrometers to be constructed, as demonstrated by Jian Ge and colleagues at Pennsylvania State University [44] and by F. Vitali et al. in Rome [45]. Techniques of grism manufacture by engraving directly onto a fused silica substrate are discussed by G. de Villele et al. [46].

Examples of recent grism spectrometers are the Faint Object Camera and Spectrograph (FOCAS) on the 8.2-m Japanese Subaru telescope on Mauna

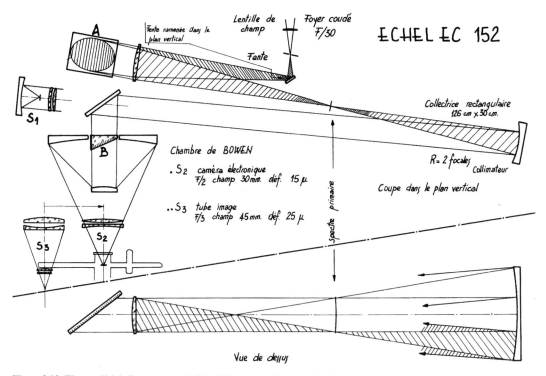

Figure 2.13. The coudé échelle spectrograph EchelEC was an early example of an instrument employing Baranne's white pupil concept. It was built for the 1.52-m telescope at Haute-Provence in 1972. A lens was placed immediately in front of the échelle (A) which served both as collimator and as the first collector element. The mirror acts as the second collector element, imaging the white pupil onto the grism (B) used for cross-dispersion. A Lallemand electronic camera is at focal position S2.

Kea [47], and the Advanced Camera for Surveys on the Hubble Space Telescope [48].

2.9   SOME MISCELLANEOUS ASPECTS OF SPECTROGRAPH DESIGN

2.9.1   Baranne's white pupil concept

One of the problems of spectrograph design arises from the need for the diverging monochromatic beams from the dispersing element to be accepted by the camera and imaged without vignetting to form the spectrum. Proximity of the camera to the dispersing element is not always practicable, in particular for reflection gratings, because of the requirement of unobstructed access to the incident beam. Therefore a large and unwieldy camera able to image peripheral beams without vignetting may be required, or vignetting losses at the limiting wavelengths being recorded must be accepted.

To overcome these difficulties, André Baranne at the Marseille Observatory proposed using a concave collector mirror after the grating [49]. Its purpose is to reimage onto the entrance pupil of the camera the white undispersed pupil from the collimator that falls on the grating. Therefore all these monochromatic beams are recombined at the camera, whose aperture is therefore much less. The angular spread of different wavelengths is still preserved, but not the spatial distribution at the reimaged white pupil. Hence vignetting is avoided and coma in the camera is also eliminated.

This concept has been applied by Baranne to a number of spectrographs, as discussed in his 1988 review [50]. One of these was the Cassegrain radial-velocity spectrograph on the 1.52-m telescope at the European Southern Observatory [51]. In this design light passes the collimator lens twice, before and after the grating, and a primary spectrum forms just in front of the collimator mirror. A similar arrangement was employed in the coudé échelle spectrograph known as EchelEC, on the 1.52-m telescope at Haute-Provence Observatory [52] – see Fig. 2.13. The ELODIE spectrograph at Haute-Provence also uses the white pupil design. It is discussed further in Section 8.5 and illustrated in Fig. 8.6. In this figure, mirror M3 images the white pupil at the échelle onto the cross-dispersing prism.

The white pupil concept has been used in other major spectrographs. One notable example is the new coudé échelle spectrograph at the 2.7-m (107-inch) telescope at the McDonald Observatory [53, 54, 55]. Here a single special collector (or pupil) mirror reimages the white pupil on the échelle onto the flat mirror of the folded Schmidt camera. The Hobby-Eberly telescope high resolution échelle spectrograph also uses a white pupil design by Bob Tull [56].

2.10 SOME GENERAL PRINCIPLES OF SPECTROGRAPH DESIGN

The designer of an astronomical spectrograph has a large number of variables to quantify, which together specify the design parameters of a given instrument. Bingham [3] has identified sixteen parameters for the spectrograph and telescope combination. They are R, P, θ_s, $(\alpha + \beta)$, $(\alpha - \beta)$, n, D, f_{tel}, f_{coll}, f_{cam}, L, A_{coll}, A_{cam}, w', λ and d. He also gives seven relationships between these variables from the general design of spectrographs and the theory of gratings (these equations are discussed further in Section 3.3), leaving nine free variables. Of these nine, some are fixed by the telescope available (D, f_{tel}) and the science to be done (R, λ), others by the typical seeing (θ_s) and the detector characteristics (w').

In general we can state that the designer will have a telescope to which a spectrograph must be matched for optimum efficiency. And the science to be done will determine the minimum resolving power necessary and the range of wavelengths required. Some typical examples are that a resolving power of only 500 may be sufficient for some types of survey work, such as finding emission-line objects, 1000 for measuring redshifts for faint quasars, 2500 for stellar spectral classifications, 10 000 for measuring line profiles in broad-lined stars (such as Balmer lines in A stars), 25 000 for element-abundance analysis for a range of elements in solar-type stars, and up to 100 000 for work on very fine interstellar lines where there are multiple components that need to be resolved. The wavelength range in a single exposure is also important. For some work, such as abundance analysis, it may be advantageous to record over 600 nm of wavelength in a single exposure, so as to maximize the number of lines observed, while for other applications only one narrow spectral feature may be of interest.

The site where the telescope is located will have a typical seeing $\theta_\star$ and the angular size of the slit θ_s should be comparable to this, so as to give a high throughput. The detector to be used will have a given pixel size, and the Nyquist condition requires that the slit image w' covers two (or slightly more than two) pixels. Overall detector dimensions and the camera focal length fix the wavelength range. Grating efficiencies depend on α and β for different blaze angles and for each of two different polarization states. The combination of wavelengths required, maximizing efficiency, having a grating large enough to deliver the required resolving power and possible physical constraints (especially for instruments mounted on a moving telescope) will determine the properties of the grating chosen.

Generally the slit size will be less than the seeing, and a significant fraction of the light can be lost at the slit as a result. The technique of improving spectrograph performance by overfilling the grating was first pointed out by Tull [22] and applied by Tull for the coudé échelle on the McDonald 2.7-m telescope and also by Diego and Walker for the coudé échelle UCLES on the Anglo-Australian telescope [57]. The principle of the technique is that a longer collimator focal length (f_{coll}) will result in a larger collimated beam size (A_{coll}). If this exceeds the grating dimension $L \cos \alpha$, then light is lost from overfilling the grating. But a longer f_{coll} also results in a smaller angular slit size $\delta\alpha$ ($= w/f_{coll}$) subtended by the collimator, which in turn results in a higher resolving power. Alternatively, the slit width w can be increased by the same factor as f_{coll} so as to keep the resolving power R the same, but the wider slit will admit more light as a result. The throughput gain can exceed the loss from overfilling. Gains of 20 to 40 per cent in overall efficiency have been calculated by Tull and by Diego and Walker under certain seeing conditions.

A general principle for maximizing throughput is that light should not be lost at just one single point in the optical train. To balance light loss at a slit and at a grating is likely to be more efficient than losing a larger amount just at the slit, because generally changes in throughput are highly non-linear with the parameter being varied – thus decreasing slit size loses a little light initially ($\theta_s \leq \theta_\star$) but a lot later on ($\theta_s \ll \theta_\star$). In this connection, Diego has discussed the throughput for Lorentzian stellar seeing profiles, and either circular apertures (such as the entrance to an optical fibre) or rectangular slits [58].

The same principle can be applied to a spectrograph that is linked to a telescope through an optical fibre feed. If point sources are being observed, and if the output from the fibre is reimaged onto the slit, then there may be light losses at the fibre input, at the slit and from overfilling of the grating. As in all spectrographs, there may also be vignetting losses in the camera. All these losses need to be carefully considered together, as they are usually not independent. Light losses can be minimized, and hence the overall figure of merit maximized, by seeking an optimum solution by adjusting several variables simultaneously.

REFERENCES

[1] Jacquinot, P., *J. Opt. Soc. Amer.* **44**, 761 (1954)

[2] Rayleigh, Lord, *Phil. Mag.* (5) **8**, 261 (1879)

[3] Bingham, R. G., *Quarterly J. R. astron. Soc.* **20**, 395 (1979)

[4] Newton, I., *Phil. Trans. R. Soc.* **6**(80), 3075 (1672)

[5] Fraunhofer, J., *Denkschr. der königl. Acad. zu München für 1821 und 1822*, **8**, 1 (1822)

[6] Herschel, J. W. F., 'Light', *Encyclopaedia Metropolitana, 2nd Division: Mixed Sciences*, vol. 2, pp. 341–586 (1830)

[7] Mousson, A., *Poggendorfs Ann. der Phys.* **112**, 428 (1861)

[8] Thollon, L., *Comptes Rendus de l'Acad. Sci., Paris* **89**, 93 (1879)

[9] Cauchy, A. L., *Mémoire sur la dispersion de la lumière*, Prague: Prague Sci. Soc., 235pp. (1836)

[10] Wadsworth, F. L. O., *Phil. Mag.* (6) **5**, 355 (1903)

[11] Pickering, E. C., *Phil. Mag.* (4) **36**, 39 (1868)

[12] Wadsworth, F. L. O., *Astrophys. J.* **2**, 264 (1895)

[13] Boutry, G. A., *Instrumental Optics*, London: Hilger and Watts (1961). Translated by R. Auerbach from the French edition of *Optique instrumentale* of 1946.

[14] *Encyclopedia Britannica*, 9th edn., **22**, 373 (1884)

[15] Merrill, P. W., *Astrophys. J.* **74**, 188 (1931)

[16] Adams, W. S., *Astrophys. J.* **93**, 11 (1941)

[17] Wood, R. W., *Phil. Mag.* (6) **20**, 770 (1910)

[18] von Littrow, O., *Sitzungsber. der kaiserl. Akad. der Wiss. zu Wien* **47** (part 2), 26 (1863)

[19] Harrison, G. R., *J. Opt. Soc. Amer.* **39**, 522 (1949)

[20] Hutley, M. C., *Diffraction Gratings*, London, New York: Academic Press (1982)

[21] Gray, D. F., *The Observation and Analysis of Stellar Photospheres*, 2nd edn., Cambridge University Press (1992)

[22] Tull, R. G., *ESO-CERN Conference on Auxiliary Instrumentation for Large Telescopes*, ed. A. Laustsen and A. Reiz, p. 259 (1972)

[23] Chaffee, F. H. and Schroeder, D. J., *Ann. Rev. Astron. Astrophys.* **14**, 23 (1976)

[24] Diego, F. and Walker, D. D., *Mon. Not. R. astron. Soc.* **217**, 347 (1985)

[25] Schuster, A., *Astrophys. J.* **21**, 197 (1905)

[26] Loewen, E. G., Nevière, M. and Maystre, D., *Appl. Optics* **16**, 2711 (1977)

[27] Nyquist, H., *Trans. Amer. Inst. Elec. Engineers* **47**, 617 (1928). Reprinted in *Proc. IEEE* **90**, 280 (2002)

[28] Shannon, C. E., *Proc. Inst. Radio Engineers* **37**, 10 (1949). Reprinted in *Proc. IEEE* **86**, 447 (1998)

[29] Lüke, H. D., *IEEE Comm. Mag.*, p. 106, April (1999)

[30] Bowen, I. S., in *Astronomical Techniques, Stars and Stellar Systems*, ed. W. A. Hiltner, University of Chicago Press, p. 34 (1962)

[31] Schroeder, D. J. and Hilliard, R. L., *Appl. Optics* **19**, 2833 (1980)

[32] Schweizer, F., *Publ. astron. Soc. Pacific* **91**, 149 (1979)

[33] Wynne, C. G. and Worswick, S. P., *Observ.* **103**, 12 (1983)

[34] Burton, W. M. and Reay, N. K., *Appl. Optics* **9**, 1227 (1970)

[35] Schroeder, D. J., *Publ. astron. Soc. Pacific* **82**, 1253 (1970)

[36] Walker, D. D. and Diego, F., *Mon. Not. R. astron. Soc.* **217**, 355 (1985)

[37] Hearnshaw, J. B., *Intl. Astron. Union Symp.* **118**, 371 (1986)

[38] Bowen, I. S. and Vaughan, A. H., *Publ. astron. Soc. Pacific* **85**, 174 (1973)

[39] Beckers, J. and Gatley, I., *ESO-CERN Conf. no. 30: Very Large Telescopes and Their Instrumentation*, ed. M.-H. Ulrich, vol. II, p. 1093 (1988)

[40] Käufl, H. U., Kühl, K. and Vogel, S., in 'Infrared astronomical instrumentation', ed. A. M. Fowler, *Proc. Soc. Photo-instrumentation Engineers (SPIE)* **3354**, 151 (1998)

[41] Jaffe, D. T., Ershov, O. and Marsh, J. P., *Bull. Amer. Astron. Soc.* **32**, 1430 (2000)

[42] Ge, J., McDavitt, D. L., Bernecker, J. L. *et al.*, in 'Optical spectroscopic techniques, remote sensing, and instrumentation for atmospheric and space research: IV', ed. A. M. Larar and M. G. Mlynczak, *Proc. Soc. Photo-instrumentation Engineers (SPIE)* **4485**, 393 (2002)

[43] Nevière, M., *Appl. Optics* **31**, 427 (1992)

[44] Ge, J., McDavitt, D. L., Chakraborty, A., Bernecker, J. L. and Miller, S., in 'Adaptive optical system technologies, II', ed. P. L. Wizinowich and D. Bonaccini, *Proc. Soc. Photo-instrumentation Engineers (SPIE)* **4839**, 1124 (2003)

[45] Vitali, F., Cianci, E., Foglietti, V. and Lorenzetti, D., *Mem. Soc. Astron. Italiana* **74**, 197 (2003)

[46] de Villele, G., Bonnemason, F., Brach, C. *et al.*, in 'Optical fabrication, testing and metrology', ed. R. Geyl, D. Rimmer and L. Wang, *Proc. Soc. Photo-instrumentation Engineers (SPIE)* **5252**, 183 (2004)

[47] Ebizuka, N., Kobayashi, H., Hirahara, Y. *et al.* in 'Imaging the universe in three dimensions: astrophysics with advanced multi-wavelength imaging devices', ed. W. van Breugel and J. Bland-Hawthorn, *Astron. Soc. Pacific Conf. Ser.* **195**, 564 (2000)

[48] Pasquali, A., Pirzkal, N., Larsen, S., Walsh, J. R. and Kümmel, M., *Publ. astron. Soc. Pacific* **118**, 270 (2006)

[49] Baranne, A., *Comptes Rendus de l'Acad. Sci., Paris* **260**, 3283 (1965), also published as *Publ. de l'Observatoire de Haute Provence* **7**(43), 1 (1965)

[50] Baranne, A., *ESO-CERN Conf. no. 30: Very Large Telescopes and Their Instrumentation*, ed. M.-H. Ulrich, vol. II, p. 1195 (1988)

[51] Baranne, A., Maurice, E. and Prévot, L., *European South. Observ. Bull.* **7**, 11 (1969)

[52] Baranne, A. and Duchesne, M., *ESO-CERN Conference on Auxiliary Instrumentation for Large Telescopes*, ed. A. Laustsen and A. Reiz, p. 241 (1972)

[53] Tull, R. G. and MacQueen, P. J., *ESO-CERN Conf. no. 30: Very Large Telescopes and Their Instrumentation*, ed. M.-H. Ulrich, vol. II, p. 1235 (1988)

[54] MacQueen, P. J. and Tull, R. G., *Instrumentation for Ground-based Optical Astronomy*, ed. L. B. Robinson, Springer-Verlag, p. 52 (1988)

[55] Tull, R. G., MacQueen, P. J., Sneden, C. and Lambert, D. L., *Publ. astron. Soc. Pacific* **107**, 251 (1995)

[56] Tull, R. G., in Optical astronomical instrumentation, ed. S. D'Odorico, *Proc. Soc. Photo-instrumentation Engineers (SPIE)* **3355**, 387 (1998)

[57] Diego, F. and Walker, D. D., *Mon. Not. R. astron. Soc.* **217**, 347 (1985)

[58] Diego, F., *Publ. astron. Soc. Pacific* **97**, 1209 (1985)

3 · High resolution spectrographs

3.1 THEORY OF THE COUDÉ SPECTROGRAPH

Any spectrograph, whether slit or diffraction limited, has a resolving power that is determined by the size of the collimated beam A and of the dispersing element, whether this be a prism or a grating. Thus

$$R_{\text{diff}} = A\frac{\mathrm{d}\theta}{\mathrm{d}\lambda} \qquad (3.1)$$

and

$$R_{\text{slit}} = A\frac{\mathrm{d}\theta}{\mathrm{d}\lambda}\frac{\lambda}{\theta_s D} \qquad (3.2)$$

are the relevant expressions, regardless of the type of dispersing element employed (see Section 2.3). Here $\frac{\mathrm{d}\theta}{\mathrm{d}\lambda}$ is the angular dispersion, θ_s is the angular width of the slit (projected onto the sky) and D is the diameter of the telescope's objective lens or primary mirror.

In practice, it is the slit-limited case that normally applies in astronomy. But it is still true, as shown by the equations above, that large resolving powers require large dispersing elements, if the slit size, θ_s, is to be comparable to the typical seeing (~ 1 arc second).

The size of prisms in prism spectrographs is limited by the expense of procuring large blocks of optical glass free of bubbles, striae and other imperfections, and also by the absorption in the glass, especially at shorter wavelengths, which results in a diminishing return for large collimated apertures. However, for Cassegrain instruments or refractors, the fundamental limit is set by flexure in the collimator for focal lengths in excess of about 1000 mm. For the older generation of telescopes working at about f/18, this limited the beam size A to little more than 50 mm. The coudé spectrograph was designed to overcome this flexure limitation and it allowed a higher resolving power to be attained using larger dispersing elements. At the same time, it was easier to control the temperature of a coudé instrument, an important factor for prism spectrographs.

When blazed gratings first became available in the mid 1930s, the largest were under 150 mm in dimensions. That installed in the 100-inch coudé at Mt Wilson by Adams and Dunham was 111×140 mm, which was comparable in size to the large Littrow prism spectrograph used earlier at that focus [1]. Larger gratings would permit either higher resolving power or wider slits, without the concomitant light loss associated with a large prism. At Palomar, a mosaic grating was installed in the coudé spectrograph of the 5-m telescope using four smaller gratings to fill the 30-cm beam [2].

The coudé arrangement generally requires four reflections to bring the light to a fixed focal plane, a common arrangement being two 45° plane mirrors, one at the intersection of the telescope's optical axis with the declination axis, and the second at the intersection of the declination and polar axes. The benefits of higher resolving power and mechanical and thermal stability therefore have to be offset against light loss associated with these reflections. Another advantage of the coudé arrangement is the ease with which different cameras can be mounted for different resolving powers in a large spectrograph, as has been exploited, for example, on the 100-inch coudé at Mt Wilson.

The resolving power of a slit-limited grating spectrograph is given by Equation 2.100:

$$R \simeq \frac{2L\sin\theta_B}{\theta_s D}. \qquad (3.3)$$

For a given grating (L, θ_B), one changes the angular size of the slit $(\theta_s < \theta_\star)$ to obtain different resolving powers. The camera focal length to satisfy the Nyquist condition is

$$f_{\text{cam}} = \Delta s_{\text{pix}} R \cot\theta_B \qquad (3.4)$$

and hence the reciprocal dispersion, or plate factor, is

$$P \simeq \frac{d\cos\theta_B}{nf_{\text{cam}}} = \frac{\lambda}{2\Delta s_{\text{pix}} R}. \qquad (3.5)$$

Clearly one should choose $f_{cam} \propto R \propto 1/\theta_s$ and hence one obtains $P \propto 1/R \propto \theta_s$. Section 2.5 gives more details.

Thus narrow slits require long focal length cameras and they give high dispersion as well as high resolving power on bright stars. The coudé arrangement provides an ideal spacious laboratory where such cameras can readily be selected when required.

3.2 HISTORY OF THE COUDÉ SPECTROGRAPH

3.2.1 Early coudé refractors

The first coudé telescope was devised by Maurice Loewy (1833–1907) at the Paris Observatory in 1882 [3]. It was a 27-cm refractor in which the 'bent' (coudé) tube contained two 45° plane mirrors to send the converging beam up the polar axle so as to image a star at a fixed focal point on the polar axis (see the book by Danjon and Couder: *Lunettes et Télescopes* for a discussion of the 'équatorial coudé' [4]).

The first equatorial coudé was followed in 1889 by a larger 60-cm instrument with an f/30 focal ratio, and Loewy, as observatory director from 1896, initiated the construction of the first coudé spectrograph on this telescope. The work was undertaken by the firm of P. Gautier and completed in 1907 [5]. It was a Littrow spectrograph using either a 30° silvered half-prism or one or two whole prisms with the beam being returned by means of a silvered mirror. This arrangement gave dispersions of 20, 8 or 4 Å/mm. The collimator was a cemented triplet of focal length 2.5 m, and this also served as the spectrograph camera. The beam size in this spectrograph was 8 cm, which was significantly larger than that of most other spectrographs mounted directly on refractors.

Another instrument of this type was Adolf Hnatek's (1876–1960) coudé spectrograph on the 38-cm Rothschild coudé refractor at the Vienna Observatory [6] – see Fig. 3.1. Here the f/24 beam was reflected down the polar axle into a one-prism spectrograph equipped with a choice of two cameras. The collimator was just over a metre in focal length.

Neither of these early coudé spectrographs were widely used in stellar spectrography. Maurice Hamy (1861–1936) in Paris certainly stressed the benefits of a stationary and fixed spectrograph for radial-velocity work, but not many results were forthcoming.

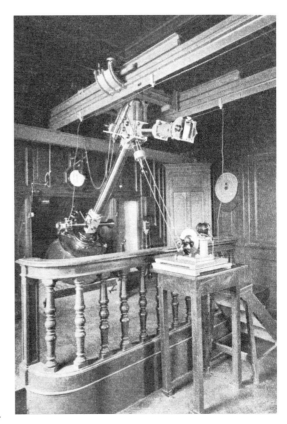

Figure 3.1. The Vienna Observatory coudé spectrograph on the 38-cm Rothschild coudé equatorial telescope, as presented by Adolf Hnatek in 1913.

3.2.2 Coudé prism spectrographs at Mt Wilson

The first coudé spectrograph on a reflector was designed by George Ellery Hale and Walter Adams on the 60-inch telescope at Mt Wilson [7]. It was a Littrow prism instrument, like that in Paris, with the single 63° flat glass prism being traversed twice after reflection. The spectrograph came into operation early in 1911, and represented Hale's vision of a large spectrograph to be employed in photographing the spectra of some of the brighter stars.

The f/30 coudé beam in this telescope was arranged to be in a vertical orientation, requiring five reflections before the slit. The collimated beam size of the Mt Wilson spectrograph was 152 mm, and the collimator focal length was 5.5 m. Both were huge by the standards of any other stellar spectrograph of that time. No doubt the ambition was to acquire spectroscopic data for bright stars of the same quality that Rowland and Hale had already obtained for the Sun.

The Mt Wilson 60-inch telescope and coudé spectrograph was a pioneering instrument for high resolution spectroscopy. An early paper studied the spectra of Sirius, Procyon and Arcturus [7]. A dispersion as high as 1.4 Å/mm was available at Hγ. The success of this instrument led the way for the 100-in Hooker telescope and coudé spectrograph, which became the standard instrument against which all other high resolving power spectrographs were judged for the next four decades.

Originally the 100-inch coudé was a Littrow prism instrument of similar design to that on the 60-inch, but with the spectrograph axis being aligned with the polar axis of the telescope [8]. It could give a resolving power as high as 70 000 at Hγ [1], which was about ten times higher than commonly attained by Cassegrain prism instruments or by prism spectrographs on refractors.

Two Littrow collimator-camera lenses were provided, with focal lengths of 4570 mm and (from 1930) of 2740 mm, so as to give a choice of two dispersions (2.9 or 5.0 Å/mm at Hγ) and hence resolving powers. However, the Littrow format was not an ideal arrangement, as it prevented changing the collimator-to-camera focal-length ratio according to the slit width and resolving power desired.

3.2.3   The coudé grating spectrograph with Schmidt cameras

The transformation of the Mt Wilson coudé spectrograph into a blazed grating instrument with a reflecting collimator and Schmidt cameras was a process that took place over several years in the mid 1930s, under the direction of Adams and Dunham. The whole concept was inspired by the work of Bernhard Schmidt, who had devised his wide-field spherical mirror telescope in Hamburg in 1932 [9], as well as by the availability of the first blazed gratings ruled in aluminium by Robert Wood in Baltimore (see Section 1.6).

The work entailed a complete rebuild of the spectrograph in 1935 and was probably the most significant development in the history of stellar spectrograph design since the pioneering days of Huggins and Vogel in stellar spectrography, about half a century earlier. In the words of Ira Bowen (1898–1973):

Suddenly, however, the picture was completely changed by the advent of the Schmidt camera and

its various modifications and by the development of the blazed grating, which permits the concentration of 60–70 per cent of the light in one order [10].

The first experiments were reported by Dunham in 1934:

A correcting plate at the center of curvature of a spherical mirror, as used by Schmidt for telescopes, has been found useful for spectrograph cameras. Such a camera, of 780 mm focus and 110 mm aperture, has been constructed with the plateholder outside the incident beam so as not to obstruct light. 3000 Å of the first order can be photographed at once in excellent definition. With a collimating mirror the spectrograph will be nearly chromatic and any region may be photographed by merely rotating the grating [11].[1]

For the first time astronomers had a truly achromatic camera that gave excellent definition over a wide angular field.

Two years later, Dunham reported on several faster Schmidt cameras that had been employed or were under construction, including one of focal ratio f/0.57 with a 2° field [12], and by 1938 several such cameras were available on a rotating turntable so as to enable rapid selection [13]. High dispersion ultraviolet spectra at 10.3 Å/mm were recorded with the 32-inch Schmidt camera in conjunction with the Wood blazed grating (first used in 1936) and an aluminized parabolic collimator. Good definition was simultaneously achieved from 310 to 440 nm [14].

Dunham later described how the new camera instantly gave excellent results following the news of Schmidt's discovery:

The ordinary lenses of those days, for short focus cameras were perfectly horrible compromises. So we rubbed up the aspherical plate ... and put it up there in a wooden mounting ... And we got some gorgeous spectra out of it instantly [15].

---

[1]   These first experiments with a grating on the 100-inch coudé used an Anderson plane grating ruled on speculum and then aluminized. The Wood grating, ruled directly on a vacuum-deposited aluminium layer, was used from 1936 and achieved a higher blaze efficiency.

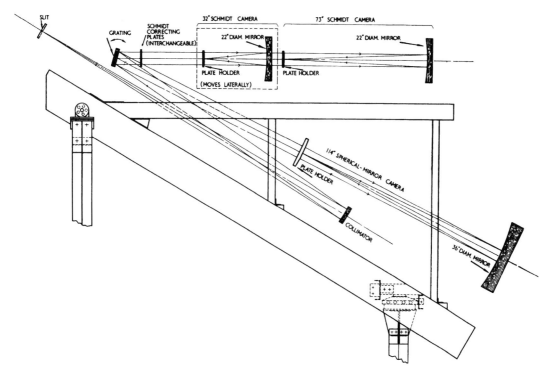

Figure 3.2. Optical and mechanical layout of the Mt Wilson 100-inch coudé spectrograph, in its final form of 1939, as described by Walter Adams in 1941 and Ted Dunham in 1956.

In its final form (in 1939), the Mt Wilson coudé had a 2.74-m off-axis parabolic aluminized collimator mirror, the Wood blazed 590 groove/mm plane grating, used in first or second order, and a choice of three spherical camera mirrors (813 mm, 1.85 m and 2.90 m – the last without a Schmidt corrector plate). Spectra with a resolving power of 80 000 were recorded with the largest camera. The various optical arrangements were discussed by Adams [8] and by Dunham [1]. The whole spectrograph was mounted on a heavy steel triangular frame some 18 feet (5.4 m) long. It was in a wooden housing in a concrete room whose temperature was controlled to 0.1 °C. Figure 3.2 shows the optical and mechanical layout of the Mt Wilson coudé spectrograph, as it was from 1939, with the choice of three cameras.

Various accessory items of equipment were added to the spectrograph, for wavelength comparison spectra, for the calibration of photographic plates (so as to convert photographic densities to intensity at any wavelength), for the uniform widening of spectra using a quartz rocking plate, for automatic guiding [16] and

for monitoring the exposure photoelectrically using an integrating exposure meter [17]. These devices were added by Dunham or by Horace Babcock over the ensuing decade (1939–49) and were described by Dunham in 1956 [1].

3.2.4 New coudé spectrographs

The success of the Mt Wilson 100-inch coudé in its late 1930s arrangement led to a considerable number of coudé spectrographs being constructed, notably during the 1960s and 1970s. These all had certain features in common: they were blazed plane reflection grating spectrographs with reflecting collimators and usually several Schmidt cameras, giving reciprocal dispersions on photographic plates in most cases between about 1 and 40 Å/mm (but values between 4 and 20 Å/mm were the most common).

Table 3.1 gives a summary of selected coudé spectrographs described in the published literature. Two major instruments were delayed by World War II, namely the McDonald 82-inch coudé (completed in

Table 3.1. *Table of selected coudé spectrographs*

Observatory	Telescope	Year coudé installed	Notes	Reference(s)
Paris	60-cm (refr.)	1907	prism spectrograph	[5]
Mt Wilson	60-inch	1911	prism spectrograph	[7]
Vienna	38-cm (refr.)	1913	prism spectrograph	[6]
Mt Wilson	100-inch	1925	prism spectrograph	[8]
Mt Wilson	100-inch	1936–9	blazed gratings (see Fig. 3.2)	[1]
McDonald	82-inch	1949	3 Schmidt cameras	[20] [10]
Mt Palomar	200-inch Hale	1952	4-grating mosaic; 5 cameras	[2]
Haute-Provence	1.93-m	1959	REOSC: 2 gratings, 5 cameras (see Fig. 3.4)	[21, 22]
Okayama	1.89-m	1960	Hilger & Watts; 2 Schmidt cameras	[23]
Lick	120-in	1961	4 cameras	[18]
Radcliffe	74-inch	1961		[24]
Mt Stromlo	74-inch	1961	3 gratings; 3 cameras	[25]
DAO	48-inch	1962	horizontal coudé 32″ and 96″ cameras (see Fig. 3.5)	[19]
RGO	30-inch Thompson	1964	vertical coudé; (see Fig. 3.3) 1 Schmidt camera	[26]
Kitt Peak	84-inch	1965	3 cameras	[27]
Crimean AO	2.6-m Shajn	1967	3 cameras	[28]
Tautenburg	2-m Schmidt	1968	horizontal coudé	[29]
McDonald	107-inch	1969	horizontal coudé; 2 gratings, 2 cameras	[30]
Ondrejov	2-m	1969	1 grating; 1 camera; 6 Å/mm	[31]
Mauna Kea	88-inch	1970		
Western Ontario	48-inch	1971	2 cameras	[32]
Lowell	42-inch	1972		[33]
ESO	1.52-m	1973		[34]
Calar Alto	2.2-m	1974	vertical coudé; 2 cameras	[35]
CFHT	3.6-m	1981		[36]
Yunnan	1-m	1984	1 grating; 1 camera; 4.2 Å/mm	[37]
Xinglong Station	2.16-m	1988		[38]

DAO: Dominion Astrophysical Observatory, BC, Canada; RGO: Royal Greenwich Observatory, UK; ESO: European Southern Observatory, La Silla, Chile; CFHT: Canada-France-Hawaii Telescope, Mauna Kea, Hawaii; Xinglong Station, Beijing Astronomical Observatory.

1949) and that at Palomar on the 5-m (200-inch) telescope (completed in 1952). Thereafter about 20 spectrographs were completed by 1980, mainly on telescopes in the 1- to 3-m class. These were fed by a variety of optical arrangements, using between three and five mirrors before the slit; that on the Lick 3-m telescope had just three mirrors for declinations $\delta < +55°$ [18]. More often four mirrors were used, with the spectrograph axis being aligned with the polar axis of the telescope. Sometimes a four-mirror vertical coudé spectrograph was constructed, with the fourth mirror being fixed and directly above the slit to the north of a symmetrically mounted telescope (as at the Royal Greenwich Observatory, Herstmonceux on the 30-inch – see Fig. 3.3 – and at Calar Alto on the 2.2-m telescopes), while others chose a horizontal coudé with five mirrors in front of the slit (for example, this was the arrangement at Victoria on the 1.2-m DAO telescope, and at the McDonald Observatory on the 2.7-m (107-inch) telescope). A vertical system was claimed to give superior protection against convection, while a horizontal coudé permitted ease of access and new optical arrangements. With a horizontal dispersion it was also relatively immune to convection currents. Focal ratios were between about f/25 and f/46.

Large replica gratings were installed in most of these instruments; often a choice of two or three gratings with different blaze wavelengths in low orders ($n = 1$ to 3) was available, in sizes rarely exceeding 30 cm. Such large gratings with slow coudé beams required very long collimators – for example, the f/38 coudé beam at Lick (on the 3-m telescope) feeds a collimator of focal length 11.6 m. Mosaic gratings were used in some cases, which increases the beam size and hence the slit-limited resolving power, but not the diffraction-limited value, as the grooves were not aligned in phase. The grating mosaic technique was pioneered by Bowen at Palomar [2] with four gratings in a 2×2 mosaic to give a 30-cm beam. A similar arrangement was also used on the 1.2-m telescope at Victoria with the 96-inch camera [19].

Collimators were generally aluminized paraboloids. However, these were sometimes figured so as to be off-axis paraboloids, in which the pole of the paraboloid is not the centre of the mirror, which allows them to operate off-axis without introducing astigmatism into the beam. This was the case, for example, on the 2.7-m coudé at McDonald Observatory [30]. The off-axis

illumination ensures that the collimated beam does not return directly to the slit, but to the grating.

More complex asphericity of collimators has been proposed [39] so as to eliminate the need for a Schmidt corrector plate in a Schmidt camera. In effect, the spherical aberration is removed before the grating, thereby avoiding the light losses caused by the only transmissive element in a conventional telescope-spectrograph system. Such an arrangement has been successfully used at Victoria for cameras slower in focal ratio than f/5 [19].

No corrector plate is in any case required for the longest focal length cameras (focal ratios of about f/10 or slower) used for the highest resolving powers with the narrowest slits in coudé spectrographs, as the spherical aberration becomes negligible in such cases.

Another innovative design aspect of coudé spectrographs involves overfilling the grating with the collimated beam, by having a longer focal length collimator than the grating size times the telescope focal ratio would predict. This loses some light, but gains either resolving power (θ_s is smaller) or, by using a wider slit for given θ_s, gains more light than is lost by overfilling (see Section 2.3.2). The suggestion was first made by Robert Tull in 1972 [40] and was used in the design of the 2.7-m coudé at McDonald Observatory.

A coudé spectrograph and telescope will generally have six or more mirrors and a reflection grating. With six not freshly silvered mirrors, the reflectivity might be only 80 per cent per surface, while a blazed grating typically delivers 70 per cent at the blaze maximum. Only 18 per cent of the light would then reach the detector, without taking losses at the slit into account. This figure rises to 33 per cent, however, if the reflectivity is 0.88, which is about the maximum for a freshly coated aluminium mirror in the visual region. At Victoria, on the 1.2-m telescope, Harvey Richardson (b. 1927) undertook pioneering developments to improve the otherwise poor throughput of the coudé configuration by using high reflectivity coatings, which gave 97.5 per cent reflectivity for all surfaces except for the primary mirror. Differently coated mirrors are however required for three different broad wavelength regions [41]. The coudé spectrograph on this telescope has a horizontal orientation, and the layout of the spectrograph in the coudé room is shown in Fig. 3.5.

In addition, an ingenious reflecting image slicer [42] was also developed. With these two improvements

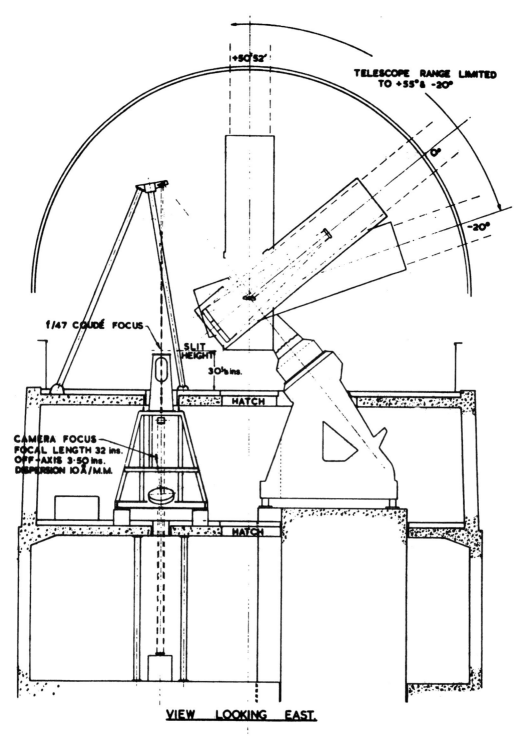

Figure 3.3. The four-mirror coudé system on the 30-inch telescope at the Royal Greenwich Observatory. The f/47 light beam was sent to a vertically mounted coudé spectrograph.

$$\epsilon = \frac{4w'}{\pi \theta_{\star} f_{\mathrm{tel}}} \qquad (3.6)$$

would be as low as 5 per cent on the Hale 5-m telescope in seeing of $\theta_{\star} = 1$ arc second, provided that the image of the coudé slit width w' on the detector is matched to the detector, which is taken to be the 30-μm grain size.

To overcome this problem, Bowen devised the first image slicer, in which the stellar image of diameter $\theta_{\star} f_{\mathrm{tel}}$ is divided into a number $\frac{\theta_{\star} f_{\mathrm{tel}}}{w}$ of parallel slices, which are placed end on end by means of reflections from a stack of plane mirrors, so as to form a linear slit [43]. Bowen proposed a stack of suitably angled (at $45°$ to the slit plane) and staggered thin glass plates with aluminized faces to achieve the slicing – see Fig. 3.6. The device was placed immediately in front of a long narrow slit, near the coudé focus. In order to avoid excessive widening of the final spectrum, caused by the slices being placed end to end on the slit, Bowen proposed using a cylindrical lens in front of the photographic plate.

Efficiencies of 75 to 90 per cent at the slit were achievable with the Bowen image slicer, instead of just 5 per cent without it. Keith Pierce has discussed the practical implementation of the Bowen devices at Kitt Peak, with either 13 or 20 slices being used [44].

The Bowen image slicer was adapted by Theodore Walraven (b. 1916) who, in 1972, described an image slicer based on multiple internal reflections within a thin glass plate [45]. The plate was cemented to a $45°$ prism in which a thin groove was cut, which allowed light in successive thin slices to pass on into the prism instead of being reflected in the plate. The result was an image in some eight or nine slices arranged end to end and giving an effectively long and narrow spectrograph slit. This device is usually known as the Bowen–Walraven image slicer.

Yet another adaptation of the Bowen slicer was devised by Hiroshi Suto and Hideki Takami in Japan in 1997 [46]. Their device comprised a stack of thin plates, with successive slices of the image suffering a single reflection from a silvered layer between the plate contacts.

A second type of image slicer, called an image transformer, was proposed by William Benesch (b. 1922) and John Strong at Johns Hopkins University in 1951 [47]. Their instrument sliced an image into three sections, using plane mirrors on each side of

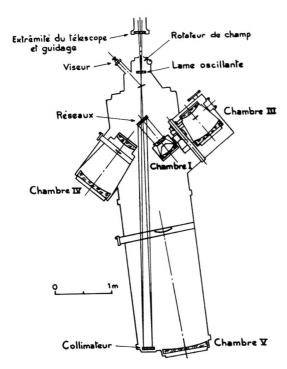

Figure 3.4. The coudé spectrograph on the 1.93-m telescope at the Observatoire de Haute-Provence was made by the firm of REOSC and installed in 1959. The instrument is entirely enclosed in a tank to keep out dust and stray light, and is aligned with the telescope's polar axis.

to light throughput, the performance of the DAO 1.2-m coudé spectrograph was comparable to that of the 5-m Palomar telescope, in spite of the more than four times larger telescope aperture and twice the spectrograph collimated beam size at Palomar. The DAO performance was several times better than either of the coudé spectrographs on the Kitt Peak 84-inch or the Mt Stromlo 74-inch telescopes at similar dispersions. This work clearly demonstrated the benefit and importance of carefully conserving the valuable photons in a long coudé light path.

3.2.5 The image slicer in high resolution spectroscopy

High resolution spectrographs often necessitate the use of narrow slits whose angular size θ_s is substantially less than the seeing disk diameter, $\theta_{\star}$. This can lead to great inefficiency of the spectrograph as a result of light loss at the slit. This problem was discussed by Bowen in 1938 [43], who pointed out that the efficiency

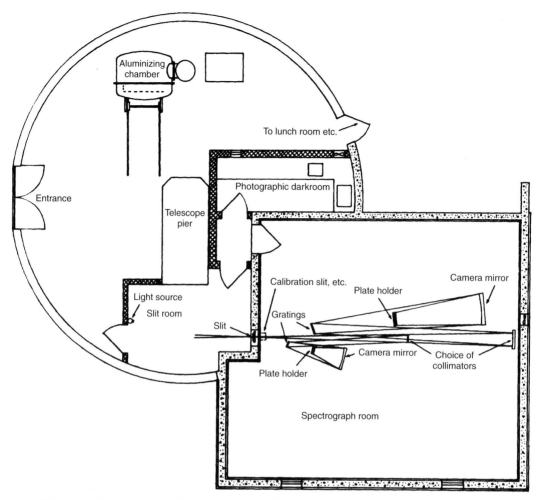

Figure 3.5. Layout of the horizontal coudé spectrograph on the 1.2-m telescope at the Dominion Astrophysical Observatory near Victoria, BC, Canada.

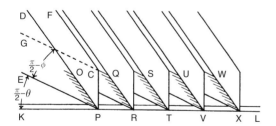

Figure 3.6. Construction of Bowen's image slicer of 1938. KL is the slit; P, R, T etc. are successive glass plates, whose aluminized faces (shaded) act as small mirrors that slice the image and reflect successive slices into the narrow slit.

the slit. These mirrors sent the side lobes to concave mirrors which, after a third plane mirror reflection,

reimaged the side lobes onto the slit above and below the central section. Another adaptation of this concept involved five reflections for the side lobes of the image, but still three slices. Although more reflections are required than for a Bowen image slicer, all parts of the image are exactly in focus on the slit, which is not the case for Bowen's device.

A later adaptation of the Benesch and Strong image transformer was devised by John Strong and Frederic Stauffer in 1963. Their instrument consisted of a slit whose jaws were two small sapphire prisms, which refracted the side sections of an extended disk. The refracted rays are reflected back by concave mirrors on each side of the direct beam to small plane mirrors where the side lobes are imaged and reflected so as to locate them above and below the central image slice that passed directly through the slit jaws [48]. This therefore formed a three-slice linear slit. This form of image slicer was limited to three slices, and the use of refracting slit jaws also limited the wavelength range over which it could operate.

Another configuration for an image slicer was described by Harvey Richardson at Victoria in 1966 [42, 19]. This device comprises a cylindrical lens followed by four concave spherical mirrors, two of which define an entrance slot whose width equals the stellar image diameter, and two of which define the spectrograph slit. The two pairs of carefully adjusted mirrors face each other and are separated by their radius of curvature. Light not passing straight through the slit is reflected back to the aperture mirrors and thence towards the slit again where, as the next slice, it may pass now or after several such double reflections which comprise successive slices towards the edge of the stellar image.

In the Richardson slicer, all slices eventually pass through the same part of the slit, but they travel in slightly different directions and illuminate different parts of the collimator. One advantage is that the spectrum is not widened any further than it would otherwise be, as the spectrum of each slice is coincident on the detector. The fullest description of the Richardson slicer is in an article by Richardson in 1972 [49]. The device was first implemented on the 1.2-m coudé spectrograph at the Dominion Astrophysical Observatory in Victoria, and has been copied elsewhere (see for example [50]).

A fourth arrangement of image slicer was described by Narinder Kapany (at the Illinois Institute of Technology, Chicago) in 1958 [51]. This device, which was called a light funnel, consisted of some 250 optical fibres of 50 μm core diameter in a bundle about one millimetre in diameter. The fibres were reformatted into a linear slit of length 12.5 mm. The length of the fibres was 60 mm. The grainy nature of the slit with its individual fibres was overcome by having a rocking plate that oscillated the slit image along its length. The final spectrum on the photographic plate was then compressed in the cross-dispersion dimension by means of a cylindrical lens – see Fig. 3.7.

The Kapany device is interesting because it was designed and built in the early days of fibre optics some

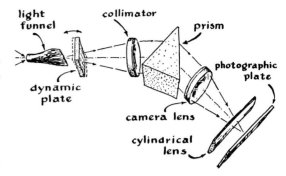

Figure 3.7. Narinder Kapany's light funnel (1958) comprised some 250 optical fibres which were reformatted into a linear spectrograph slit, so as to act as an image slicer.

two decades before low-loss fused-silica fibres were available. It can be regraded as an early prototype of the integral field unit devices in astronomy using fibre bundles.

A practical example of reformatting an image with optical fibres was built by C. Vanderriest at the Meudon Observatory in 1980 [52] (see Section 7.5). The Vanderriest arrangement reformatted a hexagonal array of 169 fibres into a 23-mm long linear slit. Devices of this type suffer from about 25 per cent light loss in the interstitial spaces between the fibres, as well as from reflection and absorption losses and focal ratio degradation in the fibres. Nonetheless, improvements in fibre technology since 1980 have made such instruments popular solutions for image slicing, multi-object spectroscopy and as integral field units (see Section 7.5).

3.3 THEORY OF THE ÉCHELLE SPECTROGRAPH

The échelle grating is a high blaze angle (θ_B) grating devised and named by George Harrison at MIT [53]. Harrison pointed out that the angular dispersion of a grating *at a given wavelength*, λ, depended only on the angle of diffraction, and not at all on the groove spacing. A coarsely ruled grating with a blaze angle of over 45° could therefore give high angular dispersion ($\frac{\mathrm{d}\beta}{\mathrm{d}\lambda}$) and hence high resolving power $R = \frac{\lambda}{\delta\beta}\frac{\mathrm{d}\beta}{\mathrm{d}\lambda}$, where $\delta\beta$ is the angular spread of monochromatic diffracted rays leaving the grating, whether arising from diffraction or from a finite slit width.

Table 3.2. *Order numbers for échelle gratings at optical wavelengths*

échelle blaze	λ (nm)	31.6	79.	316.
			$1/d$ (gr/mm)	
$\tan\theta_B = 2$	350	162	65	16
	550	103	41	10
	750	75	30	8
	950	60	24	6
$\tan\theta_B = 4$	350	175	70	
	550	112	45	
	750	82	33	
	950	65	26	

Table 3.3. *Summary of échelle order properties*

Free spectral range	$\Delta\lambda_{FSR} = \lambda/n \simeq \frac{2s}{n^2} \simeq \frac{\lambda^2}{2s}$
Angular dispersion	$\frac{d\beta}{d\lambda} = \frac{n}{d\cos\beta} \simeq \frac{2\tan\theta_B}{\lambda}$
Plate factor	$P = \frac{d\cos\beta}{nf_{cam}} \simeq \frac{\lambda}{2f_{cam}\tan\theta_B}$
Length of order	$\Delta S_{order} = f_{cam}\frac{\lambda}{d\cos\beta} \simeq \frac{2f_{cam}\tan\beta}{n}$

This property of the échelle arises directly from the grating equation,

$$n\lambda = d(\sin\alpha + \sin\beta),\qquad(3.7)$$

which results in

$$n\frac{d\lambda}{d\beta} = d\cos\beta\qquad(3.8)$$

and hence

$$\frac{d\beta}{d\lambda} = \frac{\sin\alpha + \sin\beta}{\lambda\cos\beta}.\qquad(3.9)$$

In the near-Littrow mode, where $\alpha \simeq \beta$, this can be written

$$\frac{d\beta}{d\lambda} \simeq \frac{2\tan\beta}{\lambda}.\qquad(3.10)$$

In terms of the blaze angle, we can write $\alpha = \theta_B + \theta$ and, in the centre of an order, $\overline{\beta} = \theta_B - \theta$. The angle θ in the quasi-Littrow condition is a small angle ($\theta \ll \theta_B$), termed here the Littrow angle. Using this notation, one obtains for the angular dispersion

$$\frac{d\beta}{d\lambda} = \frac{2\tan\theta_B}{\lambda(1 + \tan\theta_B\tan\theta)}.\qquad(3.11)$$

For example, if $\tan\theta_B = 2$ and $\theta = 5°$ (typical values) then $\lambda\frac{d\beta}{d\lambda} = 3.4$ radians, or $97.5°$. This compares with a conventional grating say $\theta_B = 10°$, $\theta = 15°$, giving $\lambda\frac{d\beta}{d\lambda} = 0.34$ radians, fully a factor of ten less.[2]

The échelle grating is a coarsely ruled grating operating simultaneously in many high orders. Common groove spacings have been 31.6 and 79.0 gr/mm, although some échelles have been made at 52 or 316 gr/mm. Table 3.2 gives the order numbers for the optical wavelength region, valid for the perfect Littrow case ($\theta = 0°$ and $\gamma = 0°$).

---

[2] A grating with a blaze angle of $10°$ illuminated at a Littrow angle of $15°$ and having 600 gr/mm would have $\lambda_B = 559$ nm in first order. As will be discussed, small values of θ are critical for échelles but not for conventional gratings, where a larger value of the angle θ may facilitate the layout of the spectrograph.

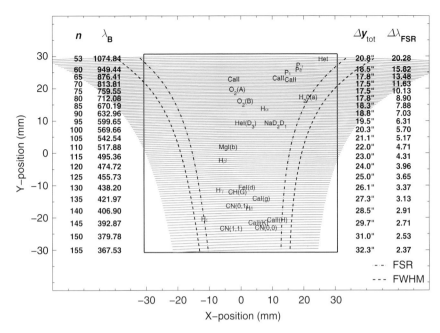

Figure 3.8. A computer-generated map of the échelle format, in this case for the Hercules spectrograph at Mt John Observatory – see Section 8.10. The free spectral range of the orders is between the dash-dot curves, while the half-maximum intensity of the blaze function is at the dashed curves. The ends of the orders as drawn are where the intensity reaches its first minimum. The approximate locations of a few prominent spectral lines are shown. The 60 mm by 60 mm square represents the boundary of the 4096 × 4096-pixel CCD detector.

As expected, the order numbers depend slightly on the blaze angle, and are proportional to the groove spacing d. The coarser gratings (smaller $1/d$) have higher order numbers but a larger number of shorter orders, which may make them more suitable for the format of CCD detectors. For this reason, finely ruled échelles (316 gr/mm) have rarely been used, even though they suffer less from a rapid variation of intensity with wavelength along an order. The spectrograph built at Queen's University, Belfast, is one exception, using a 316 gr/mm grating [54, 55].

3.3.1 Line tilt in échelles

Clearly, with an échelle spectrograph one has the possibility of simultaneously recording many orders – as many as about a hundred for a 31.6 gr/mm échelle or about 40 for 79 gr/mm within the entire optical range. A typical échelle format is shown in Fig. 3.8, which is a map of the format for the Hercules R2 spectrograph with a 31.6 gr/mm échelle and prism cross-dispersion – see Section 8.10.

Order separation can be effected by means of a prism, grism or conventional grating, giving a low dispersion in a direction orthogonal to the high échelle dispersion. If the angular cross-dispersion is $(\frac{d\beta}{d\lambda})_\times$ then the resultant angular dispersion along the orders will be marginally increased to $(\frac{d\beta}{d\lambda})_{total}$ given by

$$\left(\frac{d\beta}{d\lambda}\right)_{total} = \left(\left(\frac{d\beta}{d\lambda}\right)_{échelle}^2 + \left(\frac{d\beta}{d\lambda}\right)_\times^2\right)^{1/2}$$
$$\simeq \left(\frac{d\beta}{d\lambda}\right)_{échelle}\left\{1 + \frac{1}{2}\left(\frac{d\beta}{d\lambda}\right)_\times^2 \bigg/ \left(\frac{d\beta}{d\lambda}\right)_{échelle}^2\right\}$$
$$(3.12)$$

to first order. Moreover, the lines will show a tilt, relative to the cross-dispersed orders, given by $\tan\phi_\times = (\frac{d\beta}{d\lambda})_\times/(\frac{d\beta}{d\lambda})_{échelle}$.

There are in fact three possible causes of line tilt in échelles, relative to the orders. One is caused by the cross-dispersion, as discussed above, and is always present. Then there may be the effect of illuminating the grating at a small angle γ to the normal plane, in which case

$$\tan\phi_\gamma = 2\tan\theta_B\tan\gamma \qquad (3.13)$$

(see Section 2.7). This effect can have either positive or negative sign for ϕ_γ, depending on the sign of γ, so can be arranged so as partially to cancel the tilt due to the cross-dispersion, at least in some orders.

Thirdly, the slit can be rotated by an angle, ψ, about an axis normal to the slit. Light from different points along the slit height then has different values of the angle of incidence, α, on the échelle, and the slit image will as a result be rotated through an angle $\phi_s = \psi \frac{\cos \alpha}{\cos \beta}$. Hence the rotation angle is demagnified by the anamorphic demagnification factor $(\cos \alpha / \cos \beta)$, which is normally less than one. This rotation can also be arranged to compensate tilt due to the cross-dispersion, but at the expense of a slight loss in resolving power. It is noted that the first two effects are respectively absent or negligible in spectrographs with conventional gratings with small blaze angles.

Echelle spectrographs normally have a cross-dispersing element after the échelle, but sometimes it is before the échelle, or the light is cross-dispersed twice by a prism in a double-pass mode, both before and after. Cross-dispersion before the échelle, however, in effect causes γ to be a function of wavelength or order number. In this case the line tilt can be considerable and vary markedly from order to order, which may complicate the reductions. A severe example of this is illustrated in the variable line tilt of $16°$ to $18°$ in the MIT Cassegrain échelle spectrograph, which uses pre-échelle prism cross-dispersion [56].

3.3.2 Efficiency of échelles as a function of θ and γ

The efficiency of échelle gratings along an order is crucial to overall spectrograph performance. Echelle gratings show the related effects of shadowing and of facet obscuration far more severely than do conventional gratings, as has been discussed by Schroeder and Hilliard [57]. These authors have identified three possible modes of operating an échelle:

Case A	$\alpha > \overline{\beta}$	$\gamma = 0$
Case B	$\alpha < \overline{\beta}$	$\gamma = 0$
Case C	$\alpha = \theta_B = \overline{\beta}$	$\gamma \neq 0$

Here $\overline{\beta}$ is the angle of diffraction in the order centre at the blaze peak.

In case A, the intensity distribution over an order is the diffraction pattern of the illuminated part of a facet of width $f' = f(1 - \tan \theta \tan \theta_B)$. This is given by the square of a sinc function, namely

$$I(\delta) = \left(\frac{\sin \delta}{\delta} \right)^2 \tag{3.14}$$

where

$$\delta = \frac{2\pi}{\lambda} f' \sin \left(\frac{\beta - \overline{\beta}}{2} \right) \cos \left(\theta - \left[\frac{\beta - \overline{\beta}}{2} \right] \right) \tag{3.15}$$

is the phase difference between the edge and centre of the effective facet width, f'. In the Littrow case of $\theta = 0°$, the blaze wavelengths of order centres almost coincide with the first minima in the function $I(\delta)$, which occur when $\delta = \pm 2\pi$ or when

$$\sin(\beta - \overline{\beta}) = \pm \frac{\lambda}{f} = \pm \frac{\lambda}{d \cos \theta_B}, \tag{3.16}$$

giving

$$\Delta\beta = \beta - \overline{\beta} \simeq \pm \frac{\lambda}{d \cos \theta_B} \tag{3.17}$$

for small angles $\Delta\beta$. On the other hand, the values of β in order n that correspond to wavelengths, which in adjacent orders $(n \pm 1)$ occur at blaze maxima, are given by

$$\sin \beta - \sin \overline{\beta} = \pm \frac{2}{n} \sin \theta_B \cos \theta = \pm \frac{\lambda}{d}, \tag{3.18}$$

which results in

$$\Delta\beta = \beta - \overline{\beta} = \pm \frac{\lambda}{d \cos \theta_B} \tag{3.19}$$

as $\sin \beta - \sin \overline{\beta} \simeq \cos \theta_B . \Delta\beta$. Hence, in the Littrow condition, the first minima of the blaze function occur at wavelengths that are close to the centres of adjacent orders. Therefore it follows that photons at the order centre wavelengths given by

$$\lambda = \frac{2d(\sin \alpha + \sin \overline{\beta})}{n} \tag{3.20}$$

are not diminished in intensity in order n by photons going into the adjacent orders. This statement is equivalent to saying that the full-width of an order at zero intensity is, for the Littrow condition of $\theta = 0°$, equal to twice the free spectral range, or that photons of a given wavelength can appear in the main peaks of no more than two adjacent orders, and not in three or more.

However, for non-zero θ, the reduction in the effective facet width due to facet obscuration, or groove shadowing, means that the whole diffraction pattern is broadened by a factor $1/(1 - \tan\theta \tan\theta_B)$, which can be substantial for échelle gratings. This lowers the peak intensity in any given order by the same factor. In addition, the minima no longer occur at the principal maxima of other orders. Instead, light is removed from the peak by the wings in other orders, thereby further degrading échelle efficiency. Schroeder and Hilliard [57] find, from numerical calculations, that the fall-off in échelle efficiency with increasing θ in case A approximately goes as

$$r = \frac{\cos\alpha}{\cos\beta} = \frac{(1 - \tan\theta \tan\theta_B)}{(1 + \tan\theta \tan\theta_B)}, \quad (3.21)$$

the anamorphic demagnification of the échelle. This factor is close to unity for conventional gratings, but for example, equals 0.55 for an échelle grating having $\tan\theta_B = 2$ and operated at $\theta = 5°$. Similar conclusions were reached by Walker and Diego in discussing the relative merits of case A (in-plane mounting) and case C (which they term the quasi-Littrow mounting) [58].

In case B, the effective facet width is also reduced, as light falling on the deepest part of the facet is reflected off the step of the groove back to the collimator, and hence is lost (see Section 2.6 and Fig. 2.10). The facet width is then reduced to f' given by

$$f' = f(1 + \tan\theta \tan\theta_B) \quad (\theta < 0) \quad (3.22)$$

and hence the diffraction pattern from the facets is once again broadened, as in case A, by about the same factor for a given value of $|\theta|$. However, in case B some of the light is lost by shadowing, the fraction *not* lost being

$$\phi = \frac{f'}{2f - f'} = \frac{(1 + \tan\theta \tan\theta_B)}{(1 - \tan\theta \tan\theta_B)} = \frac{1}{r}. \quad (3.23)$$

(Note that $r = \frac{(1 - \tan\theta \tan\theta_B)}{(1 + \tan\theta \tan\theta_B)} > 1$ for $\theta < 0$.) Thus in case B, the efficiency declines both because of the broadening of the diffraction pattern (resulting in a mismatch between first minima and the blaze angles of adjacent orders), and also by shadowing, which causes the loss of some of the light altogether. Case B is therefore always less efficient than case A, and it is generally not used in échelle spectrograph design.

In case C ($\theta = 0°; \gamma \neq 0°$) there is no shadowing or obscuration, and so the full facet width f is used. The orders are therefore narrower and more peaked than in

cases A or B. The first minima occur at the same angular positions as for the true Littrow case of $\gamma = \theta = 0°$, that is, at $\Delta\beta = \pm \frac{\lambda}{d \cos\theta_B}$, but the wavelengths at these minima are slightly closer to the central blaze wavelength, because of the higher angular dispersion when $\gamma \neq 0$. The angular dispersion is

$$\frac{d\beta}{d\lambda} = \frac{n}{d \cos\gamma \cos\beta} \simeq \frac{n}{d \cos\gamma \cos\theta_B} \quad (3.24)$$

for case C, but

$$\frac{d\beta}{d\lambda} = \frac{n}{d \cos\theta \cos\theta_B (1 + \tan\theta \tan\theta_B)} \quad (3.25)$$

for case A. The result in case C is that the first minima occur at a wavelength $\Delta\lambda = \lambda_0 \pm \frac{\lambda_0 \cos\gamma}{n}$ from the blaze wavelength λ_0 in order n, whereas the free spectral range is still $\frac{\lambda_0}{n}$, thereby causing a small mismatch between first minima and the peaks of adjacent orders. According to Schroeder and Hilliard [57], échelle efficiency declines slightly, going as $\cos\gamma$.

Even in the true Littrow case of $\gamma = \theta = 0°$, there is a mismatch in low orders. The minima in order n occur at $\Delta\lambda = \lambda_0 \pm \frac{\lambda_0}{n}$, whereas the maxima of the adjacent orders occur at $\Delta\lambda = \lambda_0 + \frac{\lambda_0}{(n-1)}$ and $\Delta\lambda = \lambda_0 - \frac{\lambda_0}{(n+1)}$ respectively. The wavelength difference between successive maxima is no longer, for small n, equal to the difference between a maximum and its adjacent minimum. Efficiency therefore declines in low orders (at large wavelengths), a result that is common to all cases (A, B and C) of échelle usage.

Two further points are relevant. In case C, if the slit is rotated so as to avoid line tilt, then the slit has to be narrower to maintain the same resolving power. This reduces throughput, and more or less cancels the apparent advantage of case C over case A [57].

The second point is that the problems of shadowing or facet obscuration in cases A and B and of line tilt in case C are far more severe for R4 ($\tan\theta_B = 4$) échelles than for R2. In comparing R4 with R2 échelles, we can conclude that for R4:

1. the effective facet width f' in case A (due to facet obscuration) decreases twice as fast with increasing θ;

2. the angular spread of the orders increases twice as fast in cases A or B with increasing θ, leading to low efficiencies for moderate values of θ;

3. in case C, R4 échelles have a line tilt of 8γ, twice as much as for the R2 échelle;

4. the beam size of the R4 spectrograph is halved for a given grating size L, and hence optical components are only half as large, reducing costs;

5. the resolving power of an échelle of given size L goes as $\sin\theta_B$; therefore this is about 8.5 per cent greater for R4 than for R2 échelles.

The advantages of points (4) and (5) in favour of the R4 grating must be offset against the disadvantages of points (1) to (3). The availability of the R4 échelle grating [59] has led to its use in some recent échelle instruments, such as in the ELODIE spectrograph [60] – see Section 8.5.

3.4 HISTORICAL DEVELOPMENT OF ÉCHELLE SPECTROGRAPHS

Following Harrison's paper describing the échelle grating in 1949 [53], échelle gratings were first used for laboratory and for solar spectroscopy. The first échelle gratings from Bausch and Lomb generally had 200 gr/mm, and this gave a relatively large free spectral range in low order numbers. A 3×6-inch échelle of this type, with a $63°$ blaze angle (R2), and crossed with a concave grating was used by Harrison and colleagues at MIT in 1952 to build a laboratory spectrograph able to achieve resolving powers of 2 to 5×10^5, and reciprocal dispersions of less than 0.5 Å/mm [61]. Wavelengths were measured to a precision of about 0.001 Å with such an instrument [62]. At the same time, Keith Pierce, Robert McMath and Orren Mohler (1908–85) at the University of Michigan used one of the Bausch and Lomb échelle gratings for the first astronomical échelle spectroscopy [63]. Their instrument was used for solar spectroscopy with a resolving power of 250 000.

Further astronomical applications were clearly envisaged in Harrison's review article of 1955 [64]. Here he described a Bausch and Lomb Littrow échelle using a prism cross-disperser in a double-pass mode, and also a giant auroral échelle spectrograph built for the US Air Force, and comprising a mosaic of four échelles. Harrison noted: 'The compactness, speed, and high intrinsic resolving power and dispersion of the échelle should make it extremely useful in many astronomical applications' [64].

It is noted here in passing that the concept of cross-dispersion, essential for order separation in an échelle, was not entirely new. As early as 1876 prism cross-dispersion had been used by Charles Young to separate relatively high orders ($n = 7$ to 10) in a solar grating spectroscope [65]. Much later, but still before the échelle grating's introduction, Robert Wood was using a coarse grating as a cross-disperser for the separation of higher orders from another grating [66].

In the early 1950s, work on an échelle spectrograph for solar ultraviolet spectroscopy from an Aerobee rocket was begun by Richard Tousey (1908–97) and J. D. Purcell at the US Naval Research Laboratory. This instrument comprised a replica R2 échelle grating from Bausch and Lomb (79 gr/mm) with a double-pass prism for order separation [67]. The spectrograph made several flights between 1957 and 1964, and more than 7000 solar lines in the region 200–300 nm were photographed and catalogued [68].

3.4.1 Coudé échelle spectrographs

The first stellar échelle spectrographs were devised in the mid 1960s. Early examples were to be found at the coudé focus at the Crimean Astrophysical Observatory [69], at the Okayama Observatory in Japan [70] and at Harvard's Oak Ridge Observatory in Massachusetts [71]. All used R2 échelles with a grating cross-disperser (at Okayama) or a prism in double-pass mode (in the Crimea and at Oak Ridge). Another example of such an instrument was assembled at Mt Stromlo Observatory in Australia in 1971, and this employed a grating as cross-disperser [72]. All these instruments involved the adaptation of existing conventional coudé grating spectrographs so as to accommodate an échelle grating, to give a higher resolving power or to produce a more compact spectral image format than otherwise would be possible. Table 3.4 gives details of a selection of coudé échelle spectrographs described in the literature.

An example of a modern coudé échelle is the University College London spectrograph on the Anglo-Australian telescope, known as UCLES. This instrument has three large cross-dispersion prisms ahead of the échelle, and a folded Schmidt camera [58]. The optical layout is in Fig. 3.9.

Table 3.4. *Table of selected coudé or Nasmyth échelle spectrographs*

Observatory or instrument	Telescope	Year échelle completed	Échelle grating	Cross-disperser	Reference(s)
Crimean AO	2.6-m	1965		double-pass prism	[69]
Okayama AO	1.89-m	*c.* 1965	R2(79)	grating CD	[70]
Oak Ridge	Wyeth 1.55-m	1970	R2(73.25)	prism in double pass	[71]
Mt Stromlo	1.88-m	1971	R2(79)	grating after échelle	[72, 73]
Lick (Hamilton)	3-m Shane	1986	R2(31.6)	2 prisms after échelle	[74, 75, 76]
Mauna Kea (HIRES)	Keck I 10-m Nasmyth		$3\times$ R2(46. 5) mosaic	$4\times$ grating mosaic after échelle	[77]
CFHT	3.6-m		R2(31.6)	$3.6°$ grism	[78, 79]
ARCES	ARC 3.5-m Nasmyth		R2(31.6)	2 prisms after échelle	[80]
AAO UCLES	3.9-m AAT	1988	R2(31.6 or 79)	3 prisms before échelle	[81, 82]
McDonald	2.7-m	1992	R2(79 or 52.6)	2 prisms in double pass	[83, 84, 85]
ESO Paranal: UVES	8-m VLT (Kueyen)	2000	R4(31.6) (blue) R4(41.6) (red) 2-grating mosaics	grating after échelle	[86]

CD: cross-disperser; R2(79) refers to an R2 échelle grating with 79 gr/mm

ARC: Astronomical Research Consortium, comprising University of Chicago, NM State University, Princeton, Washington University and Washington State University and operating the Apache Point Observatory, NM

UCLES: University College London échelle spectrograph; CFHT: Canada-France-Hawaii Telescope

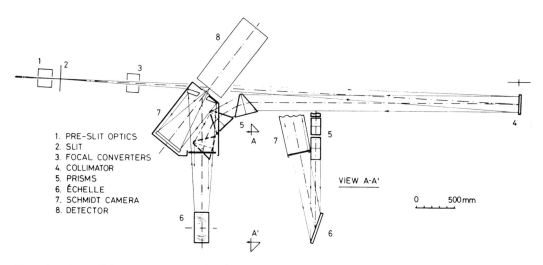

1. PRE-SLIT OPTICS
2. SLIT
3. FOCAL CONVERTERS
4. COLLIMATOR
5. PRISMS
6. ÉCHELLE
7. SCHMIDT CAMERA
8. DETECTOR

VIEW A-A'

0   500 mm

Figure 3.9. The coudé échelle spectrograph UCLES on the Anglo-Australian telescope.

3.4.2 Cassegrain échelle spectrographs

Probably the most dramatic benefits of the échelle in stellar spectroscopy were in Cassegrain échelle spectrographs, in many cases on relatively small telescopes. For the first time high dispersion studies, albeit on rather bright stars, could be undertaken on 1-m class instruments. The pioneer in this field was Daniel Schroeder [87, 88] with Cassegrain instruments on the 0.91-m Pine Bluff reflector at the University of Wisconsin.

Further instruments of this type followed, including those at the 1.5-m telescope of the Smithsonian's Whipple Observatory on Mt Hopkins, Arizona [89], which was frequently used in conjunction with the Kron electronic camera, in spite of the small 40-mm diameter cathode size of this detector. The Smithsonian échelle spectrograph was duplicated elsewhere, including at Mt John Observatory in New Zealand (0.61-m and 1.0-m reflectors) [90, 91], at the Ritter Observatory, Toledo, Ohio (1-m reflector), at Harvard's Oak Ridge Observatory (1.55-m reflector), and at the David Dunlap Observatory in Toronto (1.88-m reflector). Figure 3.10 shows the Mt John instrument on the 0.61-m telescope in 1986, with a Reticon diode array detector, and Fig. 3.11 shows a ray diagram for this instrument. Figure 3.12 is an example of an échelle spectrum recorded at Mt John Observatory of the bright star Canopus.

In 1956 the distinguished American stellar spectroscopist Dean McLaughlin (1901–65) had written: 'Detailed analysis of stellar spectra with high dispersion is strictly the province of the largest telescopes with coudé spectrographs. It seems hopeless to dream of doing such work with lesser instruments' [92]. Little more than a decade later the échelle grating was making McLaughlin's 'hopeless dream' a reality!

Cassegrain échelle spectrographs continued to be popular throughout the 1970s and 1980s, on telescopes of all sizes, including large reflectors such as the 4-m Mayall telescope at Kitt Peak [93, 94]. The introduction of CCD detectors into astronomy from about 1975, initially with small formats, made the échelle spectrograph especially suitable for obtaining a reasonable wavelength coverage over several orders.

Table 3.5 summarizes the properties of a selection of Cassegrain échelle instruments. The predominance of R2 instruments with 79 or occasionally 31.6 gr/mm is noted. The resolving power obtained was typically 30 to 40 thousand in many cases, but rarely as high as 60 000. The McDonald Observatory Cassegrain échelle for the 2.1-m reflector (known as the Sandiford échelle) [105] is an example where $R = 60\,000$ is achieved in an unusually compact Littrow design with high through-put efficiency (of over 10 per cent). This instrument featured a cross-dispersed prism in double pass and a five-element all-refracting collimator and camera.

The fundamental limitation to Cassegrain échelle spectroscopy has been the flexure of a large instrument. Despite extreme care in the mechanical design, as shown in the Sandiford échelle at McDonald, flexure can be greatly reduced but not eliminated. A spectral image shift of half a pixel in long four-hour exposures still can occur, thereby rendering the precise measurement of radial velocity impossible.

3.4.3 Fibre-fed échelle spectrographs

From the mid 1980s, the alternative approach of the fibre-fed échelle spectrograph has been an interesting solution to the problem of flexure found in Cassegrain instruments. The technique was pioneered by Laurence Ramsey and his colleagues at the Pennsylvania State University [106, 107, 108, 109]. An optical fibre coupled the Cassegrain focus to a stationary bench-mounted spectrograph, and this also allowed the spectrograph to be located in a thermally stabilized environment. Moreover the use of the fibre, which scrambles the starlight, makes such spectrographs immune to telescope guiding errors when precise radial velocities are the goal.

The Pennsylvania State fibre-coupled échelle was soon followed by others, for example at the Landessternwarte in Heidelberg in 1987 [110], at Kitt Peak in 1987 [111] and at Mt John in New Zealand in 1988 [112, 113].

In recent years some complex fibre-fed échelle spectrographs have been designed and built, which exploit the intrinsic suitability for precise radial-velocity studies of this type of instrument. The so-called ELODIE spectrograph (see Section 8.5) is one such example [60]. An R4 ($\theta_B = 76°$) grating and a white pupil design, which reimages the échelle illumination onto the prism-grism cross-disperser, has kept the size of optical elements to a minimum. In ELODIE, as well as, for example, in the 5-m Hale telescope

Figure 3.10. The Cassegrain échelle spectrograph on the 61-cm telescope at Mt John, New Zealand. The instrument came into operation in 1977, the first Cassegrain échelle in the southern hemisphere. It is seen here in 1986 with a Reticon diode array detector.

fibre-fed échelle [114], the camera comprises a multi-element all-refracting system able to record from the near ultraviolet to the far red in a single exposure. Such cameras have been pioneered by Harland Epps (b. 1936) at the University of California Santa Cruz. They are especially suited to fibre-fed échelles with CCD detectors, because of the wide wavelength range and small size of the detector. In addition, the central

obscuration (from the telescope's secondary mirror) found in coudé and Cassegrain designs is absent in fibre-fed spectrographs, so it is not possible to use a folded Schmidt spectrograph camera without losing a significant amount of light at some wavelengths through the hole in the flat fold mirror. Other fibre-fed échelles have relied on commercial telephoto camera lenses. Thus the MUSICOS instrument made use of

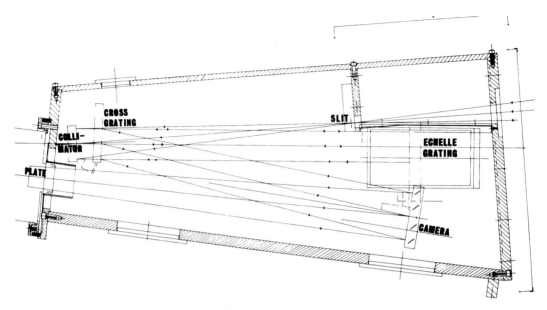

Figure 3.11. Ray diagram of the Cassegrain échelle spectrograph at Mt John Observatory, New Zealand. The échelle grating is R2 and the cross-dispersion is from a first-order grating after the échelle. The camera mirror is spherical.

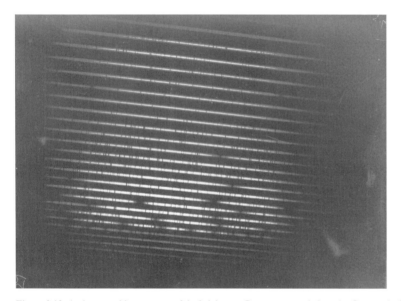

Figure 3.12. A photographic spectrum of the bright star Canopus, recorded on the Cassegrain échelle spectrograph and 61-cm telescope at Mt John in October 1978. This is an early example of the échelle spectrum of a southern star. The original was on Kodak IIIa-J emulsion and this positive image is from a contact print with the glass plate. The red orders are at the top.

an off-the-shelf 100-mm aperture Canon camera lens operating at f/4 [115, 116].

The use of CCD detectors has also favoured prism cross-dispersion over gratings. There are two reasons for this: the non-uniform dispersion of a prism can result in more nearly uniform order spacing on a CCD, thereby maximizing the wavelength coverage. Secondly, the efficiency of a blazed cross-dispersion grating, even in first order, is not maintained over such a wide wavelength range as 380–900 nm that

Table 3.5. *Table of selected Cassegrain échelle spectrographs*

Observatory or instrument	Telescope	Year completed	Échelle grating	Cross-disperser	Reference(s)
Pine Bluff, Wisconsin	0.91-m	1967	R2(30)	grating after échelle	[87]
Pine Bluff, Wisconsin	0.91-m	1971	R2(73.5)	grating after échelle	[88, 95]
Smithsonian AO Mt Hopkins, AZ	1.5-m	1973	R2(31.6)	grating after échelle	[89]
Kitt Peak Obs.	4.0-m	1974	R2(31, 58 or 79)	grating after échelle	[93, 94, 96]
Harvard Coll. Obs. Oak Ridge, MA	1.55-m	1975	R2(79)	grating after échelle	[97]
McGraw-Hill Obs., MIT (now MDM) Kitt Peak, AZ	1.32-m	1975	R0.5(54.5)	quartz prism before échelle	[56]
Johns Hopkins Univ. Goddard SFC	0.91-m	1976	R2(79)	prism before échelle	[98]
Univ. Canterbury Mt John Obs.	1.0 & 0.6-m	1977	R2(79)	grating after échelle	[90, 91]
Queen's Univ. Belfast	1.0-m (La Palma) 0.9-m (RGO)	1978	R2(31.6)	double-pass flint prism	[54, 55, 99, 100]
Siding Spring Obs., ANU	1.0-m	1980	R2(79)	grating after échelle	[101]
Figl Obs., Vienna	1.5-m	1982	R2(79)	grating after échelle	[102]
Catalina Obs. LPL, AZ	1.55-m	1982	R3.2(79)	filter	
ESO, La Silla CASPEC	3.6-m	1983	R2(31.6)	2-grating mosaic after échelle	[103]
Osmania Univ. Japal-Rangapur Obs.	1.22-m	1986	R2(79)	grating after échelle	[104]
McDonald Obs., TX Sandiford échelle	2.1-m	1991	R2(23.2)	40° prism in double pass	[105]

R2(79) refers to an R2 échelle grating with 79 gr/mm

Table 3.6. *Table of selected fibre-fed échelle spectrographs*

Observatory or instrument	Telescope	Year échelle completed	Échelle grating	Cross-disperser	Reference(s)
Penn. State Univ.	1.6-m	1984	R2(79)	prism in double pass	[106, 107, 108, 109]
Landessternwarte, Heidelberg: FLASH	0.75-m	1987	R2(31.6)	grating after échelle	[110]
Kitt Peak	2.1-m	1987	R2(79)	prism or grating after échelle	[111]
Mt John Observ.	1.0-m	1988	R2(79)	grating after échelle	[112, 113]
Pic du Midi etc. MUSICOS	2.0-m	1991	R2(31.6)	prism after échelle	[115, 116]
Geneva Obs., Marseille LAM ELODIE	1.93-m (OHP)	1993	R4(31)	prism & grism after échelle	[60]
Caltech, Mt Palomar Obs.	5.0-m Hale	1994	R2(79)	double-pass prism	[114]
High Altitude Obs. Harvard Coll. Obs. AFOE	1.5-m (Mt Hopkins)	1994	R2(59)	grating after échelle	[117]
Tautenburg Obs. TRAFICOS	2-m Schmidt	1996	R2(25)	prism in double pass	[118]
ESO, La Silla: FEROS	1.52 & 2.2-m	1998	R2(79)	prism	[119]
McDonald Obs. Hobby–Eberly HRS	9.2-m HET	2001	R4(31.6) 2–grating mosaic	grating after échelle at white pupil	[120]
Mt John Obs. Hercules	1.0-m	2001	R2(31.6)	prism in double pass	[121]
ESO, La Silla: HARPS	3.6-m	2003	R4(31.6) 2–grating mosaic	FK5 grism at white pupil	[122, 123]

R2(79) refers to an R2 échelle grating (i.e. $\tan\theta_B = 2$) with 79 gr/mm some fibre-fed échelles are capable of. A uniform order spacing for a prism, given that the central order wavelengths have $\lambda_c \propto 1/m$ and given that the intervals between these wavelengths go as $\Delta\lambda_c \propto 1/m^2 \propto \lambda_c^2$, requires that the refractive index of the prism glass should obey $\frac{dn}{d\lambda} \propto \frac{1}{\lambda^2}$. A glass obeying the Cauchy formula $n = A + B/\lambda^2$ gives $\frac{dn}{d\lambda} = \frac{-2B}{\lambda^3}$, thereby slightly overcompensating for the required cross-dispersion at shorter wavelengths.

On the other hand, the nearly uniform cross-dispersion of a grating results in order separation going as $1/m^2$. The ultraviolet orders are then either too crowded to record individually, or the red ones are too far separated and may fall off the CCD. In either case, total wavelength coverage is reduced.

Glasses used for cross-dispersion prisms have been Schott UBK7, BK7, LF5 or fused silica. The last is expensive and suitable for coudé instruments able to reach the ultraviolet, but is not necessary for fibre-fed instruments, where $\lambda < 370$ nm is in any case unattainable because of high ultraviolet absorption losses in fibres. Fused silica is the option chosen on the UCLES coudé échelle of the Anglo-Australian telescope [81].

In spite of these comments, some CCD échelle spectrographs have still opted for grating cross-dispersion. Gratings are cheaper, and they can still deliver efficiently around 250 nm of wavelength coverage. The AFOE instrument [117], which was designed for precise radial-velocity studies, is a case in point. Gratings can also deliver more cross-dispersion for those cases where interleaved star and sky spectra (or star and calibration spectra) are required. And for the very largest coudé or Nasmyth échelle spectrographs, the physical size of the prisms required may rule out the option of prism cross-dispersion, either on the grounds of expense or because of absorption losses in the glass. Thus the HIRES instrument at the Keck telescope uses a mosaic of four cross-gratings after the échelle, in a 40×60 cm array [75, 76].

A selection of fibre-fed échelle spectrographs described in the literature is listed in Table 3.6.

REFERENCES

[1] Dunham, T., *Vistas Astron.* **2**, 1223 (1956)

[2] Bowen, I. S., *Astrophys. J.* **116**, 1 (1952)

[3] Loewy, M., *Comptes Rendus de l'Acad. Sci., Paris* **96**, 735 (1883)

[4] Danjon, A. and Couder, A., *Lunettes et Télescopes.* Editions de la Revue d'Optique Théorique et Instrumentale, Paris, section **93**, p. 361 (1935)

[5] Hamy, M., *Ann. de l'Observ. de Paris* **32**, B1 (1924)

[6] Hnatek, A., *Ann. der königl. Univ.-Sternwarte in Wien* **25**, (Nr 1), 1 (1913)

[7] Adams, W. S., *Astrophys. J.* **33**, 11 (1911)

[8] Adams, W. S., *Astrophys. J.* **93**, 11 (1941)

[9] Schmidt, B., *Mitt. der Hamburger Sternw.* **7**(36) (1932)

[10] Bowen, I. S., In *Astronomical Techniques: Stars and Stellar Systems*, vol. 2, p. 34, ed. W. A. Hiltner, University of Chicago Press (1962)

[11] Dunham, T., *Phys. Rev.* **46**(2), 326 (1934)

[12] Dunham, T., *Publ. Amer. Astron. Soc.* **8**, 110 (1936)

[13] Dunham, T., *Publ. astron. Soc. Pacific* **50**, 220 (1938)

[14] Adams, W. S. and Dunham, T., *Astrophys. J.* **87**, 102 (1938)

[15] Dunham, T., Oral history interviews, American Institute of Physics, p. 49 (1977)

[16] Babcock, H. W., *Astrophys. J.* **107**, 73 (1948)

[17] Babcock, H. W., *J. Opt. Soc. Amer.* **40**, 409 (1950)

[18] Baustian, W. W., In *Telescopes: Stars and Stellar Systems*, vol. 1, p. 16, ed. G. P. Kuiper and B. M. Middlehurst, University of Chicago Press (1960)

[19] Richardson, E. H., *J. R. astron. Soc. Canada* **62**, 313 (1968)

[20] van Biesbroeck, G., *Contrib. McDonald Observ.* **1**, 103 (1943)

[21] Fehrenbach, C., *Bull. Soc. Astron. France* **74**, 193 (1960)

[22] Fehrenbach, C., *J. des Observateurs* **43**, 85 (1960)

[23] Maehara, H., private communication (1997). There are no published papers describing the Okayama coudé.

[24] Thackeray, A. D., *Quarterly J. R. astron. Soc.* **2**, 188 (1961)

[25] Dunham, T., *Astron. J.* **67**, 114 (1962)

[26] Harding, G. A., Palmer, D. R. and Pope, J. D., *R. Observ. Bull. (Greenwich)* **145**, E367 (1968)

[27] Mayall, N. U., *Astron. J.* **72**, 262 (1967)

[28] Vasiliev, A. S., *Izvest. Crimean Astrophys. Observ.* **55**, 224 (1976)

[29] Bartl, E., *Jenaer Rundschau* **15**, 335 (1970)

[30] Tull, R. G., *Sky and Tel.* **38**, 156 (1969)

[31] Veth, C., *Bull. Astron. Inst. Netherlands* **20**, 312 (1969)

[32] Wehlau, W. H., *Bull. Amer. Astron. Soc.* **4**, 201 (1972)

[33] Boyce, P. B., White, N. M., Albrecht, R. and Slettebak, A., *Publ. astron. Soc. Pacific* **85**, 91 (1973)

[34] Wood, H. J., Wolf, B. and Maurice, E., *European South. Observ. Bull.* **11**, 5 (1975)

[35] Bahner, K. and Solf, J., *ESO-CERN Conference on Auxiliary Instrumentation for Large Telescopes*, ed. A. Laustsen and A. Reiz, p. 247 (1972)

[36] Richardson, E. H., *Proc. Soc. Photo-instrumentation Engineers (SPIE)* **445**, 530 (1984)

[37] Qin, S.-N. and Li, J.-K., *Acta Astron. Sinica* **25**, 212 (1984)

[38] Jiang, S. Y., *Liège Int. Astrophys. Coll.* **27**, 241 (1987)

[39] Simon, J. M., de Novarini, L. R. and Platzeck, R., *Optica Acta* **18**, 829 (1971)

[40] Tull, R. G., *ESO-CERN Conference on Auxiliary Instrumentation for Large Telescopes*, ed. A. Laustsen and A. Reiz, p. 259 (1972)

[41] Richardson, E. H., Brealey, G. A. and Dancey, R., *Publ. Dominion Astrophys. Observ. Victoria* **14**, 1 (1971)

[42] Richardson, E. H., *Publ. astron. Soc. Pacific* **78**, 436 (1966)

[43] Bowen, I. S., *Astrophys. J.* **88**, 113 (1938)

[44] Pierce, A. K., *Publ. astron. Soc. Pacific* **77**, 216 (1965)

[45] Walraven, T. and Walraven, J. H., *ESO-CERN Conference on Auxiliary Instrumentation for Large Telescopes*, ed. A. Laustsen and A. Reiz, p. 175 (1972)

[46] Suto, H. and Takami, H., *Appl. Optics* **36**, 4582 (1997)

[47] Benesch, W. and Strong, J., *J. Opt. Soc. Amer.* **41**, 252 (1951)

[48] Strong, J. and Stauffer, F., *Appl. Optics* **3**, 761 (1964)

[49] Richardson, E. H., *ESO-CERN conference on auxiliary instrumentation for Large Telescopes*, ed. A. Laustsen and A. Reiz, p. 275 (1972)

[50] Gray, D. F., *Intl. Astron. Union Symp.* **118**, 401 (1986)

[51] Kapany, N. S., in *Concepts of Classical Optics*, by J. Strong, Appendix N, p. **553**, W. H. Freeman & Co. (1958)

[52] Vanderriest, C., *Publ. astron. Soc. Pacific* **92**, 858 (1980)

[53] Harrison, G. R., *J. Opt. Soc. Amer.* **39**, 522 (1949)

[54] McKeith, C. D., Dufton, P. L. and Kane, L., *Observ.* **98**, 263 (1978)

[55] Bates, B., McKeith, C. D., Jorden, P. R. and van Breda, I. G., *Astron. & Astrophys.* **145**, 321 (1985)

[56] Bardas, D., *Publ. astron. Soc. Pacific* **89**, 104 (1977)

[57] Schroeder, D. J. and Hilliard, R. L., *Appl. Optics* **19**, 2833 (1980)

[58] Walker, D. D. and Diego, F., *Mon. Not. R. astron. Soc.* **217**, 355 (1985)

[59] Dekker, H. and Hoose, J., *ESO Workshop on High Resolution Spectroscopy with the VLT*, ed. M.-H. Ulrich, p. 261(1992)

[60] Baranne, A., Queloz, D. and Mayor, M., *Astron. & Astrophys. Suppl.* **119**, 1 (1961)

[61] Harrison, G. R., Archer, J. E. and Camus, J., *J. Opt. Soc. Amer.* **42**, 706 (1952)

[62] Harrison, G. R., Davis, S. P. and Robertson, H. J., *J. Opt. Soc. Amer.* **43**, 853 (1953)

[63] Pierce, A. K., McMath, R. R. and Mohler, O., *Astron. J.* **56**, 137 (1951)

[64] Harrison, G. R., *Vistas Astron.* **1**, 405 (1955)

[65] Young, C. A., *Amer. J. Sci. Arts* (3) **11**, 429 (1876)

[66] Wood, R. W., *J. Opt. Soc. Amer.* **37**, 733 (1947)

[67] Tousey, R., Purcell, J. D. and Garrett, D. L., *Appl. Optics* **6**, 365 (1967)

[68] Tousey, R., *Astrophys. J.* **149**, 239 (1967)

[69] Kopylov, I. M. and Steshenko, N. V., *Proc. Crimean Astrophys. Observ.* **33**, 308 (1965)

[70] Fujita, Y., *Vistas Astron.* **7**, 71 (1966)

[71] Liller, W., *Appl. Optics* **9**, 2332 (1970)

[72] Butcher, H. R., *Publ. Astron. Soc. Australia* **2**, 21 (1971)

[73] Crawford, I. A., Rees, P. C. T. and Diego, F., *Observ.* **107**, 147 (1987)

[74] Vogt, S. S., *Publ. astron. Soc. Pacific* **99**, 1214 (1987)

[75] Vogt, S. S., *Intl. Astron. Union Symp.* **132**, 1 (1988)

[76] Vogt, S. S., in *Instrumentation for Ground-based Optical Astronomy*, ed. L. B. Robinson, Springer-Verlag, p. 33 (1988)

[77] Vogt, S. S. and Penrod, G. D., in *Instrumentation for Ground-based Optical Astronomy*, ed. L. B. Robinson, Springer-Verlag, p. 68 (1988)

[78] Richardson, E. H. and Morbey, C. L., in *Instrumentation for Ground-based Optical Astronomy*, ed. L. B. Robinson, Springer-Verlag, p. 47 (1988)

[79] Grundmann, W. A., Moore, F. A. and Richardson, E. H., *Proc. Soc. Photo-instrumentation Engineers (SPIE)* **1235**, 577 (1990)

[80] Schroeder, D. J., in *Instrumentation for Ground-based Optical Astronomy*, ed. L. B. Robinson, Springer-Verlag, p. 39 (1988)

[81] Diego, F., in *Instrumentation for Ground-based Optical Astronomy*, ed. L. B. Robinson, Springer-Verlag, p. 6 (1988)

[82] Diego, F., Charalambous, A., Fish, A. C. and Walker, D. D., in 'Instrumentation in astronomy VII', *Proc. Soc. Photo-instrumentation Engineers (SPIE)* **1235**, 562 (1990)

[83] Tull, R. G., MacQueen, P. J., Sneden, C. and Lambert, D. L., *Publ. astron. Soc. Pacific* **107**, 251 (1995)

[84] MacQueen, P. J. and Tull, R. G., in *Instrumentation for Ground-based Optical Astronomy*, ed. L. B. Robinson, Springer-Verlag, p. 52 (1988)

[85] Tull, R. G. and MacQueen, P. J., *Very Large Telescopes and their Instrumentation*, ESO conference and workshop proceedings no. 30, ed. M.-H. Ulrich, p. 1235 (1988)

[86] Dekker, H., D'Odorico, S., Kaufer, A., Delabre, B. and Kotzlowski, H., *Proc. Soc. Photo-instrumentation Engineers (SPIE)* **4008**, 534 (2000)

[87] Schroeder, D. J., *Appl. Optics* **6**, 1976 (1967)

[88] Schroeder, D. J. and Anderson, C. M., *Publ. astron. Soc. Pacific* **83**, 438 (1971)

[89] Chaffee, F. H. and Schroeder, D. J., *Ann. Rev. Astron. Astrophys.* **14**, 23 (1976)

[90] Hearnshaw, J. B., *Publ. Astron. Soc. Australia* **3**, 102 (1977)

[91] Hearnshaw, J. B., *Sky and Tel.* **56**, 6 (1978)

[92] McLaughlin, D. B., in *The Present and Future of the Telescope of Moderate Size*, ed. F. B. Wood, p. 205, University of Pennsylvania Press (1958), from University of Pennsylvania Symposium of June 1956

[93] Gull, T. R., *Quart. Bull., Kitt Peak Nat. Observ. for Oct–Dec 1974*, p. 20 (1974)

[94] Goldberg, L., *Bull. Amer. Astron. Soc.* **10**, 152 (1978), see p. 160

[95] McNall, J. F., Michalski, D. E. and Miedaner, T. L., *Publ. astron. Soc. Pacific* **84**, 176 (1972)

[96] Goad, J. W., *KPNO Facilities Book*, pp. 3–37 (1982)

[97] Latham, D. W., *Intl. Astron. Union Coll.* **40**, 45 (1977)

[98] McClintock, W., *Publ. astron. Soc. Pacific* **91**, 712 (1979)

[99] McKeith, C. D., *Irish Astron. J.* **17**, 487 (1986)

[100] McKeith, C. D., Conal, D. and Barnett, E. W., *Proc. Liège Internat. Astrophys. Coll.* **27**, 21 (1987)

[101] Weiss, W. W., Barylak, M., Hron, J. and Schmiedmayer, J., *Publ. astron. Soc. Pacific* **93**, 787 (1981)

[102] Hunten, D. M., Wells, W. K., Brown, R. A., Schneider, N. M. and Hilliard, R. L., *Publ. astron. Soc. Pacific* **103**, 1187 (1991)

[103] D'Odorico, S., Enard, D., Lizon, J. L. *et al.*, *European South. Observ. Messenger* **33**, 2 (1983)

[104] Bhatia, R. K., Swaminathan, R. and Vijas, M. L., *Bull. Astron. Soc. India* **12**, 79 (1984)

[105] McCarthy, J. K., Sandiford, B. A., Boyd, D. and Booth, J., *Publ. astron. Soc. Pacific* **105**, 881 (1993)

[106] Ramsey, L. W., Brungardt, C., Huenemoerder, D. P. and Rosenthal, S., *Bull. Amer. Astron. Soc.* **17**, 574 (1985)

[107] Ramsey, L. W. and Huenemoerder, D. P., Fourth Cambridge Workshop on Cool Stars, Stellar Systems and the Sun, in *Lecture Notes in Physics* **254**, 238 (1986)

[108] Ramsey, L. W. and Huenemoerder, D. P., *Proc. Soc. Photo-instrumentation Engineers (SPIE)* **627**, 282 (1986)

[109] Ramsey, L. W., SHIRSOG conference, *Proceedings of a Conference for a New Synoptic High Resolution Spectroscopic Observing Facility*, ed. M. S. Giampapa, Tucson, Arizona: NOAO, p. 80 (1986)

[110] Mandel, H., *Intl. Astron. Union Symp.* **132**, 9 (1988)

[111] Barden, S. C., in *Instrumentation for Ground-based Optical Astronomy*, ed. L. B. Robinson, Springer-Verlag, p. 250 (1988)

[112] Kershaw, G. M. and Hearnshaw, J. B., *Southern Stars* **33**, 89 (1989)

[113] Murdoch, K. A. and Hearnshaw, J. B., *Astrophys. & Space Sci.* **186**, 169 (1991)

[114] Libbrecht, K. G. and Peri, M. L., *Publ. astron. Soc. Pacific* **107**, 62 (1995)

[115] Baudrand, J. and Böhm, T., *Astron. & Astrophys.* **259**, 711 (1992)

[116] Catala, C., Baudrand, J., Böhm, T. and Foing, B., *Intl. Astron. Union Coll.* **137**, 662 (Astron. Soc. Pacific Conf. Ser. 40) ed. W. Weiss and A. Baglin (1993)

[117] Brown, T. M., Noyes, R. W., Nisenson, P., Korzenik, S. G. and Horner, S., *Publ. astron. Soc. Pacific* **106**, 1285 (1994)

[118] Hildebrandt, G., Scholz, G., Rendtel, J., Woche, M. and Lehmann, H., *Astron. Nachrichten* **318**, 291 (1997)

[119] Kaufer, A., Stahl, O., Tubbesing, S. *et al.*, *European South. Observ. Messenger* **95**, 8 (1999)

[120] Tull, R. G., *Proc. Soc. Photo-instrumentation Engineers (SPIE)* **3355**, 387 (1998)

[121] Hearnshaw, J. B., Barnes, S. I., Kershaw, G. M. *et al.*, *Exper. Astron.* **13**, 59 (2002)

[122] Pepe, F., Mayor, M. and Rupprecht, G., *European South. Observ. Messenger* **110**, 9 (2002)

[123] Mayor, M., Pepe, F., Queloz, D. *et al.*, *European South. Observ. Messenger* **114**, 20 (2003)

4 · Solar spectrographs and the history of solar spectroscopy

4.1 SPECTROSCOPY OF THE SOLAR CHROMOSPHERE

4.1.1 The discovery of helium

A new era in solar spectroscopy was launched in the late 1860s when a series of solar eclipses provided an opportunity to study the solar chromosphere.[1] This is the hot tenuous layer lying above the photosphere, which is the region of the Sun that emits the vast proportion of the visible light. At the time of a total solar eclipse, it was well known by the 1860s that the light from the thin chromospheric layer (with a height of $\sim$ 10 to 12 arc seconds or 8000 to 10 000 km) became briefly visible. Moreover, the solar prominences are large structures of chromospheric material extending out from the limb, often a minute of arc or more in height. They too were seen at times of eclipse, and the question of the physical conditions in the chromosphere and prominences arose. The spectroscope was the natural tool to settle the issue. If they were comprised of hot low density gas, then bright emission lines would be expected, as William Huggins pointed out [2], and hence they should have spectra similar to those of the gaseous nebulae, such as the Orion nebula. Huggins also believed that if bright lines were present, then perhaps they could be observed with suitable glass filters to isolate the line radiation, even outside of eclipse, but attempts to do so did not come to fruition [3, 4, 5, 6].

Meanwhile, several expeditions from England and France were organized to observe the especially favourable eclipse of 18 August 1868, which was observable from India and the Malayan peninsula. These included expeditions by Jules Janssen (to Guntoor, India) and Georges Rayet (1839–1906) (from Paris Observatory, to Malaya), by Lt John Herschel (1837–1921) (son of Sir John) (on a Royal Society expedition) and by J. F. Tennant (1829–1916) (who was sent by the Royal Astronomical Society to Guntoor, India). All these observers employed simple prism spectroscopes. The instrument used by Lt Herschel is shown in Fig. 4.1. All observers found bright lines in the spectra of prominences at the time of the eclipse. Thus Herschel found three bright lines, one being yellow and apparently the sodium D line [7]; Tennant saw five bright lines, including C, D and b, and others near F and G [8], and Rayet in Malaya, using a direct vision spectroscope, saw as many as nine lines [9]. Telescopes were needed, given the weakness of the chromospheric and prominence light and the extended nature of the source, as well as to produce an image to be focussed onto the slit. The nature of the prominences as hot low density gaseous structures, as Huggins had hypothesized, was thus established.

Most interesting, however, was the report of Janssen, also in Guntoor, who devised a technique of observing the prominence and chromospheric spectra even outside the time of the eclipse. The method relied on carefully aligning the slit of the spectroscope on the chromosphere and tangentially to the solar limb. He wrote:

> But the most important result of these observations is the discovery of a method, whose principle was conceived during the eclipse itself, and which allows one to study the prominences and regions around the Sun all the time, without it being necessary to resource to the placing of an opaque object in front of the Sun's disk. This method is based on the spectral properties of the light of the prominences, which light is resolved into a number of very bright rays which correspond to the dark lines of the solar spectrum.

---

[1] The word 'chromosphere' was introduced by Lockyer in 1869 [1] and so named for the reddish colour of this region of the Sun. Before that expressions such as 'regions around the Sun' were used.

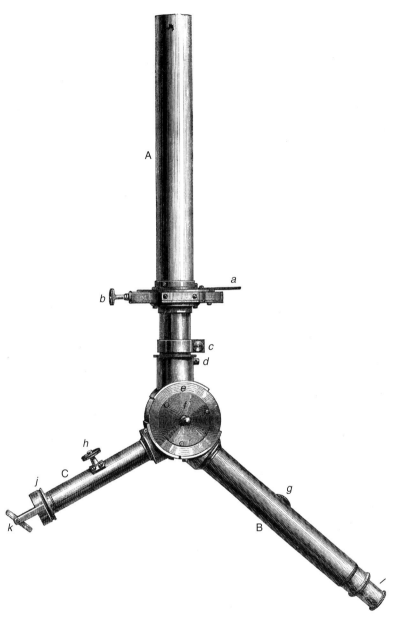

Figure 4.1. Lt John Herschel's single-prism spectroscope used on the 18 August 1868 solar eclipse expedition to India. The instrument was made by Troughton and Simms.

From the day after the eclipse, the method was applied with success and I was able to be present at the phenomena presented by a new eclipse which lasted the whole day. The prominences from the previous day were considerably altered. There remained hardly any traces of the big prominence

and the distribution of the gaseous matter was quite different [10].

This report of Janssen's technique did not reach the Académie des Sciences in Paris until 24 October that year (see [11]), by chance the same day that

Norman Lockyer also communicated the same independent discovery of the method to the Academy (reported by Warren de la Rue (1815–89) [12]). Lockyer's first successful observations were made on 20 October from England, though he had attempted the method as early as 1867, but using a spectroscope of inadequate resolving power. According to Faye:

> ... If therefore the priority for the idea [of the method] belongs without contest to Mr Norman Lockyer, that for the success and fruitful application goes of right to Mr Janssen, for, if he were not so far away, we would have had two months earlier, and by him alone, these marvellous discoveries on the regions surrounding the Sun [13].

Janssen's main report on the spectroscopy undertaken during the eclipse was sent to Paris from Calcutta on 3 October 1868 [14]. He observed five or six very bright emission lines, which he believed corresponded with those seen in absorption in the Fraunhofer spectrum. He summarized his three main conclusions as follows:

1. the gaseous nature of the prominences (bright spectral lines);
2. the general similarity of their chemical composition (spectra correspond line for line);
3. their chemical species (the red and blue lines of their spectrum were none other than the C and F lines of the solar spectrum, which are characteristic, as is well known, of hydrogen gas) [14].

Lockyer's most detailed report of the chromospheric spectrum was sent to the Royal Society and showed how the bright-line spectrum was rendered visible by carefully placing the spectroscope slit tangentially to the limb of the Sun [1]. Moreover, with a slit instead radial to the solar limb, encroaching partly on the photosphere and partly into a prominence, the bright-lined prominence spectrum could readily be compared with the dark Fraunhofer lines. The important discovery that the bright yellow line in prominences was near the Fraunhofer D line but not coincident with it, but instead at a slightly shorter or more refrangible wavelength, was a major result. At first Lockyer concluded that the line, which he named D_3 (at 5876 Å) was 'most probably proceeding from the

same substance which gives off the light at C (Hα) and F (Hβ)' – namely hydrogen [1].

This conclusion radically changed a quarter century later, when William Ramsay (1852–1916) identified in the laboratory a new gaseous element, helium, which emanated from the uranium-containing mineral clèveite, and which also produced the D_3 line [15]. Ramsay noted that Lockyer had named the element helium 'many years ago'; but in fact Lockyer rarely if ever used this word before Ramsay had popularized it. It appears just once in a paper of his in 1895 [16]. And also in 1895, Lockyer discussed the presence of several lines for the new element that had been observed both in the chromosphere and in the spectra of stars. In addition to D_3 (5876 Å) there were lines at 4472 Å, 7665.5 Å and others [17]. Lockyer did not refer at all to the word 'helium' in his landmark book *Contributions to Solar Physics* of 1874 [18].

The discovery of helium and its observation in the Sun decades before its isolation in the laboratory was a major breakthrough for nineteenth century spectroscopy, and also a triumph for instrumentation and the specialized technique that made this possible.

Other observers went on to exploit the Janssen–Lockyer technique to obtain new information on the solar prominences. One such assiduous observer was Angelo Secchi, who also wrote a classical monograph *le Soleil* in the 1870s [19]. Secchi's three-prism spectroscope on the 24-cm Merz equatorial refractor at the Collegio Romano Observatory is shown in Fig. 4.2.

In his book on the Sun, his discussion of the technique for observing prominences is especially clear. With a narrow slit tangential to the limb, bright lines such as C (Hα) were clearly seen, superimposed on the photospheric light with absorption lines (see [19, 2nd edition, part 2, p. 7]). Likewise a radial slit, perpendicular to the limb, enabled bright line positions to be compared with the Fraunhofer absorption lines. That D_3 did not coincide with D was readily observed, provided the dispersion was sufficient.

> With the equipment thus arranged, I slowly came up to the edge of the solar disk, until such time as the slit came to fall on the limb, so as to be perfectly tangential. Suddenly one would then see the C line become bright, because the chromosphere was projected onto the slit so that its light filled the field of the analysing telescope. This result is shown in

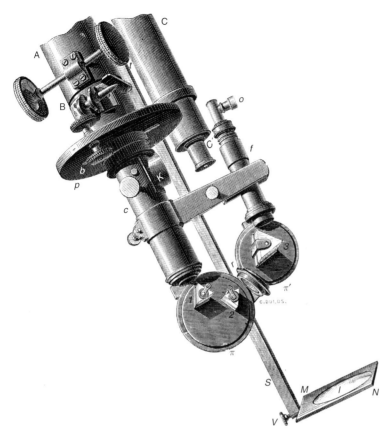

Figure 4.2. Secchi's three-prism spectroscope on the equatorial refractor at Collegio Romano, about 1870.

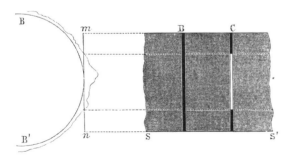

Figure 4.3. Secchi's diagram showing the chromospheric spectrum with a slit tangential to the solar limb.

figure [4.3]. BB′ is the limb of the solar disk, above which the chromosphere extends; *mn* is the position of the slit, SS′ is the spectrum in which the C line is reversed, whereas the line B is dark.

To see the prominences clearly, it is necessary that the image of the Sun is focussed with the highest precision possible onto the plane of the slit. To achieve that, I place the slit so as to be perpendicular to the solar limb, and one sees then in the field two distinct parts, one which is brightly illuminated by the direct light from the Sun, while the other comes from the much fainter light of the surrounding atmosphere, as shown in figure [4.4] ... If one sets on a prominence, one sees at once jump into view, beyond the Sun, three bright lines: one in the red on the extension of the line C, the second quite close to D, and the third being an extension of F. There may even be a fourth near G. Even if there are no prominences, these lines will nonetheless still be there, but they will be very short and consequently difficult to recognize [19, 2nd edition, part 2, pp. 6–8].

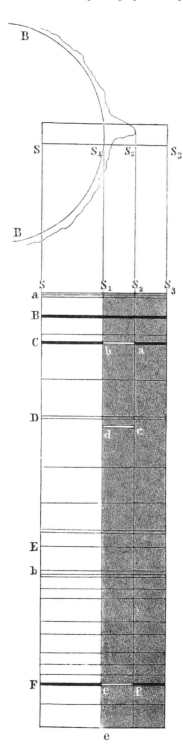

Figure 4.4. Secchi's diagram showing both photospheric and chromospheric spectra when the slit is aligned in a radial direction, perpendicular to the limb.

4.1.2   The structure of prominences revealed

Another important application of the spectroscope in solar chromospheric spectroscopy was in revealing the shape, structure and evolution of the solar prominences. Here the spectroscope was the essential tool, because the dispersion of the residual photospheric light that entered the slit effectively eliminates much of the bright background, so enabling the prominences to be seen with reasonable contrast. This is especially so, given that at Hα the prominences produce the line in emission, whereas for the photospheric light at that same wavelength there is a deep absorption line.

With a narrow tangential slit sweeping rapidly across the height of a prominence, it was possible to determine its shape or structure. Indeed, this was a technique that was the forerunner of the spectrohelioscope, and early experiments were carried out by Jules Janssen [20, 21], who devised a spectroscope with a rotating slit that repeatedly swept across the image. He wrote:

> This new method, in principle, consists in isolating one of the luminous beams in the spectral domain emitted by the prominence. This beam is deficient in solar [i.e. photospheric] light. These linear sections of prominence images are then transformed into images themselves, by means of a fairly rapid rotational movement imposed on the spectroscope.

Zöllner (1834–82) criticized Janssen's rotating spectroscope, and instead proposed an instrument with an oscillating slit:

> Apart from the mechanical difficulties of such a rotating spectroscope, for which the three bright prominence lines must lie exactly in the rotation axis, the sought for goal would be achieved more simply and completely by an oscillation of the slit at right angles to its length. It would thereby be possible to observe the same prominence simultaneously in three different coloured images, corresponding to the three different lines of its spectrum [22].

Janssen also pioneered the application of a second spectroscope slit so as further to reduce the harmful effects of scattered light, a technique very similar to that used some decades later by Hale and Deslandres in the spectroheliograph [21]. Janssen wrote:

A metal diaphragm, placed at the focus of the spectroscope, and pierced by a slit at the precise point where one of the bright lines of prominence light should be seen, allows one to completely separate this light from that of the photosphere, which lacks precisely the rays of this refrangibility, and one can then follow still further the traces of the prominence material. I will have to return to the use of the focal plane or eyepiece slit, which allows one to obtain a series of monochromatic images that a luminous body produces, when one combines this with a rotating movement applied to the spectroscope.

The bright lines which then appear on the whole rim of the disk are mainly those that we recognize as being characteristic of incandescent hydrogen gas [21].

The technique of predispersing the light using an objective prism or small direct-vision Amici prism was another method that helps overcome the scattered light problem in observations of prominences. This was used by Secchi in 1871, and is also reminiscent of modern solar spectrographs that employ a predisperser in front of the main slit [23, 24].

Several variations of the Janssen–Lockyer technique for viewing the prominences outside eclipse were proposed. Thus in 1873, Lockyer and Seabroke (1838–1918) experimented with a ring-shaped slit which exactly coincided with the position of the chromospheric image in the focal plane of the telescope, so as to be able to see the entire annulus of the chromosphere at once rather than just a small part of it selected by a linear slit [25]. Although successful observations were reported with this device, it was seldom used by other observers.

William Huggins, on the other hand, found that prominences could be viewed quite easily by opening the entrance slit of his spectroscope wide enough to encompass the entire prominence [6, 26]. No movement of the slit across the prominence was then necessary. In effect the wide slit meant that the residual photospheric light which inevitably gets scattered into the spectroscope gave an Hα line that was much broader and shallower than normal. However, the background was still just dark enough to see the structure of the prominence using the bright Hα emission line. The basis of the wide-slit method is carefully described

by Secchi in his book *le Soleil* [19, 2nd edition, part 2, pp. 13–16].

The ability to observe the size, shape and changes in the structure of the prominences outside of eclipse was a major breakthrough in solar physics of the 1870s, made possible by the spectroscope. The next step forwards came with the application of photography to solar physics and the development of the spectroheliograph – this is covered in Section 4.5. First, however, the work of Kirchhoff, Ångström and Rowland in cataloguing the Fraunhofer spectrum and measuring solar wavelengths is described, as these developments hinged on the application of the diffraction grating to solar spectroscopy, and is in itself an important part of the history of solar instrumentation.

4.2 THE MEASUREMENT OF SOLAR SPECTRUM WAVELENGTHS

4.2.1 Solar wavelengths from Fraunhofer's gratings

The first measurement of the wavelengths of lines in the solar spectrum came from the pioneering work of Joseph Fraunhofer. His first observations of the solar spectrum in 1814 had used a prism mounted in front of a small 25-mm telescope taken from a theodolite. With this simple objective prism arrangement, Fraunhofer catalogued the principal spectrum lines [27].

A few years later he used a simple transmission grating comprising a number of fine parallel wires on a framework, with the wires being stretched between two screws with fine threads. Fraunhofer's grating had 260 parallel wires, each about 0.002 inches (50 μm) thick, and spaced 0.039 inches ($\sim$1 mm) from each other. He went on to describe the spectra in several diffraction orders on each side of the white zero-order image of a slit, which was illuminated by sunlight from a heliostat. In addition Fraunhofer produced gratings with more closely spaced wires, up to 340 per inch ($\sim$13/mm), and he also experimented with gratings ruled on a gold layer deposited on a glass plate. Finally, he measured the deviations of the principal solar absorption lines produced by ten different gratings, so as to obtain their wavelengths, using the simple theory for the interference of light waves [28]. The

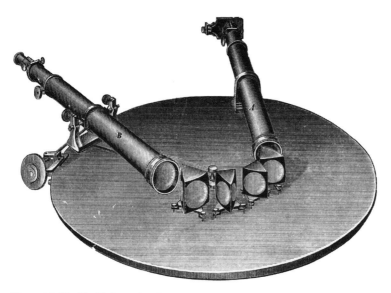

Figure 4.5. Kirchhoff's four-prism Steinheil spectroscope which he used for cataloguing solar spectral lines, 1861–3.

wire gratings had constants between 1.46 wires/mm and 18.9 wires/mm, but later gratings had as many as 130 wires/mm, enabling larger deviations and higher precision to be obtained [29].

Fraunhofer's values from the 1821 paper, after conversion to metric units, were for the D lines (mean of two lines) 5887.7 Å; Hα, 6563.1 Å; Hβ, 4841.2 Å; H, 3926.5 Å. Clearly some significant systematic errors, amounting to some 20 Å (1 part in 250), were incurred in the blue spectral region.

4.2.2   Kirchhoff's drawing of the solar spectrum

Gustav Kirchhoff produced the first detailed map of the solar spectrum in his landmark work which he published in two parts in 1861–2 and in 1863 [30, 31]. The two-year delay between parts I and II arose from the severe eye strain he had suffered in carrying out this work.

Kirchhoff's apparatus comprised a Steinheil four-prism spectroscope (three with 45° apex angles, one with 60°) with which the positions of solar lines could be compared with the spark spectra of 32 different elements produced in the laboratory. The apparatus is shown in Fig. 4.5. Line positions were given through micrometer readings that recorded the position of the spectroscope's viewing telescope, so these were on an arbitrary scale and not fundamental wavelengths as Fraunhofer had obtained. The whole apparatus must

have been cumbersome to use, as movements of the telescope required for scanning through the spectrum also entailed relocating the prisms manually on the circular platform. The dispersion of the instrument changed after each relocation of the prisms.

Although Kirchhoff's work was a landmark for advancing our knowledge of the Sun's qualitative chemical composition, it was not, however, a source of fundamental wavelengths. George Biddell Airy (1801–92) later undertook to calibrate the wavelengths of some 1600 solar lines in Kirchhoff's work [32]. However, he initially used Fraunhofer's rather imprecise wavelengths for this purpose, and when improved wavelengths from the grating observations of Leander Ditscheiner (1839–1905) and Ångström were available, he was able to make large corrections to the scale that had previously been published [33].

4.2.3   New grating measurements from Ångström and others

A new landmark in solar spectrum wavelengths resulted from a careful investigation by Anders Ångström in Uppsala, published in 1868 [34]. For this work a high quality spectrometer with collimator and viewing telescope manufactured by the Berlin firm of Pistor and Martins was used. The prism was replaced with a transmission grating ruled by F. A. Norbert, of which Ångström had two available, with respectively

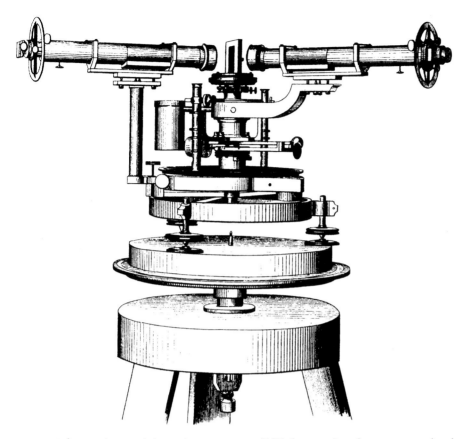

Figure 4.6. Ångström's transmission grating spectroscope of 1868, for measuring solar spectrum wavelengths.

4501 and 2701 grooves ruled in glass. Figure 4.6 shows Ångström's grating spectroscope which he used for his solar spectrum wavelength measurements.

Ångström calibrated his wavelength scale by carefully measuring nine principal solar lines, using measurements in several orders of diffraction. Thus he obtained 6562.1 Å for Hα, 5895.13 and 5889.12 Å for the D lines and 4860.72 Å for Hβ. Over 1000 other solar lines were also measured and tied to this scale.

In spite of the care with which this work was conducted, by 1872 Ångström himself suspected a major systematic error had crept into his work, as a result of an error in calibrating the overall width and hence groove spacing of his Norbert gratings. Ångström died in 1874 with this problem unresolved, but his assistant, Robert Thalén (1827–1905) in 1884 published a probable correction that should be applied to Ångström's scale, in which all wavelengths were to be multiplied by the factor of 1.00013 [35]. This amounted to nearly

a one angstrom increase for the red solar lines, but proportionately less in the blue.

The uncertainty persisted for several years until further studies in the 1880s, using various gratings by Norbert, Rutherfurd and Rowland and carried out by Gustav Müller (1851–1925) and Paul Kempf [36], by F. Kurlbaum [37] and by Louis Bell [38]. These investigations were all conducted with great care, and although they confirmed the systematic error in Ångström's work, differences of a few tenths of an angstrom persisted in the results from different workers, or in those obtained from different gratings.

4.3 HENRY ROWLAND AND THE CONCAVE GRATING

The contribution of Ångström had been a detailed drawing of the solar spectrum obtained with a visual grating spectroscope. The next major advance came from the photographic recording of the solar spectrum and the measurement of the plates in a precise measuring engine. Henry Rowland at the Johns Hopkins University in Baltimore was the person who took the art and science of solar spectrum photography to a new level. His invention of the concave grating [39] was a breakthrough that enabled him to obtain spectra of hitherto undreamed of resolving power and quality, which in particular extended from the far red right down to the ultraviolet region at 2152 Å.

Rowland developed the theory of the concave grating in 1883 [40] and he showed that for a spherical grating of radius of curvature R_{gr}, then a sharp spectral image was always located on a circle of diameter R_{gr}, provided that the slit was also on this Rowland circle. The simplicity of this arrangement, which entailed only a slit, a concave grating and a photographic plate was one of the keys to its success, for the Rowland spectrograph at once dispensed with lenses for collimator and camera, enabling ultraviolet spectra free of chromatic aberration to be obtained.

In the Rowland mounting (Fig. 1.22), which was not widely copied elsewhere, the wavelength region was changed by moving both the plate and the grating on the Rowland circle, but keeping the slit's location fixed. The relationship of plate to grating was also fixed, by mounting both on a rigid beam which moved on a diameter of the Rowland circle. The angle of diffraction β was therefore always close to zero in this type of mounting, whereas the angle of incidence α varied according to the wavelength region to be recorded.

The quality of Rowland's gratings, which were mostly reflection gratings ruled on speculum metal, was also a key to his success, and resulted from his ability to build a ruling engine with a screw of exceptional quality. Many gratings with 14 438 gr/in (573 gr/mm) were ruled, though values up to three times greater were possible. Some gratings had as many as 10^5 grooves over a 15 cm diameter, which permitted the production of spectra of superior quality to any that had preceded.

Rowland produced a *Photographic Map of the Normal Solar Spectrum* [41] with his new spectrographic equipment – here the 'normal' refers to the linear wavelength scale provided by the grating dispersion. The resolution was of the order of 0.01 Å.

From the photographic plates was derived Rowland's greatest contribution, his table of solar spectrum wavelengths, in which, with the help of Lewis Jewell, all the lines visible in the solar spectrum were recorded, their wavelength measured as well as their intensity on an arbitrary scale and, where possible, their chemical identification [42].

Before publishing this large work, Rowland produced a preliminary paper in 1893, being a table of standard solar spectrum wavelengths for the region 2152 to 7715 Å [43]. A variety of gratings of 150 mm diameter and 6.5 m radius of curvature were used for this purpose, and the best final wavelength scale took account of the earlier works that had appeared in the literature. Wavelengths to about 0.001 Å precision were recorded for about 750 lines; that for NaD_1 was 5896.154 Å, which was adopted as the primary standard relative to which other solar lines were measured. Extensive use was made of the method of coincidences, in which the coincidence of lines in different orders gave their wavelength ratio as the inverse ratio of the order numbers.

The Rowland scale of solar wavelengths and the intensity scale for line strengths was widely used for a generation, yet even at its inception a serious systematic error was suspected. This arose from the interferometric measures of Albert Michelson (1852–1931) at the University of Chicago in about 1894, with the aim of obtaining the absolute wavelength of the red cadmium line [44, 45]. Further interferometry by Charles Fabry and Alfred Pérot (1863–1925), at the Université de Marseille with their Fabry–Pérot interferometer, showed that the Fraunhofer lines have slightly smaller wavelengths, by about 34 parts in a million, than those obtained by Rowland [46]. The discrepancy amounted to about 0.212 Å at the sodium D_1 line. The whole episode that followed is reminiscent of the problems with the earlier Ångström scale, and underlines the complications and dangers of obtaining absolute wavelengths, even when scientists of the highest calibre and with the best equipped laboratories of their day were involved.

As a result of the suspected problems with Rowland's wavelengths, a vigorous debate ensued in the literature between the supporters of Rowland and the diffraction grating measurements, and the supporters of the new interferometric techniques. Louis Bell (1864–1923) in Massachusetts [47, 48] and Rowland's assistant, Lewis Jewell [49] were in the former camp. Johannes Hartmann made an extensive study of the

Rowland and the interferometric wavelengths [50] and the relationship between them. Although he gave corrections and transformations to the Rowland scale, he also recommended the new measurements. Heinrich Kayser [51] supported Fabry and Pérot and identified the likely source of systematic error in Rowland's work as being differences in spectrograph illumination between solar and arc lines, as well as errors in applying the method of coincidences.

By about 1905 the time was already ripe for a revision of the Rowland scale, but it was not till 1928 that this was finally accomplished. This is discussed in the next section.

4.4 THE REVISION OF THE ROWLAND SCALE

In spite of the high precision of the Rowland scale, the fact that it had a systematic error led to the need for a revision, especially as accurate wavelengths were desirable for radial-velocity work. By 1907 there was no doubt that the interferometric work of Michelson, using the red cadmium line, was essentially correct and indeed the wavelength of this line was adopted in that year as 6438.4696 Å as a primary standard [52].

Using the excellent facilities at Mt Wilson, Charles St John (1857–1935) and his colleagues set about remeasuring a large number of lines in the solar spectrum, using the red cadmium line's wavelength as the primary standard. Two independent series of measurements were made, one set using the grating spectrographs on the 60-foot and 150-foot tower telescopes at Mt Wilson, while another set used an interferometer on the Snow telescope and spectrograph at Mt Wilson as well as laboratory measurements in Pasadena [53]. Corrections to the Rowland scale were approximately $0.212\lambda/5900$, but with some significant departures from this linearity at different wavelengths.

As many as 21 835 solar lines were listed in the new work, together with their wavelengths and Rowland intensities, which included 1808 lines in the near infrared region to 1.02 μm not explored by Rowland. Some 57 per cent of these had element identifications.

The St John revision of the Rowland scale largely settled the issue of the solar wavelengths, which were now known within a few milli-angstroms from ultraviolet to the far red. After the production of the *Utrecht Solar Atlas* by Marcel Minnaert (1893–1970)

and his colleagues in 1940, which was a microphotometric tracing based on Mt Wilson photographic plates of the solar spectrum, a second revision of the wavelengths of solar lines was produced by Charlotte Moore (1898–1990) with Minnaert and J. Houtgast (1908–82) in 1966 [54]. The corrections to the wavelengths advocated by William Meggers (1888–1966) in 1950 brought the solar scale into conformity with that recommended by the International Astronomical Union in 1928. The IAU scale was essentially the 1907 International Solar Union scale based on the red cadmium line, and with secondary standards for iron arc lines also given throughout the spectrum [55].

4.5 HALE AND DESLANDRES AND THE SPECTROHELIOGRAPH

A new golden age of large solar instrumentation commenced in the 1890s. The new developments were introduced almost simultaneously and independently by George Ellery Hale (1868–1938) in the United States and by Henri Deslandres (1853–1948) in Paris.

Hale's interest in solar spectroscopy dated from his student days at MIT. In 1888 he established his private Kenwood Observatory in Chicago, and he produced solar spectra with a concave Rowland grating. By 1891 he had acquired a 12-inch Brashear refractor at Kenwood and in 1892 he successfully designed and built a spectroheliograph for this telescope with which he could photograph the chromosphere of the Sun in monochromatic light centred on a strong photospheric absorption line such as K [56]. According to Hale's own account, the idea of the instrument had come to him as early as 1889 (see discussion in [57]) and an earlier thesis report at MIT had included a description of the new instrument (see also [58, 59]). Some of Hale's early experiments were carried out at the Harvard College Observatory during 1889–90.

The principle of the spectroheliograph entailed the continuous driving of the entrance slit of a spectrograph over an image of the disk of the whole Sun, while at the same time synchronously driving a photographic plate past a second exit slit, so as to build up an image on the plate of the entire solar disk in monochromatic light. The spatial resolution of such a spectroheliogram was determined by the entrance slit width, while the

Figure 4.7. Hale's first spectroheliograph at the Kenwood Observatory, 1892.

wavelength resolution was dependent on the second or exit slit width.

The instrument completed in January 1892 at Kenwood comprised a grating spectrograph with two slits, $3\frac{1}{4}$ inches in length, which were mechanically linked so as to be driven at a uniform rate, one across the solar disk (diameter 2 inches), the other in front of the plate. The grating was a 4-inch Rowland grating (14 438 gr/in) and the collimator and camera were at a fixed 25° angle to each other, each with lenses of $42\frac{1}{2}$-inch focal length [60, 57] – see Fig. 4.7.

Hale took up a position at Yerkes Observatory in 1892, and he transferred the Kenwood instruments

there in 1895. But it was not until 1903 that a new spectroheliograph could be constructed and put into operation for the 40-inch refractor [61], thanks to a grant from the Rumford Committee. The Rumford spectroheliograph employed two 60° prisms, as they gave less scattered light and a brighter spectrum. The instrument is shown in Fig. 4.8.

Although Hale independently invented the spectroheliograph and built the first practical working instrument of this type, the idea of displaying the shape of individual prominences and structures in the Sun using a slit that scanned the image goes back a lot further. Apparently Jules Janssen at the time of the

Figure 4.8. The Rumford spectroheliograph of Hale and Ellermann on the 40-inch telescope at Yerkes Observatory, 1903.

1869 total eclipse experimented with a primitive spectrohelioscope, in which the first slit scans the image rapidly, so that in the eyepiece the observer sees a two-dimensional monochromatic image in the light of a bright spectral line [20, 21]. He wrote:

> This new method, in principle, consists in isolating one of the luminous beams in the spectral domain, emitted by a prominence. The beam is deficient in [photospheric] solar light. These linear sections of prominence images are then transformed into

images themselves, by a fairly rapid rotational movement imposed on the spectroscope [20].

Figure 4.10 shows Janssen's instrument of 1869.

The concept of replacing the eye by a photographic plate and building up an image of the Sun one strip at a time by the movement of two slits was devised by the Jesuit astronomer Carl Braun, in Bohemia in 1872 [62]. The idea of using two co-moving slits and a dark photospheric line were features of Braun's instrument, which was described in some detail, but not

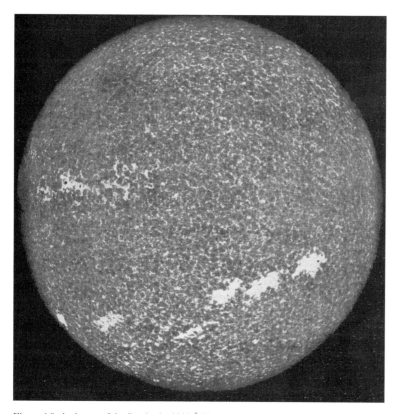

Figure 4.9. An image of the Sun in the 3933 Å H_2 line, recorded by Hale and Ellermann on 12 August 1903, using the Rumford spectroheliograph at Yerkes.

realized in practice – see Fig. 4.11. Hale was unaware of Braun's earlier description of a spectroheliograph (see [63]).

Another attempt to design and build a spectroheliograph was made by Oswald Lohse (1845–1915) at Potsdam. Here a direct vision spectroscope was repeatedly scanned through the solar image on a rotating drum, with a fixed plate being located behind a second slit that isolated the H_γ line [64]. Despite experimenting for several years, Lohse never brought this apparatus to successful operation. These various early attempts to record the chromosphere in monochromatic light were reviewed by Hale in 1893 and again in 1929 [57, 65].

Henri Deslandres at the Paris Observatory began his work to record monochromatic images of the solar chromosphere at about the same time as Hale. In 1891 he published an outline of an instrument with two slits, essentially based on the same principle as the Braun and Hale spectroheliographs [66]. He also proposed

placing a static entrance slit at some 200 equally spaced strips across the Sun's disk, and to record a narrow section of spectrum (defined by a second slit) in the vicinity of a strong line, so as to determine the changes in the velocity field on the Sun at different locations – this was the 'spectro-enregistreur de vitesses', or spectral velocity recorder – see Fig. 4.12. The second slit was much wider than in the spectroheliograph, and neither slit was given a continuous movement, as was done in the spectroheliograph. Deslandres pioneered this instrument so that, with the spectroheliograph, both the shape and the line-of-sight motions of structures in the chromosphere could be determined [66].

A key discovery by Deslandres in 1892 was greatly to influence all future work on spectroheliographs and on the velocity recorder. This was that the strong calcium H and K absorption lines in the solar spectrum had weak emission cores [67, 68]. It was soon realized that the emission was chromospheric in origin,

Figure 4.10. Jules Janssen's simple spectrohelioscope of 1869.

that is, it came from a higher layer of the Sun's atmosphere than the much brighter photospheric light. The bright H_2 and K_2 line reversals proved to be ideal for spectroheliograph observations, as the strong photospheric absorption lines (H_1 and K_1) rendered the disk quite dark at these wavelengths. Figure 4.13 shows an image from Deslandres' velocity recorder, showing the emission in the K_2 line.

Deslandres went on to build and bring into operation his own spectroheliograph in Paris during 1893 [69]. He wrote:

A spectroscope with two slits, capable of imaging these line reversals, gives an exact image of the chromosphere, such as would be seen by someone sensitive only to the extreme violet rays, the photosphere having been removed.

I have obtained this summer photographs of the chromosphere showing all the details with a siderostat, a simple 6-inch mirror and a

spectroscope with one prism, giving a separation of the H and K lines of only 2 mm …

For photographing the motions, it is convenient, on the other hand, to use a high dispersion, a very wide second slit, and a discontinuous movement produced by stops and equal steps. I use therefore two spectroscopes, one of low and one of high dispersion. For simplicity, I arrange them so that they receive light simultaneously from the same solar image, by means of a siderostat and a single objective lens [69].

In 1897 Deslandres took up a position at the Meudon Observatory, then an independent institution under the directorship of Jules Janssen. The low dispersion single-prism spectrograph and spectroheliograph, and the high dispersion spectrograph and velocity recorder were relocated to Meudon on the outskirts of Paris. New improved versions of both instruments were later brought into operation in respectively 1899 and 1903

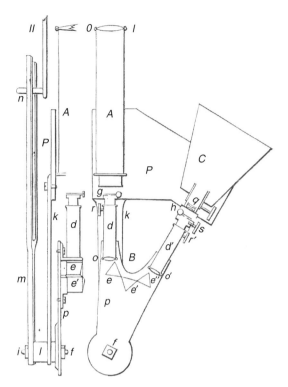

Figure 4.11. Carl Braun's spectroheliograph, which he described in 1872, but the instrument was never built.

[70, 71]. A full description of these instruments and of the solar spectroscopy programme at Paris and Meudon was given by Deslandres in 1905 [72, 73]. The low dispersion instrument used a 60° apex angle light flint prism. The beam size was $A = 6$ cm. The high dispersion instrument employed the 4-inch Rowland grating acquired by Deslandres in 1887, with 560 gr/mm (14 438 gr/in) and a 1-m focal length camera [73].

From 1906 new forms of spectroheliograph were constructed at Meudon, some with up to three prisms, some with a grating and one with three slits so as to reduce the scattered light and to accept light from several dark solar lines simultaneously – a polychromator [72, 73, 74, 75]. The spectroheliograph laboratory at Meudon is seen in Fig. 4.14. A spectroheliogram of the Sun obtained by Deslandres in K_2 light is shown in Fig. 4.15.

The question of priority for the development of the spectroheliograph, and in the interpretation of its results was never far from the mind of either Hale

or Deslandres. The latter made frequent references to Hale's publications, and generally grumbled at the lack of recognition the French work had received from Yerkes or Mt Wilson. Thus, and not with complete accuracy, Deslandres wrote:

> I am led to present here my own personal point of view on the question (of precedence), and in a full manner, in as much as Hale, after having reproduced in his journal my first notes in *Comptes Rendus* on the subject, omitted to publish the numerous notes that followed after 1892.
>
> I have not been ignorant ... of the advantages of the spectroheliograph, which I have continuously used for the recording of vapors in my first notes of August 1891 and February 1892, which Hale always fails to mention. Not only have I used the spectroheliograph ... but even the spectral velocity recorder. I have considered the instrument ... [spectroheliograph] neither as very new nor as very difficult to construct ... In addition, the spectroheliograph ... obeys certain simple optical rules which I have been the first to propound in 1893 ... and which alone assure me of a clear and complete image. These rules have been followed subsequently by all, and in particular by Hale in his last researches of 1903 [72, pp. 332–3].

Hale published his reply in the *Astrophysical Journal* in 1906 [76].

> I regret exceedingly the necessity of discussing a question of priority, but the repeated statements of M. Deslandres leave no alternative. My reply, however, will be brief.
>
> I am quite content to leave the question of priority in the use of the spectroheliograph to the judgement of those who are acquainted with the facts. In 1894 the French Academy of Sciences awarded me the Janssen medal for the construction and use of the first spectroheliograph. In the statement for the reasons for the award ... no reference is made to M. Deslandres. This might reasonably be considered to settle the matter ...
>
> Although M. Deslandres did not use the spectroheliograph until more than a year after my first successful work with this instrument at the Kenwood Observatory, his observations of the

Figure 4.12. Part of the velocity recorder of Henri Deslandres. The objective collimator lens is on the left. The second slit is also seen. Both collimator lens and slit are displaced simultaneously in discrete jumps.

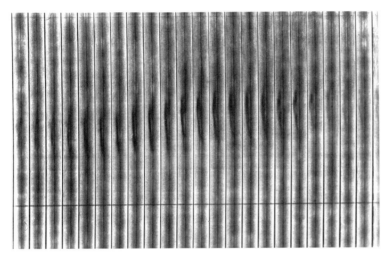

Figure 4.13. A part of the image of the Sun recorded by Deslandres in the K line using the velocity recorder at Meudon in April 1910. Each strip is a narrow band on the solar disk, showing spatial information from top to bottom, and velocity information from one side of the strip to the other. Shorter wavelengths (approaching velocity) are on the left. The black line in each strip's centre is the K_2 line in emission on these negative photographic images.

spectra of the calcium flocculi were commenced in 1891, almost simultaneously with my own investigations of these spectra. Before 1893, when he also obtained a spectroheliograph, which he has since used with marked success, M. Deslandres devoted special attention to a study of the K line in successive sections of the Sun's disk. The *spectrograph* employed for this purpose was moved a short distance between each exposure, but the exposures were made when the instrument was at rest, and the resulting photographs are photographs of *spectra*. This method is extremely

useful ... But a spectrograph thus employed is in no sense a spectroheliograph ...

M. Deslandres complains (in the *Bulletin Astronomique*) that his papers have not been published in the *Astrophysical Journal*. He alone is responsible for this, as the editors have neither been favored with his manuscripts nor informed of his desire for such publication.

The Royal Astronomical Society did redress the balance somewhat between the pioneering work of Hale and Deslandres in 1913, by awarding Deslandres the

Figure 4.14. Deslandres' spectroheliograph laboratory at Meudon, 1910. Four individual spectroheliographs are united around the same astronomical objective and same collimator. The instrument on the left is the 3-m prism spectroheliograph, while that on the right is the 3-m grating instrument. Both are ready for operating.

Society's Gold Medal for research in solar phenomena. The president, Frank Dyson (1868–1939), noted that Deslandres had not only contributed greatly to the development of the spectroheliograph

but he has also constructed 'the Spectro-Enregistreur des Vitesses' or velocity recorder, which is of equal and may prove of even greater importance. Thanks to this discovery, we can follow and study the various movements of the Solar Atmosphere, as well as the forms of the different layers [77].

With the success of the spectroheliographs of Hale and Deslandres, several other observers quickly

acquired instruments of this type. For example, Paul Kempf (1856–1920) at Potsdam had a spectroheliograph constructed for the Grubb refractor at the Potsdam Astrophysical Observatory in 1904 by the local firm of Otto Toepfer and Son [78, 79]. It was a compact grating instrument with parallel collimator and camera tubes, similar in design to one Hale himself had used to record images of the solar corona from Mt Etna a decade earlier [80]. This instrument was also made by Toepfer and was loaned to Hale by the British amateur astronomer, Arthur Cowper Ranyard (1845–94).

Another spectroheliograph was constructed in England for the Solar Physics Observatory in South Kensington. This had a single prism as dispersing element and light was fed from a siderostat [81]. In

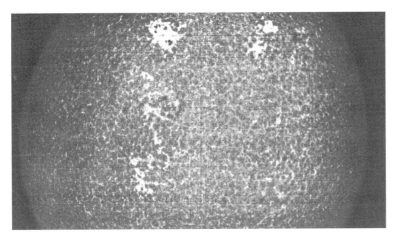

Figure 4.15. Spectroheliogram of the Sun in K_2 light, obtained by Deslandres in March 1904 with a low dispersion single-prism spectroheliograph. The exposure was 7 minutes.

this design, the whole instrument was translated side-ways, while the solar image and the photographic plate remained stationary, whereas in Hale's Rumford spectroheliograph, scanning was achieved by driving the solar image by means of the telescope's declina-tion axis. The South Kensington spectroheliograph was brought into operation in 1904 and successful K-line spectroheliograms were obtained by William Lockyer (1868–1936) (son of Sir Norman Lockyer) at that time.

When Arthur Cowper Ranyard died in December 1894, he left his astronomical instruments, including the spectroheliograph, to John Evershed (1864–1956), the English solar physicist, then near the start of his distinguished career. Evershed further improved on the Ranyard design in 1899, and installed the instrument on the 18-inch reflector at his home in Kenley, Surrey [82] – see Fig. 4.16.

In 1906 Evershed accepted a post as assistant at the Kodaikanal Observatory in India, and he constructed there a high dispersion solar spectrograph using a Row-land grating [83] (see also [82]). Evershed became the director at Kodaikanal in 1911 and stayed in close touch with Hale. As a result, Hale urged him to undertake Hα spectroheliograms in 1914. Evershed did this by building a grating spectroheliograph for the purpose, which gave sufficient resolving power for the red Hα line. He also installed a large spectroheliograph at a temporary high altitude solar observatory in Kashmir in 1915, which he operated with his wife, Mary, for some 17 months.

After retirement in 1923, Evershed returned to England, and he established a private observatory at Ewhurst, Surrey, where he continued his researches in solar physics. A spectroheliograph was used for daily observations of the Hα spectroheliograms from this site. It comprised two 45° prisms of 6 inches aperture and was used in an autocollimating mode by reflecting the light back through both prisms [84]. Light was fed to the spectroheliograph from a 15-inch coelostat and the solar image was produced by a 6-inch achromatic lens of focal length about 6.3 metres.

4.6 HALE AND THE DEDICATED SOLAR SPECTROGRAPH AND OBSERVATORY

The Rumford spectroheliograph at Yerkes Observatory was brought into operation in 1903, but Hale soon rec-ognized the limitations of doing solar spectroscopy with an instrument mounted at the end of a long refractor. So in that same year, thanks to a $10 000 gift from a benefactor, Miss Helen Snow, Hale set about build-ing a dedicated solar telescope and spectrograph. This was the Snow telescope at Yerkes Observatory, a tele-scope fed by a 76-cm coelostat. A second plane mirror then sent the beam horizontally to a concave mirror of aperture 24 inches (61 cm) (focal length 50.3 m) which produced a large image of the Sun [85]. Although this was not the earliest dedicated solar telescope (for example, the Paris Observatory was undertaking solar spectroscopy from 1892 using the Foucault siderostat),

Figure 4.16. John Evershed's first spectroheliograph of 1899, with a direct vision prism, attached to the 18-inch telescope.

the Snow telescope nevertheless set a new standard as a large-scale instrument specially designed for solar research. The Snow telescope was at Yerkes for barely a year, when Hale moved it to the new and better site of Mt Wilson in California in 1904. This was a relatively high altitude site at 1700 m and with much better climatic and seeing conditions. The Snow telescope became the first major instrument at the new location.

Hale equipped the Snow telescope with three solar instruments as well as a small prism spectrograph and a concave grating spectrograph for stellar spectroscopy. The solar instruments included two spectroheliographs: one had four flint prisms of 8 inches aperture and light was supplied by a 60-foot focal length collimator. A smaller spectroheliograph had three 5-inch aperture prisms and a 30-foot collimator, or alternatively a plane grating could be used in place of the prisms. Figure 4.17 shows the Snow telescope spectroscopic laboratory at Mt Wilson, and Fig. 4.18 shows a spectroheliograph in the laboratory under construction.

Thirdly a Littrow spectrograph used a collimating lens of 18 feet (5.49 m) focal length mounted in conjunction with a grating. It was used with the 60-foot mirror for the study of solar rotation and of sunspot spectra [85]. The Snow telescope was installed in a horizontal long low building, some 220 feet in length, which was approximately horizontal and north–south in orientation. Hale went to some trouble to overcome the effects of temperature stratification in the Snow building, using louvres and a well-insulated room in the spectroscopy laboratory. Unfortunately image definition was never satisfactory, and even before the Snow telescope became operational in 1904, Hale was searching for a better site for a new solar observatory.

Meteorological testing at Mt Wilson near Pasadena was undertaken from 1903 [86] and the solar observatory was established there in 1904 to 1905. The Snow telescope became operational at Mt Wilson in 1905. A new large spectroheliograph was built for it, using from one to four prisms, and 8-inch aperture collimator and camera lenses of 5 feet (152 cm) focal length [87].

Although Hale claimed that image quality from the Snow telescope at Mt Wilson was excellent [88], in reality he was never satisfied with the status quo. By 1907 Hale was convinced that better image quality for solar work could be achieved with a tower telescope

Figure 4.17. The Snow telescope spectroscopic laboratory at the Mt Wilson Solar Observatory, 1906. A quartz spectrograph is just above the concave mirror. To the right of this is a Hilger one-prism spectroscope. The main instrument is the Littrow spectrograph with a plane Michelson grating, 8. 3 × 4. 4 cm, with 7000 gr/cm.

Figure 4.18. Five-foot spectroheliograph for the Snow telescope at Mt Wilson, under construction in 1906.

in which the coelostat was high above the ground, and light rays travelled vertically so as to avoid the problems of temperature stratification. The solar image was to be formed near ground level by an objective lens in the tower, and the light then entered an underground laboratory housing the spectrograph [89].

He immediately set to work to build such a tower telescope and spectrograph comprising a tower 60 feet

Figure 4.19. The 60-foot solar tower telescope at Mt Wilson, with the Snow telescope in the background.

high (Fig. 4.19) with a 12-inch visual objective of 60 feet focal length producing the Sun's image. The vertical spectrograph was in a concrete well 30 feet below ground and was a grating instrument of the Littrow type with a 6-inch Brashear autocollimating lens [90] – see Fig. 4.20.

A 4-inch Rowland grating was used in the new spectrograph and 17-inch (43-cm) long plates were used to record the spectra. The whole instrument could be adapted to become a spectroheliograph, with the lens and the plate-holder both moving for this purpose.

Hale made a special study of sunspot spectra with the Snow telescope [91] and these investigations were continued with the 60-foot tower telescope. Probably his most significant discovery, that of a magnetic field in sunspots, which was detected by means of the Zeeman splitting or broadening of spectral lines, was made with this instrument in 1908 [92].

Hale claimed that the image quality from the 60-foot tower telescope was always superior to that from the horizontal Snow telescope [90], a result never accepted by Deslandres at Meudon [75, p. 43], who believed that vibrations in the tower and the chimney effect of rising air currents in the tower dissipated any apparent advantages. In spite of that opinion, tower

telescopes were copied elsewhere, including at Potsdam in 1924 [93], in Florence in 1926 [94], in Tokyo in 1926 [95], at Oxford in 1935 [96, 97] and at the McMath–Hulbert Observatory of the University of Michigan in 1936 (50-foot tower) and 1940 (70-foot tower) [98, 99]. An article on tower telescopes at solar observatories by Heinz Gollnow summarizes the situation in 1949 [100].

As early as 1909 Hale had initiated the design and construction of a much larger solar tower telescope, the 150-foot instrument. This was completed and brought into operation in 1912 [101]. The design was similar to the 60-foot tower, apart from the larger overall dimensions. In particular the focal length of a Brashear triple objective gave a solar image about 17 inches (42.5 cm) in diameter near ground level. The vertical beam then went to an autocollimating spectrograph with a Michelson grating. The autocollimating lens was either 75 feet or 30 feet in focal length. Hale described the instrument as follows:

The 75-foot spectrograph has also proved to be very satisfactory. With the Michelson grating the definition of the solar spectrum is excellent in the first three orders; in the third spectrum, where the distance between D lines is over an inch, both D_1 and D_2 are clearly double. A large number of

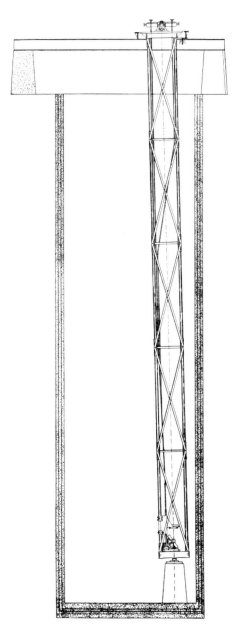

Figure 4.20. Section through the underground spectrograph chamber of Hale's 60–foot tower telescope at Mt Wilson, 1908.

from the bottom of the well to the 30-foot level is a valuable feature of the design.

Another advantage of this instrument is the possibility of changing the dispersion through a very wide range without disturbing the adjustments [101].

Figure 4.21 shows the combined spectrograph and spectroheliograph of the 150-foot tower telescope. The head of the 75-foot spectrograph is in Fig. 4.22. Fig. 4.23 shows solar spectra recorded by Charles St John with this instrument.

4.7 FURTHER DEVELOPMENTS IN SOLAR SPECTROGRAPHS AND TELESCOPES

The legacy left by Rowland, Hale and Deslandres for solar physics was immense and influenced the course of solar spectroscopy for the first half of the twentieth century: Rowland through his provision of diffraction gratings which were used at many observatories in America and Europe; Hale through the development of the spectroheliograph and the tower telescope, which were widely copied elsewhere; and Deslandres for further developing the spectroheliograph.

The tower telescope at Utrecht was one of the first to follow the Hale design. The coelostat was in fact mounted on the roof of the four-storey high Heliophysical Institute, thus avoiding the need for a separate tower, while the spectrograph was installed at ground level [102]. The objective lens was of 25 cm aperture and 13 m in focal length. The spectrograph was of the autocollimating type (see Fig. 4.24) and employed a Rowland grating (8 × 5 cm) with 568 gr/mm. The instrument could be adapted to operate as a spectroheliograph.

The famous Einstein Tower at Potsdam (designed by Erich Mendelsohn in the Jugendstil) was completed in 1924 and had an especially large 60-cm objective lens (focal length 14.5 m) [103, 93] – see Fig. 4.25. This instrument had a horizontal spectrograph and spectroheliograph in the basement, some 3 m underground. Using a Rowland grating (800 gr/mm) in first order, a resolving power of 100 000 was achieved with a dispersion of 1.4 Å/mm. A high dispersion Littrow prism spectrograph intended for stellar spectroscopy was also installed in the Einstein Tower [104].

excellent photographs of spectra have been taken with the focal length of 75 feet (22.88 m), but at present the spectrograph is used with a focal length of 30 feet (9.15 m), since the exposure times are of course much shorter in this case. The ease with which the base of the instrument can be shifted

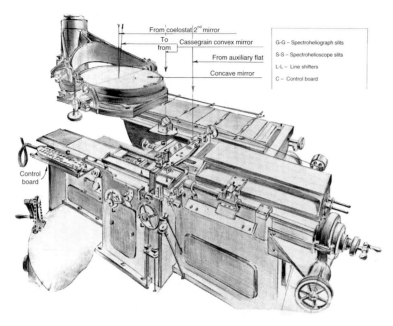

Figure 4.21. Combined spectrograph, spectroheliograph and spectrohelioscope at the 150-foot tower telescope at Mt Wilson, 1929.

Figure 4.22. Head of the 75-foot spectrograph of the 150-foot tower telescope at Mt Wilson, showing the mounting for the comparison arc.

λ 4175   λ 4270

Figure 4.23. Two spectra of the Sun (top and bottom) and of a comparison iron arc spectrum, exposed on the 75-foot spectrograph by Charles St John in 1928.

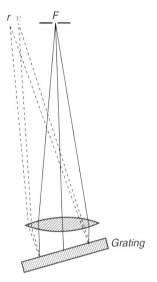

Figure 4.24. Giorgio Abetti's diagram illustrating the principle of an autocollimating grating spectrograph.

A solar tower telescope was also constructed in Tokyo with German Zeiss optics and a prism spectrograph, and completed in 1926 [95] – see Fig. 4.26. This instrument had three large prisms of height 220 mm. As at Potsdam, the spectrograph was mounted horizontally at the base of the tower. The dispersion was 0.78 mm/Å at 3950 Å. The Tokyo instrument marked the introduction of astronomical spectroscopy into Japan by Yoshio Fujita.

The instrumentation at the Arcetri Observatory in Florence was also completed in 1926. It had a 30-cm objective lens and a grating spectrograph and spectroheliograph (568 gr/mm) with Zeiss optics [94, 105] – see Fig. 4.27.

Another solar tower telescope was built in Canberra, Australia, at the Commonwealth Solar Observatory on Mt Stromlo. This instrument had a 30-cm objective lens of focal length 13 m. A 45° flat mirror sent the beam to a horizontal prism spectrograph made by Adam Hilger of London and located in the tower's basement. Three glass prisms in a Littrow double-pass mounting were used, so as to give a 'theoretical' (that is diffraction-limited) resolving power of 150 000 at 430 nm [106, 107].

Two solar tower telescopes with a somewhat different optical arrangement were built at Pasadena and at Oxford. These both used Cassegrain reflecting optics with the concave primary mirror at the bottom of the tower facing up to the coelostat, and a convex secondary reflecting the beam down again to a small 45° flat [108, 97]. Such Cassegrain telescopes were built by Hale at the California Institute of Technology in Pasadena, and by Harry Plaskett (1893–1980) in Oxford. Although five reflections ahead of the slit were required, chromatic aberrations were avoided. Such all-reflecting systems had not been advocated previously, because of the thermal effects of sunlight distorting the mirror's figure. At Oxford all mirrors were from fused silica (which has one-seventh of the expansion coefficient of Pyrex), thereby greatly reducing such thermal distortions, which have always been an issue for coelostats feeding any solar telescope. The Oxford telescope was equipped with a three-prism autocollimating spectrograph, which has been described in a further paper by Plaskett [109].

In summary, the 1920s and 1930s can be seen as a time of consolidation rather than innovation in the domain of high resolution solar spectroscopy and the use of the spectroheliograph. Many observatories were copying the basic design principles introduced by Hale at Mt Wilson for solar instrumentation. Solar physics became widely practised at dedicated solar observatories, without any new techniques being prominent. Table 4.1 lists many of the world's major dedicated solar observatories established from the 1890s and through the twentieth century and which were equipped with solar spectrographs or spectroheliographs.

The golden age of the spectroheliograph was during the first third of the twentieth century. It is true that its importance diminished somewhat with the

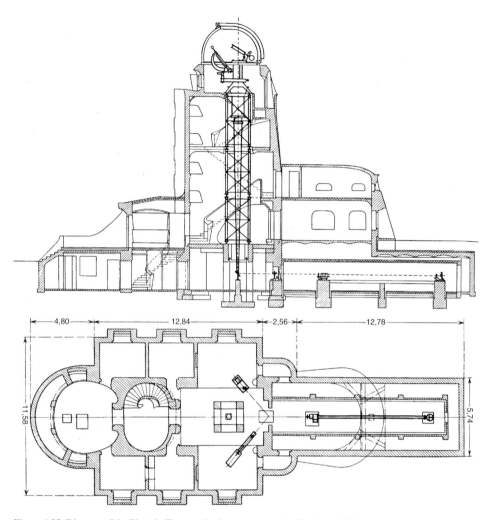

Figure 4.25. Diagram of the Einstein Tower and solar spectrograph at Potsdam, 1924.

invention of the narrow-band Lyot filter, by Bernard Lyot (1897–1952) in France in 1933 [111], and a few years later but independently by Yngve Öhman (1903–88) in Sweden [112]. The Lyot filter comprised a series of polarizing filters sandwiched between sheets of birefringent material such as quartz or Iceland spar, whose thickness was chosen to isolate essentially monochromatic radiation. Such a filter allowed wide-angle images of the solar disk in the light of one spectral line, such as Hα, to be obtained and with greater simplicity than was possible with the spectroheliograph. Lyot correctly predicted that his filter would allow '...the study either visually or by means of cine films the movements in the solar atmosphere at an accelerated rate, including those of plages, faculae, filaments and prominences' [111].

4.8 THE WORK OF FRANCIS AND ROBERT MCMATH

The pioneering contributions to solar physics of Francis and Robert McMath (respectively 1867–1938 and 1891–1962), father and son amateur astronomers in Michigan, together with their collaborator, Henry Hulbert (a Michigan county judge) deserve special mention. Their interest in astronomy dated from the early 1920s, and by 1928 they pioneered the technique of astronomical cinematography, using time-sequence

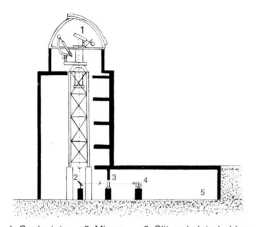

1. Coelostat.   2. Mirror.   3. Slit and plate-holder.
   4. Prism spectrograph.   5. Basement.

Figure 4.26. Solar tower telescope and spectrograph of the
Tokyo Astronomical Observatory, built for Yoshio Fujita in 1926.

images recorded on a 4-inch telescope. In 1930 they established an observatory at Lake Angelus, Michigan, with a $10\frac{1}{2}$-inch reflecting telescope and here, with assistance from Ralph Curtiss (1880–1929), Heber Curtis (1872–1942) and Robert Petrie (1906–66) from the University of Michigan, they developed an instrument they called a spectroheliokinematograph in 1932 [113]. Essentially this was a compact grating spectroheliograph mounted at the Cassegrain focus of their telescope and designed to record prominence images of the Sun in quick succession. A typical day's observing resulted in some 600 images, mainly using the Hα line, and these were later carefully registered together into a motion film strip to show the time evolution of prominences, especially those of the eruptive type [113]. The instrument is shown in Figs. 4.28 and 4.29.

With the success of the spectroheliokinematograph in 1932–3, the McMaths and Hulbert went on to build a 50-foot tower telescope in 1936, with advice and assistance from Hale [114] – see Fig. 4.30. This was also an all-mirror reflecting system for a solar telescope (as at Pasadena), and mostly it operated with a 40-cm off-axis parabola of 12 m focal length at the tower's base, with the beam being folded down again with a flat mirror. The vertically mounted spectroheliograph installed in this tower was an autocollimating grating instrument, using a 15 240 gr/in grating supplied from Mt Wilson. Later, in 1939, a new spectroheliograph (known as the Julius Stone instrument) was installed

at the 50-foot tower [115] – see Fig. 4.31. It was also an autocollimating grating spectroheliograph, and was designed especially for scanning the radial-velocity profile of solar prominences, as well as their spatial structure. This was achieved by applying successive small displacements of the second slit relative to the first, so as to make spatial scans over a prominence at several different velocity positions within the Hα line.

After the death of Francis McMath in 1938, his son Robert went on to build a larger 70-foot tower telescope (known as the McGregor telescope) at Lake Angelus, which fed a series of horizontal grating spectrographs [116] – see Fig. 4.32. The McGregor telescope became operational in 1941, and its optical system comprised an off-axis parabolic mirror (diameter 30 cm, focal length 15 m), as well as a flat mirror half way up the tower to fold the beam back down.

Several pioneering developments took place with this telescope, including the first use of échelle gratings, infrared scanning of solar spectra and the development of the first solar vacuum spectrograph.

The observatory at Lake Angelus was gifted to the University of Michigan as early as 1931, and Heber Curtis, the then director of the astronomy department, named it the McMath–Hulbert Observatory to recognize the three founders. Robert McMath became professor of solar physics at the university in 1945. University of Michigan support for the McMath–Hulbert Observatory extended through to 1979.

4.9 NEW INNOVATIONS IN SOLAR SPECTROSCOPY, 1940–1965

4.9.1 Photoelectric recording of the solar spectrum

After World War II, the impetus to technology, especially in electronics, led to new innovations in solar research. Even in the 1930s some early indications of these new advances were seen. At this time the potassium hydride (KH) photoelectric cell was being introduced into photometric astronomy, notably by Joel Stebbins (1878–1966), Albert Whitford (1905–2002) and others in Wisconsin and at Lick. Theodore Dunham used such a cell with a triode DC amplifying tube in conjunction with the 30-foot and 75-foot Littrow spectrographs on the 75-foot tower telescope at

Figure 4.27. Solar spectrograph and spectroheliograph of the Arcetri Observatory solar tower, 1926.

Mt Wilson [117] – see also [118]. The amplifier output was sent to a galvanometer from which traces of the profiles of individual solar absorption lines could be produced. Currents of only 5×10^{-14} amperes in the solar continuum near 5000 Å were recorded at a resolving power of about 150 000. This was the start of a new electronic era in astronomical spectroscopy.

The new photoelectric technique in solar spectroscopy was taken up by Hermann Brück (1905–2000) in Cambridge in 1939 using the McLean solar spectrograph and an Osram potassium on silver oxide photocell. The light spot from the galvanometer was arranged so as directly to record the spectrum on bromide paper attached to a rotating drum. The photocell moved along the spectrum at the rate of only 1 mm/min, so that 4 Å of spectrum took 10 minutes [119]. The disadvantage was a slow rate of collecting the data in one channel at a time, but the advantage was a reasonably linear detector with a dynamic range superior to the photographic plate.

4.9.2   Lead sulphide cells and the solar infrared

After the war, lead sulphide photoconductive cells, sensitive to the near infrared (to 3.6 μm), were becoming available as a result of the work of Robert

Cashman (1906–88) at Northwestern University [120]. A Cashman lead sulphide cell was acquired by Robert McMath and Orren Mohler at the McMath–Hulbert Observatory and was soon used for scans of the solar spectrum to 2.0 μm [121, 122] on the McGregor 70-foot tower telescope – see Fig. 4.33. The resolving power was 32 000 and the long wavelength limit was set by the glass objective and collimator lenses. An all-reflecting spectrometer was accordingly designed so as to extend the observable wavelengths to 3.6 μm [123]. It used the so-called Pfund optical system in which a plane grating received light from an on-axis paraboloid collimator mirror, with the light beam being folded using a perforated flat mirror [124]. The camera was also a folded paraboloid. Scanning was achieved by a slow rotation of the grating, giving a speed of 0.24 Å/s at the exit slit in front of the cell. The cell was cooled in dry ice and a mechanical chopping system (at 1.08 kHz) enabled the signal in the amplified output to be recorded [125].

The infrared spectrometer system was also taken by McMath and Mohler to Mt Wilson and used on the Snow telescope in September and October 1949. A photometric atlas of the solar spectrum 8465–25 242 Å, with a resolving power of about 30 000, was the result [126] – see Fig. 4.34. For the first time the resolving power was sufficient to show numerous infrared solar

Table 4.1. *Table of selected solar observatories with spectroheliographs or spectrographs*

Observatory	Telescope	Year installed	Instrument(s) and notes	Reference(s)
Kenwood	12″ Brashear refractor	1892	Hale's first spectroheliograph	[60, 56, 57]
Paris	12-cm Foucault siderostat	1892	low dispersion spectrograph	[67, 68]
		1894	Deslandres' first spectroheliograph	[69]
		1894	velocity recorder	[66]
Meudon	30-cm solar telescope	1899	Deslandres single-prism spectrograph	[70]
		1903	high dispersion grating spectrograph	[71]
		1906	new spectroheliographs	[72, 73]
Yerkes	40″ refractor	1903	Rumford spectroheliograph	[61]
Yerkes	Snow telescope	1904		
Potsdam	20-cm Grubb refractor	1904	Toepfer spectroheliograph	[78, 79]
South Kensington	12″ siderostat	1904	single-prism spectroheliograph	[81]
Mt Wilson	Snow telescope	1905	two spectroheliographs and Littrow spectrograph	[85, 87]
Kodaikanal	60′ tower telescope	1909	grating spectrograph and spectroheliograph	[83, 82]
Mt Wilson	150′ tower telescope	1908		[90]
		1912		[101]
Utrecht	25-cm solar telescope	1922	autocollimating grating spectrograph	[102]
Potsdam	Einstein tower and 60-cm telescope	1924	autocollimating grating spectrograph	[103, 93]
Arcetri Obs. Florence	30-cm solar telescope	1926	Zeiss grating spectrograph and spectroheliograph	[94, 105]
Tokyo		1926	Zeiss three-prism spectrograph	[95]

Table 4.1. (cont.)

Observatory	Telescope	Year installed	Instrument(s) and notes	Reference(s)
Commonwealth Solar Obs. Mt Stromlo	30-cm solar telescope	1932	horizontal Hilger three–prism spectrograph	[106, 107]
Oxford	26″ solar tower telescope	1935	three–prism spectrograph	[96, 97]
Caltech, Pasadena	50′ tower telescope	1935	75′ spectrograph and spectroheliograph	[108]
McMath–Hulbert	16″ or 12″ parabolic mirrors	1936	grating spectroheliographs Stone spectroheliograph in 1939 and Littrow grating spectrograph	[114, 115]
McMath–Hulbert	McGregor 70′ tower telescope	1940	horizontal spectrograph	[116]
Pulkovo	horizontal solar telescope	1940	CCD spectroheliograph from 1996	[110]
Kitt Peak	McMath solar telescope with heliostat and 1.6-m mirror	1962	vacuum Czerny–Turner grating spectrograph	[150]
Big Bear Lake (Caltech)	65-cm vacuum reflector	1969	spectrograph	[141]
Sacramento Peak	76-cm Dunn solar telescope	1969	horizontal spectrograph	[140]
Meudon	60-cm solar tower telescope	1968		
San Fernando (Cal.)	61-cm vacuum solar telescope	1969	coudé spectrograph and spectroheliograph	[142]
Crimean AO	90-cm solar tower telescope	1973	spectrograph with magnetograph	
	45-cm solar tower telescope		échelle spectrograph	
Huairou Solar Obs.	60-cm solar telescope	1994		

A more complete list of solar observatories is at www.bbso.njit.edu/newtelescope/large.html
Big Bear Lake Observatory has been managed by New Jersey Institute of Technology since 1997. The 65-cm solar telescope there closed in 2006, and a new 1.6-m instrument is under construction.

Figure 4.28. The McMath–Hulbert spectroheliokinematograph on the $10\frac{1}{2}$-inch reflector at the Lake Angelus observatory, Michigan, 1932.

lines, thereby opening up a whole new domain in solar spectroscopic research. Previous scans of the region beyond 1.3 µm with thermocouples and bolometers had revealed many strong telluric molecular bands, but not the mainly weaker lines of solar origin. A preliminary catalogue of 888 solar photospheric lines was published by the McMath–Hulbert team in 1953 [127] and a more detailed catalogue with over 7400 lines followed in 1955 [128]. However some 93 per cent of these features were telluric in origin.

4.9.3 First use of the échelle grating for solar spectroscopy

The introduction of the échelle grating by George Harrison in 1949 [129] led almost immediately to its use in laboratory and solar spectroscopy. The benefits of high dispersion in a compact instrument and large wavelength coverage over several orders were soon recognized. Moreover the échelle spectrograph does not require very long spectrograms, and hence it uses cameras with only a limited angular field of view [130].

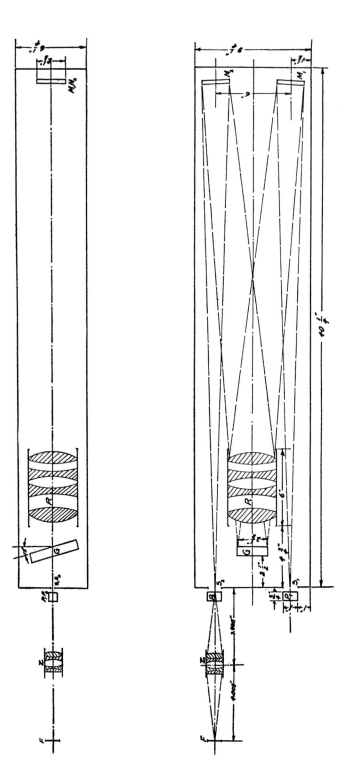

Figure 4.29. Optical system of the McMath–Hulbert spectroheliokinematograph.

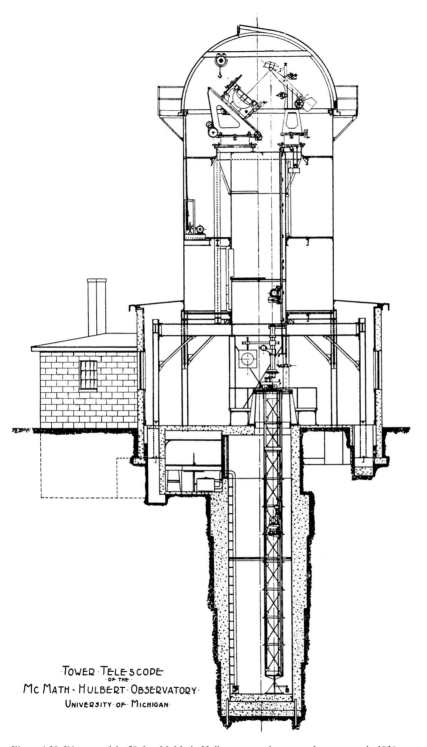

TOWER·TELESCOPE
·OF·THE·
MC MATH·HULBERT·OBSERVATORY·
UNIVERSITY·OF·MICHIGAN·

Figure 4.30. Diagram of the 50-foot McMath–Hulbert tower telescope and spectrograph, 1936.

Figure 4.31. The Julius Stone spectroheliograph at the 50-foot tower telescope at Lake Angelus, Michigan, 1939.

Soon échelle gratings were being manufactured by the Bausch and Lomb Co. [131] and an échelle grating was acquired by Keith Pierce, Robert McMath and Orren Mohler at the McMath–Hulbert Observatory, where it was tested for high resolution solar spectroscopy as early as 1951 [132]. A spectrograph with concave mirrors for collimator and camera and a 150 × 75-mm échelle grating successfully gave photographic spectra at a high resolving power of 250 000 and a dispersion of 2.9 Å/mm at 500 nm. An échelle spectrogram of a solar prominence in the far-red region is seen in Fig. 4.35.

4.9.4 Solar ultraviolet spectroscopy from rockets

The échelle grating allowed a compact spectrograph to be designed for launching in rocket nose cones.

Such instruments were therefore able to observe the the ultraviolet solar spectrum from above the Earth's ozone layer. The design of such a spectrograph was commenced in 1952 by Richard Tousey at the US Naval Research Laboratory, and the instrument was first flown in 1957 on an Aerobee rocket. It comprised a Littrow échelle (73 gr/mm) crossed with a 30° fluorite prism for order separation. Spectra were recorded on 35-mm film in the 200–300 nm wavelength range, and had a resolving power of around $R = 10^5$ [133]. A successful flight in August 1961 recorded the solar spectrum to 220 nm [134]. The instrument is shown in Fig. 4.36.

A further flight followed in 1964, and the data enabled a solar ultraviolet line catalogue to be compiled [135]. Experiments were also carried out on another échelle instrument for the extreme solar ultraviolet,

Figure 4.32. The observatory at Lake Angelus with its 50- and 70-foot solar tower telescopes and a 24-inch reflector in the foreground.

using a grating for order separation. This was designed for the region 100–130 nm, including the Lyman lines [136].

A new era in high resolution solar ultraviolet spectroscopy was introduced with the work of the Naval Research Laboratory, using the new technologies of both the échelle grating and the rocket-borne spectrograph.

4.9.5 The solar vacuum spectrograph

In the mid 1950s, Robert McMath at the McMath–Hulbert Observatory in Michigan recognized that one of the limiting factors in the performance of solar spectrographs was the air currents inside the spectrograph itself. Line positions would vary by at least ± 0.06 Å, and sometimes by as much as ± 0.20 Å on the Snow telescope as a result. On the other hand, a

vacuum spectrograph would not only eliminate these currents, but also give immunity to temperature and pressure effects on the refractive index of air. Moreover, changes in the groove spacing of diffraction gratings with temperature would be minimized in a carefully temperature-controlled environment.

The result was the first solar vacuum spectrograph, built by McMath for the McGregor 70-foot tower telescope at the McMath–Hulbert Observatory [137]. Construction began in November 1953. The instrument had two camera mirrors, one for photographic work and the other used for photoelectric scanning of solar line profiles. The whole instrument was installed in a vacuum tank at a pressure of 3 or 4 mm of mercury. The spectrograph was an all-reflecting system with a large Babcock plane grating (600 gr/mm) with a width of 203 mm.

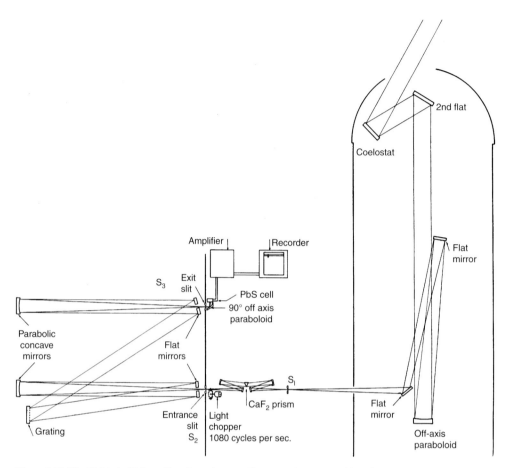

Figure 4.33. The McMath–Hulbert direct intensity recording monochromator, with coelostat, prism pre-disperser and PbS detector on the 70-foot solar telescope at Lake Angelus, Michigan, about 1948.

One innovation was the use of an iodine vapour cell in the incoming beam of sunlight. This superimposed numerous narrow iodine absorption lines on the solar spectrum, an analysis of which showed that the resolving power attained was 600 000 [137].

Preliminary results from the new instrument immediately showed a wealth of detail in the cores of the Fraunhofer lines. McMath and his colleagues noted for Hα and the K lines: ' ...as the dispersion is increased to about 5 mm/Å and greater, the cores of both lines are seen to be composed of large numbers of relatively small elements, varying in width and intensity' [138]. The spectra showed line variations over the solar surface on scales as small as 2 arc seconds, corresponding to about 1500 km on the solar surface.

A detailed discussion of the performance of the vacuum spectrograph was given by Keith Pierce (1918–2005) in 1957 [139]. A resolving power as high as 630 000 was measured for the green mercury line (5769 Å) using the grating in the fifth order. The instrumental profile was also studied and the total strength of the Rowland ghosts (1.8%) and general scattered light (0.7%) evaluated.

The vacuum spectrograph was adopted at solar observatories elsewhere, most notably at the McMath telescope of the Kitt Peak National Observatory (see Section 4.9.7). Even an evacuated solar telescope was built at the Sacramento Peak Observatory, which fed light to an evacuated spectrograph [140]. This was followed by the Caltech vacuum telescope and spectrograph at Big Bear Lake in 1970 [141]. This was the

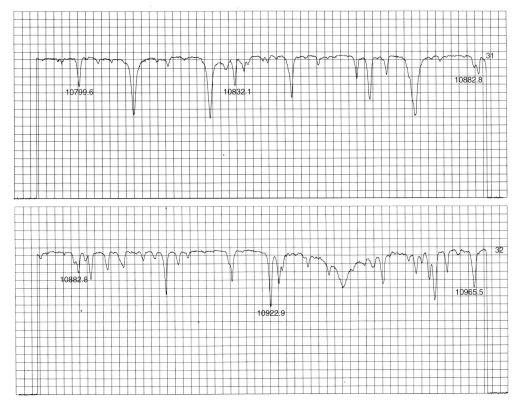

Figure 4.34. A sample from the *Photometric Atlas of the Near Infrared Solar Spectrum* by Orren Mohler and colleagues at the University of Michigan. The atlas was published in 1950, and extended the solar spectrum to about 2.5 μm with a resolving power of 30 000.

first solar telescope surrounded by water, to take advantage of good seeing conditions throughout the day. The original instrument was a 40-cm equatorially mounted Cassegrain reflector with a coudé feed to a vertical Littrow grating spectrograph. Another evacuated telescope and spectrograph was installed at San Fernando, California, also at this time. It had a 60-cm equatorial vacuum telescope and evacuated Littrow grating spectrograph with a coudé feed [142]. At Kitt Peak a 60-cm vacuum tower telescope with a coelostat feed and vertical Littrow grating spectrograph was installed in 1975 [143]. However the spectrograph is in air.

4.9.6   The double-pass solar spectrograph

As early as 1893, Henri Deslandres in Paris was considering the problem of scattered light in his solar grating spectrograph, which had the effect of partially filling in the centres of the darkest lines, such as H and K. This problem arose partly because irregularities in the ruling of diffraction gratings gave so-called Rowland ghosts – secondary peaks in the instrumental profile well displaced from the main peak. It is known that around 1900 many of the gratings of that time ruled in speculum metal had as much as 10 to 20 per cent scattered light. Some of this was light in the ghosts, arising out of periodic ruling errors, but some was a general long range scattered light in the dispersion direction which fills in the absorption lines in the solar spectrum. Further comments on this topic were made by Keith Pierce in 1957 [144].

Henry Rowland had discussed the effect of grating ruling errors on scattered light, also in 1893. He wrote, with reference to emission lines in laboratory spectra:

> The effect of small errors of ruling is to produce diffused light around the spectral lines. This diffused light is subtracted from the light

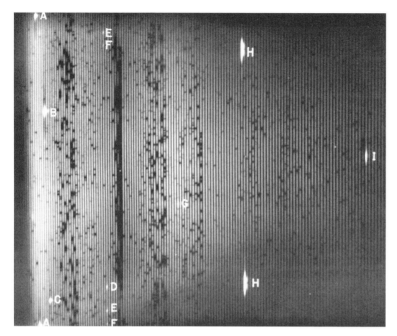

Figure 4.35. An échelle spectrum showing emission lines in a solar prominence, superimposed on a spectrum of skylight. A, B, C: CaII near infrared triplet 866.2, 854.2, 849.8 nm; D, E, F: OI 777.5, 777.4, 777.2 nm; G: HeI 706.5 nm; H: Hα 656.3 nm; I: HeI D₃ 587.6 nm. The spectrum was photographed by Robert McMath at Lake Angelus, 1951.

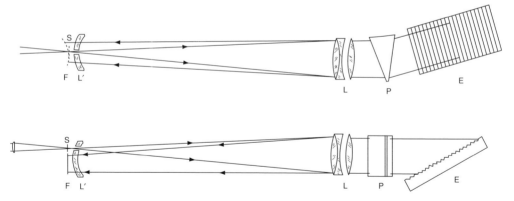

Figure 4.36. The échelle spectrograph for the solar rocket ultraviolet of Richard Tousey *et al.* of the US Naval Research Laboratory. L is an achromat of quartz and lithium fluoride, P is the fluorite cross-disperser prism in double pass, E is the 73 gr/mm échelle grating, L′ is a quartz field-flattener lens, S is the slit and F is the instrument's focal plane.

of the primary line, and its comparative amount varies as the square of the relative error of ruling and the square of the order of the spectrum [145].

Such ghosts, although weak, contribute to the diffuse light and filling in of the dark lines in solar spectra, and are especially troublesome in spectroheliographs, as diffuse or scattered light from the grating contributes to a general fogging or lack of contrast in the resulting spectroheliogram.

Deslandres' novel solution to this problem in 1893 was a spectroheliograph with three slits. In effect it was a double spectrograph, in which the exit slit of a grating spectrograph became the entrance slit of a second prism spectrograph. He wrote:

As the diffuse light, which is significant with this grating ... was a serious nuisance, I devised an arrangement called a spectrograph with three slits, which I did as early as 1893, for photography [of the Sun] using the very dark lines [73, p. 368].

The exit slit of the second spectrograph then became the third slit. He went on:

In fact, this spectrograph has just the properties which were predicted; when the second and third slits are narrow, diffuse light is completely excluded, and this advantage is valuable with the very dark solar lines, and for photographing the shape of solar structures [73, p. 368].

Unfortunately, Deslandres noted the poor light throughput of this double spectrograph, and he appears not to have persisted with the experiment.

The concept of greatly reducing scattered light using a grating in double pass was studied in the 1950s, and can be regarded as an adaptation of Deslandres' ideas, but without requiring a second spectrograph. Several authors considered double-pass grating monochromators, and pointed out that they gave twice the dispersion and resolving power yet retained the large free spectral range of low orders, in particular the first order [146, 147, 148].

The double-pass concept was introduced into solar spectroscopy by J. W. Evans and J. Waddell at the Sacramento Peak Observatory [149]. Light from the camera is sent to a reflecting prism, and from there to a second slit where the scattered light is eliminated. After being reflected by a second small prism, it passes to the collimator, grating and camera a second time. Evans and Waddell noted that, in spite of the extra reflections, throughput in double-pass spectrographs at a given resolving power may even increase, as the first and third slits can be twice as wide as in the single-pass mode. They found:

The system eliminates the Rowland ghosts and reduces residual selective scatter to an undetectable level [149].

The practice of placing a small prism monochromator so as further to reduce scattered light (or to reduce it in the first place in the case of a single-pass spectrograph) was also adopted at this time. Thus the McMath–Hulbert vacuum spectrograph, which was a single-pass instrument, had a small calcium fluoride monochromator ahead of the first main slit of the spectrograph. This not only eliminates unwanted orders from the grating, but also reduces general scattered light (but not the light in the ghosts) [139].

4.9.7   The McMath solar telescope and spectrograph at Kitt Peak

The largest solar telescope and spectrograph to be built, and at the present time still the largest, is the instrument at Kitt Peak National Observatory, which was dedicated in 1962 to Robert McMath, who had died that same year. McMath was chairman of the committee that had recommended the establishment of the National Observatory in the United States, and a large solar telescope was one of their recommendations to the National Science Foundation.

The new telescope employed a heliostat. This comprises a single large flat mirror that directs the sunlight down along the polar axis, which at Kitt Peak is at about 32° to the horizontal. The heliostat arrangement incurs image rotation, and was generally not favoured over the two-mirror coelostat used by Hale and for most other solar telescopes. However, the cost is lower, which is a significant factor for a 2-metre mirror, and moreover the noon-time shadowing problem of coelostats is avoided.

The heliostat of the McMath telescope is 30 m above the ground, and (in the original configuration of the telescope) it sent light to a 1.6-m cast aluminium concave metal mirror of focal length 88 m (f/55), which was located below ground at the end of the optical tunnel. A third flat mirror sends light into the observation room where an image of the Sun 82 cm in diameter is produced [150]. From there a vertical vacuum grating spectrograph with a Czerny–Turner mounting and operating in a double-pass mode was installed in a pit below ground level. A large aluminium-coated Babcock grating (25 cm by 15 cm; 610 gr/mm) gave a resolving power of about 500 000 in the fifth order [150]. In a later modification of the telescope the 1.6-m aluminium mirror was replaced with a 1.52-m pyrex mirror. Figure 4.37 shows the optical and mechanical layout of the McMath solar telescope at Kitt Peak. The double-pass system adopted in the McMath vacuum spectrograph at Kitt Peak is shown in Fig. 4.38.

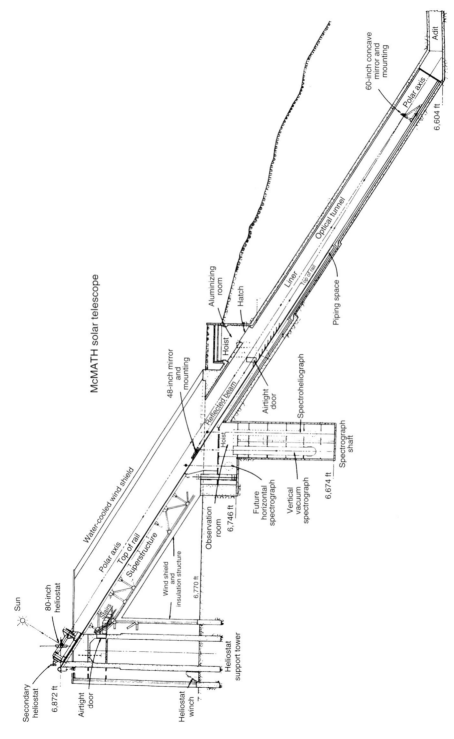

McMATH solar telescope

Sun

Secondary
heliostat

80-inch
heliostat

6,872 ft

Airtight
door

Stairs

Heliostat
winch

Heliostat
support tower

Wind shield
and
insulation structure

6,770 ft

Superstructure

Top of rail

Polar axis

Water-cooled wind shield

48-inch mirror
and
mounting

Reflected beam

Observation
room

6,746 ft

Hoist

Future
horizontal
spectrograph

Vertical
vacuum
spectrograph

6,674 ft

Spectrograph
shaft

Airtight
door

Spectroheliograph

Hoist

Aluminizing
room

Hatch

Liner

Top of rail

Optical tunnel

Piping space

60-inch concave
mirror and
mounting

Polar axis

6,604 ft

Adit

Figure 4.37. The McMath solar telescope at Kitt Peak National Observatory was dedicated in 1962 to Robert McMath. The vacuum spectrograph and spectroheliograph are mounted in the vertical well 29 m below ground level.

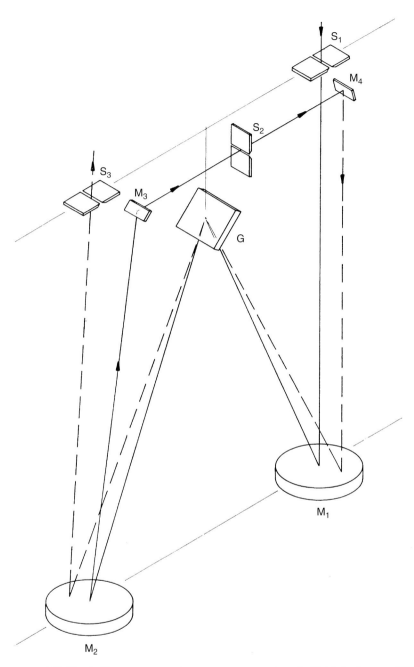

Figure 4.38. The double-pass system at the vacuum spectrograph on the McMath telescope at Kitt Peak. The spectrograph is of the Czerny–Turner arrangement. M_1 is the collimator mirror, M_2 the camera, M_3 is a small mirror sending the light to the second slit S_2 from where it goes to M_4 and through the spectrograph a second time. Finally the light emerges through slit S_3 where it is detected. Scattered light is eliminated in this double pass arrangement.

4.10 CONCLUDING REMARKS ON SOLAR SPECTROGRAPHS

It is instructive to compare the historical development of instrumentation in solar spectroscopy with that in stellar spectroscopy. Many of the developments in the former field were later introduced into the stellar domain. Thus diffraction gratings were rarely employed for stellar spectroscopy before 1930, and were only widespread after 1950, but the inefficiency of early unblazed gratings still meant they could be used for solar work from the 1890s. Likewise the échelle grating for stellar work became established in the early 1970s, some two decades after its early use for solar spectroscopy.

Ultraviolet stellar spectroscopy from rockets followed in 1965 [151], less than a decade after rocket solar spectroscopy had been successfully carried out. The benefits of vacuum spectrographs for ground-based stellar telescopes were only exploited a half century after the pioneering work of McMath at Michigan [152, 153, 154]. The photoelectric recording of stellar spectra was successfully practised only from the 1950s, first at low resolution for stellar continuous fluxes [155], and then at high resolution for line profiles [156, 157, 158]. These developments followed some 30 years after the use of photoelectric recording techniques for solar work.

In most cases the time lag between developments in solar and stellar spectroscopy reflects the low efficiency of early detectors and optical elements (especially gratings). Improved technology resulted in more efficient devices, thus permitting stellar astronomy to benefit from advances first pioneered in solar physics.

REFERENCES

[1] Lockyer, J. N., *Phil. Trans. R. Soc.* **159**, 425 (1869)

[2] Huggins, W., *Mon. Not. R. astron. Soc.* **28**, 86 (1868)

[3] Huggins, W., *Mon. Not. R. astron. Soc.* **29**, 4 (1868)

[4] Herschel, (Sir) J., *Mon. Not. R. astron. Soc.* **29**, 5 (1868)

[5] de la Rue, W., *Mon. Not. R. astron. Soc.* **30**, 22 (1870)

[6] Huggins, W., *Mon. Not. R. astron. Soc.* **30**, 36 (1870)

[7] Herschel, (Lt.) J., *Proc. R. Soc.* **17**, 116 (1869)

[8] Tennant, J. F., *Mem. R. astron. Soc.* **37**, 1 (1868)

[9] Rayet, G., *Comptes Rendus de l'Acad. Sci., Paris* **67**, 757 (1868)

[10] Janssen, J., *Comptes Rendus de l'Acad. Sci., Paris* **67**, 838 (1868)

[11] Delaunay, [C. E.], *Comptes Rendus de l'Acad. Sci., Paris* **67**, 867 (1868)

[12] de la Rue, W., *Comptes Rendus de l'Acad. Sci., Paris* **67**, 836 (1868)

[13] Faye, [C. A.-E.-A.], *Comptes Rendus de l'Acad. Sci., Paris* **67**, 840 (1868)

[14] Janssen, J., *Comptes Rendus de l'Acad. Sci., Paris* **68**, 367 (1869)

[15] Ramsay, W., *Proc. R. Soc.* **58**, 65 and **58**, 81 (1895)

[16] Lockyer, J. N., *Proc. R. Soc.* **58**, 116 (1895)

[17] Lockyer, J. N., *Proc. R. Soc.* **59**, 4 (1895)

[18] Lockyer, J. N., *Contributions to Solar Physics*, Macmillan & Co. (1874), p. 676

[19] Secchi, A., *le Soleil*, Gauthier-Villars, Paris, p. 422 (1870); 2nd edition, Gauthier-Villars, Paris, p. 428 (part 1) and p. 225 (part 2) (1875)

[20] Janssen, J., *Comptes Rendus de l'Acad. Sci., Paris* **68**, 93 (1869)

[21] Janssen, J., *Comptes Rendus de l'Acad. Sci., Paris* **68**, 713 (1869)

[22] Zöllner, J. C. F., *Astron. Nachrichten* **74**, 305 (1869)

[23] Secchi, A., *Astron. Nachrichten* **77**, 299 (1871)

[24] Secchi, A., *Comptes Rendus de l'Acad. Sci., Paris* **73**, 984 (1871)

[25] Lockyer, J. N. and Seabroke, G. M., *Proc. R. Soc.* **21**, 105 (1873); also in *Comptes Rendus de l'Acad. Sci., Paris* **76**, 363 (1873)

[26] Huggins, W., *Proc. R. Soc.* **21**, 127 (1873)

[27] Fraunhofer, J., *Denkschr. der Münch. Akad. der Wiss.* **5**, 193 (1817). Also in *Gilberts Ann. der Phys.* **56**, 264 (1817) and in *Edin. Phil. J.* **9**, 296 (1823) and **10**, 26 (1824)

[28] Fraunhofer, J., *Denkschr. der Münch. Akad. der Wiss.* **8**, 1 (1821)

[29] Fraunhofer, J., *Gilberts Ann. der Phys.* **74**, 337 (1823)

[30] Kirchhoff, G., *Untersuchungen über das Sonnenspectrum und die Spectren der Chemischen Elemente*, Part I. Abhandl. Berlin Akad., pp. 63–95 (1861); pp. 227–240 (1862)

[31] Kirchhoff, G., *Untersuchungen über das Sonnenspectrum und die Spectren der Chemischen Elemente*, Part II. Abhandl. Berlin Akad., pp. 225–240 (1863)

[32] Airy, G. B., *Phil. Trans. R. Soc.* **158** (I), 29 (1868)

[33] Airy, G. B., *Phil. Trans. R. Soc.* **162** (I), 89 (1872)

[34] Ångström, A. J., *Recherches sur le spectre normal du soleil*, Upsala: W. Schultz (1868). See also *Poggendorfs Ann. der Phys.* **123**, 489 (1864) for a preliminary account of this work.

[35] Thalén, R., *Nova acta reg. soc. sci. Ups.* (3) **12**, 1 (1884)

[36] Müller, G. and Kempf, P., *Publik. der astrophysik. Observ. Potsdam* **5** (Nr. 20), 1 (1886)

[37] Kurlbaum, F., *Wiedemanns Ann. der Phys.* **33**, 159 & 381 (1888)

[38] Bell, L., *Amer. J. Sci.* (3) **33**, 167 (1888); (3) **35**, 265 & 347 (1888)

[39] Rowland, H. A., *Phil. Mag.* (5) **13**, 469 (1882); also in *Nature* **26**, 211 (1882)

[40] Rowland, H. A., *Amer. J. Sci.* (3) **26**, 87 (1883); also in *Phil. Mag.* (5) **16**, 197 (1883)

[41] Rowland, H. A., *Photographic Map of the Normal Solar Spectrum*, Baltimore: Johns Hopkins University (1887 and 1888)

[42] Rowland, H. A., *Preliminary table of solar spectrum wavelengths*, a series of 18 papers in *Astrophys. J.* **1–5** (1895–1897); *Astrophys. J.* **1**, 29 (1895); **1**, 131 (1895); **1**, 222 (1895); **1**, 295 (1895); **1**, 377 (1895); **2**, 45 (1895); **2**, 109 (1895); **2**, 188 (1895); **2**, 306 (1895); **2**, 360 (1895); **3**, 141 (1896); **3**, 201 (1896); **3**, 365 (1896); **4**, 106 (1896); **4**, 278 (1896); **5**, 11 (1897); **5**, 109 (1897); **5**, 181 (1897)

[43] Rowland, H. A., *Astron. & Astrophys.* **12**, 321 (1893)

[44] Michelson, A. A., *Astron. & Astrophys.* **13**, 92 (1894)

[45] Michelson, A. A., *Détermination experimentale de la valeur du mètre en longueurs d'onde lumineuses* (translation by J. R. Benoît), Paris: Gauthier-Villars (1894)

[46] Fabry, C. and Pérot, A., *Ann. Chim. et Phys.* (7) **25**, 98 (1902); also in *Astrophys. J.* 15, 73 and 261 (1902)

[47] Bell, L., *Astrophys. J.* **15**, 157 (1902)

[48] Bell, L., *Astrophys. J.* **18**, 191 (1903)

[49] Jewell, L. E., *Astrophys. J.* **21**, 23 (1905)

[50] Hartmann, J., *Astrophys. J.* **18**, 167 (1903)

[51] Kayser, H., *Astrophys. J.* **19**, 157 (1904)

[52] Anon., *Trans. Int. Union for Cooperation in Solar Research* **2**, 18 (1907)

[53] St John, C. E., Moore, C. E., Ware, L. M., Adams, E. F. and Babcock, H. D., *Revision of Rowland's Preliminary Table of Solar Spectrum Wavelengths*, Washington: Carnegie Institution (1928)

[54] Moore, C. E., Minnaert, M. G. J. and Houtgast, J., *Second Revision of Rowland's Preliminary Table of Solar Spectrum Wavelengths*. NBS Monograph 61 (1966)

[55] Babcock, H. D., *Trans. Int. Astron. Union* **3**, 77 (1928)

[56] Hale, G. E., *Astron. & Astrophys.* **11**, 407 (1892)

[57] Hale, G. E., *Astron. & Astrophys.* **12**, 241 (1893)

[58] Hale, G. E., *Technology Quarterly* **3**, 310 (1890)

[59] Hale, G. E., *Astron. Nachrichten* **126** (Nr 3006), 81 (1890)

[60] Hale, G. E., *Sidereal Mess.* **10**, 257 (1891)

[61] Hale, G. E. and Ellermann, F., *Publ. Yerkes Observ.* **3**, 1 (1903)

[62] Braun, C., *Astron. Nachrichten* **80** (Nr 1899), 33 (1872)

[63] Braun, C., *Astron. Nachrichten* **126** (Nr 3014), 227 (1890)

[64] Lohse, O., *Zeitschr. für Instr. Kunde* **1**, 22 (1880)

[65] Hale, G. E., *Astrophys. J.* **70**, 265 (1929)

[66] Deslandres, H. A., *Comptes Rendus de l'Acad. Sci., Paris* **113**, 307 (1891)

[67] Deslandres, H. A., *Comptes Rendus de l'Acad. Sci., Paris* **115**, 22 (1892)

[68] Deslandres, H. A., *Comptes Rendus de l'Acad. Sci., Paris* **119**, 457 (1894)

[69] Deslandres, H. A., *Comptes Rendus de l'Acad. Sci., Paris* **117**, 716 (1893)

[70] Deslandres, H. A., *Comptes Rendus de l'Acad. Sci., Paris* **129**, 1222 (1899)

[71] Deslandres, H. A., *Comptes Rendus de l'Acad. Sci., Paris* **135**, 500 (1903)

[72] Deslandres, H. A., *Bull. Astron.* **22**, 305 (1905)

[73] Deslandres, H. A., *Bull. Astron.* **22**, 337 (1905)

[74] Deslandres, H. A., *Comptes Rendus de l'Acad. Sci., Paris* **148**, 968 (1909)

[75] Deslandres, H. A., *Ann. de l'Observ. de Meudon* **4**, 1 (1910)

[76] Hale, G. E., *Astrophys. J.* **23**, 92 (1906)

[77] Dyson, F. W., *Mon. Not. R. astron. Soc.* **73**, 317 (1913)

[78] Kempf, P., *Zeitschr. für Instr. Kunde* **24**, 317 (1904)

[79] Kempf, P., *Astrophys. J.* **21**, 49 (1905)

[80] Hale, G. E., *Astron. & Astrophys.* **13**, 681 (1894)

[81] Lockyer, W. J. S., *Mon. Not. R. astron. Soc.* **65**, 473 (1905)

[82] Evershed, J., *Vistas Astron.* **1**, 33 (1955)

[83] Evershed, J., *Mon. Not. R. astron. Soc.* **69**, 454 (1909)

[84] Evershed, J., *Mon. Not. R. astron. Soc.* **88**, 127 (1927)

[85] Hale, G. E., *Astrophys. J.* **21**, 151 (1905)

[86] Hale, G. E., *Astrophys. J.* **21**, 124 (1905)

[87] Hale, G. E. and Ellermann, F., *Astrophys. J.* **23**, 54 (1906)

[88] Hale, G. E., *Astrophys. J.* **23**, 6 (1906)

[89] Hale, G. E., *Astrophys. J.* **25**, 68 (1907)

[90] Hale, G. E., *Astrophys. J.* **27**, 204 (1908)

[91] Hale, G. E. and Adams, W. S., *Astrophys. J.* **23**, 11 (1906)

[92] Hale, G. E., *Astrophys. J.* **28**, 315 (1908)

[93] van der Pahlen, E., *Zeitschr. für Instr. Kunde* **46**, 49 (1926)

[94] Abetti, G., *Pubbl. Osserv. Astrofis. Arcetri* **43**, 11 (1926)

[95] Fujita, Y., *Proc. Phys.-Math. Soc. Japan* (3) **16**, 327 (1934); also referenced as *Ann. Tokyo Astron. Observ.* No. 41 (1934)

[96] Anon., *Observ.* **58**, 210 (1935)

[97] Plaskett, H. H., *Mon. Not. R. astron. Soc.* **99**, 219 (1939)

[98] Curtis, H. D., *Pop. Astron.* **48**, 348 (1940)

[99] Curtis, H. D., *Publ. astron. Soc. Pacific* **52**, 212 (1940)

[100] Gollnow, H., *Naturwiss.* **36**, 175 and 213 (1949)

[101] Hale, G. E., *Publ. astron. Soc. Pacific* **24**, 223 (1912); also in Hale, G. E., *Mt Wilson Solar Observatory: Annual report of the Director for 1912*, Carnegie Inst. Year Book 11, 172 (1912)

[102] Julius, W. H., *Bull. Astron. Inst. Netherlands* **1** (No. 20), 119 (1922)

[103] Freundlich, E., *Das Turmteleskop der Einstein-Stiftung*, Berlin: Springer Verlag (1927)

[104] Wurm, K., *Zeitschr. für Astrophys.* **2**, 133 (1931)

[105] Abetti, G., *Handbuch der Astrophys.* **4**, 57 (1929)

[106] Allen, C. W., *Mem. Commonwealth Solar Observ.* **5** (Pt 1), 1 (1934)

[107] Rimmer, W. B., *Mon. Not. R. astron. Soc.* **92**, 295 (1932)

[108] Hale, G. E., *Astrophys. J.* **82**, 111 (1935)

[109] Plaskett, H. H., *Mon. Not. R. astron. Soc.* **112**, 177 (1952)

[110] Parfinenko, L. D., *Solar Phys.* **213**, 291 (2003)

[111] Lyot, B., *Comptes Rendus de l'Acad. Sci., Paris* **197**, 1593 (1933)

[112] Öhman, Y., *Nature* **141**, 157 and 291 (1938)

[113] McMath, R. R. and Petrie, R. M., *Publ. Observ. Univ. Michigan* **5** (No. 8), 103 (1933)

[114] McMath, R. R., *Publ. Observ. Univ. Michigan* **7** (No. 1), 1 (1937)

[115] McMath, R. R., *Publ. Observ. Univ. Michigan* **8** (No. 11), 141 (1941)

[116] Curtis, H. D., *Pop. Astron.* **48**, 348 (1940)

[117] Dunham, T., *Phys. Rev.* **44**, 329 (1933)

[118] Adams, W. S., *Mt Wilson Observatory Annual Report for 1933*

[119] Brück, H. A., *Mon. Not. R. astron. Soc.* **99**, 607 (1939)

[120] Cashman, R. J., *J. Opt. Soc. Amer.* **36**, 356 (1946)

[121] McMath, R. R. and Mohler, O., *Publ. astron. Soc. Pacific* **59**, 267 (1947)

[122] McMath, R. R. and Mohler, O., *Astron. J.* **53**, 114 (1948)

[123] McMath, R. R. and Mohler, O., *Astron. J.* **53**, 200 (1948)

[124] Pfund, A. H., *J. Opt. Soc. Amer.* **14**, 337 (1927)

[125] McMath, R. R. and Mohler, O. C., *J. Opt. Soc. Amer.* **39**, 903 (1949)

[126] Mohler, O. C., Pierce, A. K., McMath, R. R. and Goldberg, L., *Photometric Atlas of the Near Infrared Solar Spectrum λ8465 to λ25,242*. McMath-Hulbert Observatory, Ann Arbor: University of Michigan Press (1950)

[127] Mohler, O. C., Pierce, A. K., McMath, R. R. and Goldberg, L., *Astrophys. J.* **117**, 41 (1953)

[128] Mohler, O. C., *Table of the Solar Spectrum Wave Lengths 11 984 Å to 25 578 Å*. Ann Arbor: University of Michigan Press (1955)

[129] Harrison, G. R., *J. Opt. Soc. Amer.* **39**, 522 (1949)

[130] Harrison, G. R., *Vistas Astron.* **1**, 405 (1955)

[131] Bausch and Lomb Optical Co., Rochester, *Catalogue of Echelle Spectrographs*, Rochester, NY (1954)

[132] Pierce, A. K., McMath, R. R. and Mohler, O. C., *Astron. J.* **56**, 137 (1951)

[133] Tousey, R., Purcell, J. D. and Garrett, D. L., *Appl. Optics* **6**, 365 (1967)

[134] Garrett, D. L., Michels, D. J., Purcell, J. D. and Tousey, R., *J. Opt. Soc. Amer.* **52**, 597 (1962)

[135] Moore, C. E., Tousey, R. and Brown, C. M., *The Solar Spectrum 3069 Å–2095 Å*. Washington, DC: Naval Research Laboratory (1982)

[136] Detweiler, C. R. and Purcell, J. D., *J. Opt. Soc. Amer.* **52**, 597 (1962)

[137] McMath, R. R., *Astrophys. J.* **123**, 1 (1956)

[138] McMath, R. R., Mohler, O. C., Pierce, A. K. and Goldberg, L., *Astrophys. J.* **124**, 1 (1956)

[139] Pierce, A. K., *J. Opt. Soc. Amer.* **47**, 6 (1957)

[140] Dunn, R. B., *Appl. Optics* **3**, 1353 (1964)

[141] Zirin, H., *Sky and Tel.* **39**, 215 (1970)

[142] Mayfield, E. B., Vrabel, D., Rogers, E., Janssens, T. and Becker, R. A., *Sky and Tel.* **37**, 208 (1969)

[143] Livingston, W. C., Harvey, J., Pierce, A. K. *et al.*, *Appl. Optics* **15**, 33 (1976)

[144] Pierce, A. K., *J. Opt. Soc. Amer.* **47**, 14 (1957)

[145] Rowland, H. A., *Astron. & Astrophys.* **12**, 129 (1893)

[146] Jenkins, F. A. and Alvarez, L. W., *Phys. Rev.* **85**, 763 (1952)

[147] Fastie, W. G. and Sinton, W. M., *J. Opt. Soc. Amer.* **42**, 283 (1952)

[148] Rank, D. H. and Wiggins, T. A., *J. Opt. Soc. Amer.* **42**, 963 (1952)

[149] Evans, J. W. and Waddell, J., *Appl. Optics* **1**, 111 (1962)

[150] Pierce, A. K., *Appl. Optics* **3**, 1337 (1964)

[151] Morton, D. C. and Spitzer, L., *Astrophys. J.* **144**, 1 (1966)

[152] Bouchy, F., Connes, P. and Bertaux, J. L., *Intl. Astron. Union Coll.* **170**, 22 (1999); in 'Precise stellar radial velocities', *Astron. Soc. Pacific Conf. Ser.* **185**, ed. J. B. Hearnshaw and C. D. Scarfe

[153] Hearnshaw, J. B., Barnes, S. I., Kershaw, G. M. *et al.*, *Exper. Astron.* **13**, 59 (2002)

[154] Mayor, M., Pepe, F., Queloz, D. *et al.*, *European South. Observ. Messenger* **114**, 20 (2003)

[155] Guérin, P. and Laffineur, M., *Comptes Rendus de l'Acad. Sci., Paris* **238**, 1692 (1954)

[156] Rogerson, J. D., Spitzer, L. and Bahng, J. D., *Astrophys. J.* **130**, 991 (1959)

[157] Oke, J. B. and Greenstein, J. L., *Astrophys. J.* **133**, 349 (1961)

[158] Grainger, J. F. and Ring, J., *Mon. Not. R. astron. Soc.* **125**, 93 (1963)

5 · Objective prism spectrographs

5.1 THEORY OF THE OBJECTIVE PRISM SPECTROGRAPH

The apparent simplicity of the objective prism instrument is possibly the reason that its performance has rarely been analysed in the literature prior to the 1960s, in spite of its long and illustrious use as a tool of stellar spectroscopy.

The objective prism spectrograph comprises a prism of small apex angle, typically a few degrees, mounted at the aperture of either a refracting astrograph, a Schmidt telescope or other wide-field catadioptric camera. The detector is either a photographic plate, or, in recent times, a CCD camera (preferably of a large format).

The advantages and disadvantages of the objective prism arrangement have been discussed by several authors, including by C. B. Stephenson (1929–2001) [1] and by William Bidelman (b. 1918) [2]. In summary, the advantages are:

1. many spectra can be simultaneously recorded;
2. a generally higher efficiency is achieved than that for slit spectrographs,

while the disadvantages are:

1. the resolving power is generally lower than for slit instruments;
2. the resolving power is often seeing-dependent;
3. the limiting magnitude is relatively bright;
4. an absolute wavelength scale, necessary for radial-velocity work, can be determined only with considerable difficulty and moderate precision;
5. in crowded fields, spectra may overlap.

Karl Schwarzschild (1873–1916) was amongst the first to analyse the performance of the objective prism theoretically [3]. His especial interest was the determination of stellar radial velocities from objective prism plates, mainly by using the reversion method, in which two exposures with reversed dispersions are recorded on one plate. The question of field distortions arising from the dependence of objective prism dispersion with field angle was therefore one he addressed in some detail.

In more recent times, the performance of objective prism cameras has been considered by Haffner [4], by Bowen [5], by Geyer [6], by Fehrenbach [7] and by Stock and Upgren [8].

Hans Haffner (1912–77) [4] considered the limiting magnitudes that arise from overlapping spectra, but he did not take the sky background light into account. His conclusion was that the limiting magnitude from this cause for Milky Way fields ranged from $m = 11$ for a $10°$ prism to $m = 16$ for a $1°$ prism. He gave a useful table of data for some objective prism Schmidt cameras in use in the 1960s. Bowen on the other hand takes the bright sky background into account as the principal limiting factor. He noted that this was generally about $m_{pg} = 12$ for reciprocal dispersions of 200–300 Å/mm in the blue, presumably for a dark sky with no moonlight.

A similar analysis is given by Edward Geyer [6] who, like Bowen, noted that the limiting magnitude set by the sky increased as $\Delta m \sim 5 \log f$ for short focal length cameras ($f_{tel} \sim 4$ m), where $f_{tel}\theta_\star$ is less than the photographic grain size (a condition that is generally fulfilled, except in poor seeing). Charles Fehrenbach [7] estimated a limiting magnitude of 14.35 for a 400 mm focal length (the value for the Tautenburg Schmidt) at a low dispersion of 800 Å/mm, but this became as bright as 8.35 for a 1-m focal length instrument operating at 50 Å/mm.

Finally Jürgen Stock (1923–2004) and Arthur Upgren (b. 1933) discussed prism mounting angles, and showed how the dispersion increases slightly for spectra formed near the edge of a wide field [8]. They also discussed the origin of ghost spectra that result from unwanted multiple reflections in the prism, in the corrector plate or between these two elements, the effects of prism flexure as a function of telescope position (causing astigmatism), and the ultraviolet transmission of prisms of different dimension and of different glass absorption coefficient.

Given that most objective prism instruments use prisms of small apex angle ($\alpha \sim 10°$), the prism equation simplifies to the approximate form

$$\delta = (n - 1)\alpha, \qquad (5.1)$$

where δ is the deviation and n is the refractive index. The dispersion is therefore

$$\left(\frac{d\delta}{d\lambda}\right) = \alpha\left(\frac{dn}{d\lambda}\right) \qquad (5.2)$$

and the reciprocal dispersion becomes

$$P = \left(\frac{d\lambda}{dx}\right) = \left(\frac{1}{f_{tel}}\frac{d\lambda}{d\delta}\right) \qquad (5.3)$$
$$= \frac{1}{f_{tel}\alpha\left(\frac{dn}{d\lambda}\right)}.$$

If the photographic grain size (or for a CCD detector, the pixel size) is Δx_{pix} and the seeing is $\theta_\star$ (radians), then the wavelength resolution is

$$\delta\lambda = P\Delta x_{pix} = \frac{\Delta x_{pix}}{f_{tel}\alpha\left(\frac{dn}{d\lambda}\right)} \qquad (5.4)$$

for the detector-limited case, or

$$\delta\lambda = \theta_\star \Big/ \left(\frac{d\delta}{d\lambda}\right) = \theta_\star \Big/ \left(\alpha\frac{dn}{d\lambda}\right) \qquad (5.5)$$

for the seeing-limited case, depending on whether $\theta_\star f_{tel} < \Delta x_{pix}$ (detector limited) or $> \Delta x_{pix}$ (seeing limited).

The resolving power is then

$$R = \frac{\lambda}{\delta\lambda} = \frac{\lambda f_{tel}\alpha\left(\frac{dn}{d\lambda}\right)}{\Delta x_{pix}} \qquad (5.6)$$

(detector limited) or

$$R = \frac{\lambda\alpha\left(\frac{dn}{d\lambda}\right)}{\theta_\star} \qquad (5.7)$$

(seeing limited).

For $\lambda = 450$ nm, $(\frac{dn}{d\lambda}) = 10^{-4}$ nm^{-1}, $\alpha = 5.7°$ ($= 0.1$ rad) and $\theta_\star = 2''$ ($\simeq 10^{-5}$ rad), then the last expression becomes $R \sim 450$, which is low. The wavelength resolution is about 1 nm. Higher resolving

powers require a higher dispersion (by increasing the prism apex angle or using a more dispersive glass) or observing with better seeing, if seeing is the limiting factor. If the plate is the limiting factor, then a larger focal length is advantageous for improving the resolving power as well as the limiting magnitude, but this will increase the exposure time (for a given aperture).

The determination of the limiting magnitude due to the sky brightness using an objective prism comes from finding the intensity of the starlight in a spectrum per unit area of plate, and comparing this to the effectively undispersed sky background, as has been done, for example, by Haffner [4], Geyer [6], and Bowen [5]. Here the following notation is adopted:

$f_\star$ mean flux level of starlight over interval $\Delta\lambda$ within which the instrument records a spectrum (J m^{-2} s^{-1} nm^{-1})

f_s mean sky brightness in the same wavelength interval $\Delta\lambda$ (J m^{-2} s^{-1} nm^{-1} per steradian of sky)

D telescope aperture

f_{tel} telescope focal length

α prism apex angle (radians)

$\frac{dn}{d\lambda}$ mean prism dispersion (nm^{-1})

L length of spectrum over wavelength range $\Delta\lambda$ (mm)

W width of spectrum (mm).

It is assumed, following Haffner [4], that in photographic work the greater the dispersion, the greater the widening, and hence that $W = kL$ (with $k \sim \frac{1}{15}$). Widening is necessary for photographic work, because of the small dynamic range of the photographic emulsion, but not for CCDs, where an unwidened spectrum will have $W = \theta_\star f_{tel}$ or Δx_{pix} (whichever is larger).

For a circular telescope aperture, the intensity of the spectral image on unit area of the detector is

$$I_\star = \frac{(\frac{\pi D^2}{4})f_\star \Delta\lambda}{k\Delta\lambda^2\alpha^2 f_{tel}^2(\frac{dn}{d\lambda})^2} \qquad (5.8)$$

(J m^{-1} s^{-1}) as the length of the spectrum is $L = (\Delta\lambda)\alpha f_{tel}(\frac{dn}{d\lambda})$, and the area covered by one spectrum is $A = kL^2$.

The intensity due to the sky background is

$$I_s = \frac{(\frac{\pi D^2}{4})f_s \Delta\lambda}{f_{tel}^2} \qquad (5.9)$$

and the ratio of these is

$$r = \frac{I_\star}{I_s} = \frac{f_\star}{f_s k (\Delta\lambda)^2 \alpha^2 (\frac{dn}{d\lambda})^2}. \qquad (5.10)$$

For spectra at the sky background limit, $r \simeq 1$ and

$$f_\star(\text{lim}) \simeq f_s k (\Delta\lambda)^2 \alpha^2 \left(\frac{dn}{d\lambda}\right)^2. \qquad (5.11)$$

Note that the limiting magnitude ($m_{lim} = \text{const.} - 2.5 \log_{10} f_\star(\text{lim})$) does not depend on either f_{tel} or D, but it does depend on the prism properties (α and $\frac{dn}{d\lambda}$). If this is expressed in terms of the reciprocal dispersion $P = 1/(f_{tel}\alpha(\frac{dn}{d\lambda}))$, then $f_\star(\text{lim}) \propto 1/(P^2 f_{tel}^2)$ and there is then a dependence of limiting magnitude on f_{tel} only for fixed values of the dispersion, as noted for example by Fehrenbach [7].

Haffner [4] has considered the faintest stellar magnitude visible on an objective prism plate that has not been exposed to reach the sky background limit. In this case, $I_\star$ should exceed a plate's minimum threshold and

$$f_\star(\text{lim}) \propto I_\star(\text{threshold})k(\Delta\lambda)^2\alpha^2\left(\frac{dn}{d\lambda}\right)^2 (f_{tel}/D)^2, \qquad (5.12)$$

so now for a given prism and exposure time, a faster telescope (f_{tel}/D smaller) will go fainter.

Finally, there is the question of overlapping spectra. Consider two prisms of angles α_1 and α_2. As $I_\star \propto \frac{1}{\alpha^2}$, the larger prism will give a greater dispersion, but a brighter limiting magnitude, given by $m_{lim} = \text{const.} - 5\log\alpha$. If the density of stars were everywhere uniform and space transparent, then the number of stars visible per square degree, N, to a given magnitude limit, m_{lim}, would increase fourfold per extra magnitude, or $N \propto 10^{0.6m_{lim}}$. In practice, star numbers increase more slowly than this, more nearly as $N \propto 10^{0.4m_{lim}}$. Adopting this canonical value results in $N \propto \alpha^{-2}$. The area of each spectrum is $A = kL^2 \propto \alpha^2$, so the area of each plate per square degree covered by spectra (if there are no overlaps) is NA, which is independent of α, a result noted by Haffner [4].

In practice star numbers increase a little faster than $N \propto 10^{0.4m_{lim}}$, as noted by Fehrenbach [7], in fact

$N \propto 10^{0.456m_{lim}}$, so the smaller prism angle produces a slightly larger value of NA.

Fehrenbach has also shown that if the N spectra per square degree are distributed at random, then, if each spectrum covers an angular area $a \propto \frac{A}{f_{tel}^2}$ (in square degrees), the total area covered will be $1 - (1 - a)^N$. This differs from Na by about $\frac{1}{2}N^2 a^2$, so the fractional loss in total spectrum area is $\frac{1}{2}Na$. Since each overlap eliminates two otherwise useful spectra, the number of useful spectra remaining per square degree is reduced by Na, from N to $N' = N(1 - a)$.

Haffner has calculated the effect of overlaps for fields in the Milky Way using different prism apex angles α, with $a \propto \alpha^2$ and widening to $W = L/15$. If Na is limited to 0.2, then the limiting magnitude as a result of overlaps would range from $m = 16$ for $\alpha = 1°$ to $m = 11$ for $\alpha = 10°$. These values are apparently about a magnitude fainter than given by Fehrenbach for the limits due to sky background, so even in the Milky Way, where overlaps are high, sky background is still likely to limit the magnitude before the overlaps become serious. This is a statistical argument, and there is no doubt that overlapping spectra do limit the magnitude attainable in some regions of the sky, especially as the star distribution is not in practice random.

Many objective prism spectrographs use a crown glass such as Schott BK7 for the prism, whose transmission is about 90 per cent at 350 nm for a thickness of 25 mm. A flint glass such as F2 has the same transmission at 365 nm for this same thickness, though in practice the higher dispersion allows for a smaller prism apex angle and hence thinner prism, partially compensating for the loss of the ultraviolet. Values of n and $(\frac{dn}{d\lambda})$ at Hγ for these glasses are:

glass	$n(\text{H}\gamma)$	$(\frac{dn}{d\lambda})(\text{H}\gamma)(\text{nm}^{-1})$
Schott BK7	1.528	1.12×10^{-4}
Schott F2	1.644	2.46×10^{-4}

The data are from the paper by Geyer [6].

One advantage of a flint glass prism is higher dispersion in the yellow and red regions, for example, for surveys of stars with Hα in emission. Alternatively, for surveys in the ultraviolet, a quartz prism and optics can be employed.

5.2 THE HISTORY OF OBJECTIVE PRISM SPECTROSCOPY

5.2.1 Objective prism spectroscopy in the nineteenth century

The history of the objective prism instrument in astronomy is as old as the history of astronomical spectroscopy. For the early visual observations of spectral lines in the Sun's spectrum by Joseph Fraunhofer in 1817, he simply made use of a 60° flint glass prism mounted in front of a small theodolite telescope of 25-mm aperture [9].

Fraunhofer used the same type of spectroscope again when he turned to stellar spectroscopy in 1823. This time a 10-cm telescope had a prism angle of 37° 40′ mounted in front of its object glass [10]. The first observations of spectral lines in the light from the Moon, Mars, Venus and half a dozen of the brightest stars were made. A century after Fraunhofer, Hugh Newall, the professor of astrophysics at Cambridge University, described Fraunhofer's objective prism as follows:

> The star takes the place of the slit in ordinary spectroscopes and the universe is the collimator, rendering the rays parallel to one another, in consequence of the distance of the star [11, p. 74].

Another of the pioneers of stellar spectroscopy, Angelo Secchi in Italy, also used an objective prism for visual observations on the 24-cm Merz refractor at the Collegio Romano observatory in Rome [12] – see Fig. 5.1. The flint glass prism of Secchi's instrument had a 12° apex angle. Although this was not his only instrument for stellar spectroscopy, some of Secchi's finest drawings of the high dispersion spectra of bright stars were produced in this way. However, the prism had an aperture of 16 cm, so it considerably reduced the light entering the telescope. No doubt the objective prism offered some advantages to the visual observer of stellar spectra. Although the telescope is more difficult to set (it is not pointed directly at the star), once a spectrum is in view it is less troublesome to maintain the observation of a spectrum over a certain time interval, given that a stellar image does not have to be guided on a slit.

However, as a tool in stellar spectroscopy, the objective prism excelled only after the introduction of

FIG. 1.

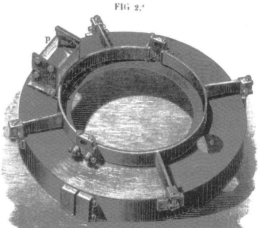

FIG 2.

FIG. 3.

Figure 5.1. Angelo Secchi's objective prism and its mounting cell for the 24-cm Merz refractor at the Collegio Romano, 1872.

astronomical photography with dry plates. The unrivalled pioneer in objective prism spectrography was Edward Pickering (1846–1919) at Harvard. After his first experiments in 1885, he wrote:

> The most striking results have been obtained with stellar spectra. Replacing the slit spectroscope by a

large prism in front of the lens, photographs have been obtained of stars as faint as eighth magnitude in which lines are shown with sufficient distinctness to be clearly seen in a paper positive. As all the stars in a large region are thus photographed, more than a hundred spectra have been obtained on a single plate [13].

The instrument used by Pickering was the 20-cm Bache f/5.6 astrograph. Equipped with a 13° dense flint glass prism, it could record the spectra of stars to $m_{pg} = 6$ in 5 minutes, whereas those of eighth magnitude took an hour. With 20×25 cm plates, a 10° field of view was recorded [14].

In 1912 Pickering lectured to the American Philosophical Society meeting on the merits of the objective prism instrument.

The principal advantages of the objective prism are the small loss of light, and the large number of stars which may be photographed simultaneously. Also, that it is not necessary to follow, as when photographing star charts. The best authorities claim that of the entire light entering the telescope less than one per cent reaches the photographic plate, when a slit spectroscope is used. The proportion of light transmitted by the objective prism must be at least 50 times as great. In fact, the principal loss of light is from absorption of the objective. Consequently, far fainter stars can be photographed with an objective prism than with a slit spectroscope, the difference amounting to several magnitudes. Another great advantage of the objective prism is that the spectra of all the stars in the field of the telescope can be photographed simultaneously, while with a slit spectroscope only one star can be taken at a time. With the Harvard 8-inch doublet as many as three or four hundred spectra are often photographed on a plate including all stars of the ninth magnitude and brighter, in a region ten degrees square [15].

The Bache telescope was installed at Arequipa in Peru in 1889. By that time over 28 000 spectra had been recorded for classification of over 10 000 stars. This material formed the basis of the *Draper Memorial Catalogue of Stellar Spectra* [16], which was published by Pickering in 1890 from the classifications of Williamina Fleming (1857–1911). It was the most extensive stellar classification to be published in the nineteenth century, and a most significant milestone, not only for Harvard College Observatory, but also for the new technique of spectral classification.

In 1889 the Bache telescope was used for the discovery of the first double-lined spectroscopic binary stars, first by Pickering (ζ UMa), and soon afterwards by Antonia Maury (1866–1952) (β Aur). Pickering described this new result from the objective prism work as follows:

One of the most remarkable results derived from the study of the spectra of the brighter stars is the detection of very close binaries. In 1886, a photograph was obtained of ζ Ursae Majoris, in which the K line was distinctly double. Later, photographs of the same star showed the line single. During the autumn of 1889, the K line in the spectrum of β Aurigae was found on successive nights to be alternately single and double. The change was so rapid that it was perceptible in successive photographs . . . The cause of this curious phenomenon appears to be that the star is a very close binary, having a period of revolution slightly less than four days, and that the two components are nearly equal in brightness and have identical spectra. When one component is approaching, the other receding, the wave-length of each line will be diminished for one component and increased, for the other star. Each line will therefore appear double [14, pp. xvii–xviii].

With the Bache telescope in Peru, there was a need for an instrument with similar capabilities in Cambridge. Such an instrument was funded by Mrs Anna Palmer Draper (1839–1914), the widow of Henry Draper, and installed at Harvard in 1891. The 20-cm (8-inch) f/6.25 Draper telescope was equipped with a 5° objective prism, which allowed stars about one magnitude fainter to be recorded than with the 13° prism on the Bache instrument [14].

Other astrographic telescopes at Harvard were used for higher dispersion spectra of brighter stars. Thus the 11-inch f/14 Draper telescope was equipped with a 15° objective prism which gave a dispersion of 45 Å/mm at Hγ. Later three further 15° prisms were acquired – see Fig. 5.2. When all four were mounted, the dispersion was only 11 Å/mm and the limit was as bright as second magnitude. Nevertheless, some

Figure 5.2. Objective prism cell being mounted on the 11-inch Draper telescope at Harvard, 1892. The cell contains two prisms and the weight is about 100 pounds.

detailed studies of the spectra of bright stars were made with the 11-inch Draper telescope, notably by Antonia Maury, who introduced her own classification scheme at Harvard with a second luminosity-sensitive line-width parameter in addition to the main parameter of line strength [17].

The same telescope (with one prism) was also used by Annie Cannon (1863–1941) for further classifications of bright stars in 1911 [18], a work that was a predecessor of her major opus, the Henry Draper Catalogue.

At the Boyden Station in Peru, a further astrograph was installed in 1891 for work on the brighter southern stars. This was the 13-inch Boyden telescope. It could work with up to three prisms, and gave dispersions of 40 Å/mm (one prism) to 12 Å/mm with three. The plate material with this telescope was used, for example, by Annie Cannon in 1901 to classify bright southern stars [19] in a programme that developed the earlier classification scheme devised by Williamina Fleming.

The objective prism spectrographs at Harvard and Boyden observatories were not the only instruments of this type used in the late nineteenth century, but they were by far the most productive. At the Paris Observatory, Ernest Mouchez (1821–92) reported on early objective prism experiments in 1885 using the 33-cm astrograph equipped with a 5° prism of 21 cm diameter, which could record the spectra of ninth magnitude stars. 'Using this spectroscopic apparatus, whose merits

are well known, we have obtained with a clarity which has hitherto not been surpassed, the spectra of a certain number of stars . . . ' The year was the same as Pickering's early experiments at Harvard, but Mouchez did not pursue spectroscopic survey work with the same vigour as Pickering. The Carte du Ciel project consumed all his energies, and few mentions of objective prism spectroscopy feature in reports of subsequent years. By 1890 Mouchez noted that two further prisms of angle 23° and 45° had been procured and that 80 stellar spectra were recorded [20].

A few other observers used objective prisms in the nineteenth century, notably Norman Lockyer (1836–1920) from about 1890 at the Solar Physics Observatory in South Kensington, London, where he was the first director. Most of his observations for classifying stellar spectra (using Lockyer's own idiosyncratic system) were with an objective prism on the 15-cm refractor [21].

Another objective prism observer was the English amateur, Francis McLean (1837–1904), who installed a 12-inch Grubb astrograph with a 20° prism in 1895 at his private observatory in Kent. He classified stars on his own system, which however resembled that of Vogel at Potsdam. McLean found several stars containing helium lines [22, 23]. In 1897 McLean visited the Royal Observatory at the Cape, taking his prism with him and using it on the astrograph there [24] – see Fig.5.3. McLean had in that year made the munificent donation of a 24-inch Grubb photographic refractor to the Cape

Figure 5.3. Francis McLean's objective prism mounted on the Cape Observatory astrograph, 1898. The prism aperture is 12 inches.

Observatory, but it was not operational until 1901. Sir David Gill (1843–1914), the director, reported some objective prism observations of η Carinae (η Argus) with an $8\frac{1}{4}^\circ$ objective prism to test the telescope's performance [25].

5.2.2   Objective prism astrographs in the twentieth century

The huge programme of objective prism spectroscopy at Harvard and its Boyden Observatory outstation in Peru continued into the twentieth century, culminating with the production of the Henry Draper Catalogue. The four astrographs already used for earlier studies, the 8-inch Bache and 13-inch Boyden telescopes

in Peru and the 8-inch Draper and 11-inch Draper telescopes in Massachusetts, were also the principal ones used for the new catalogue. The classifications were undertaken by Annie Cannon between 1911 and 1915 and the catalogue comprised the spectral types for 225 300 stars. The northern stars were complete to about $m_{\mathrm{pg}} \simeq 8.0$, although some as faint as 9.5 were included. In the south, the limiting magnitude was about a magnitude fainter [26]. This was partly the result of better seeing (image quality) in Peru, and partly because the fainter southern stars on the 8-inch Bache telescope were recorded at 400 Å/mm with a 5° prism. On the other hand faint northern stars were recorded at 160 Å/mm with the 8-inch Draper

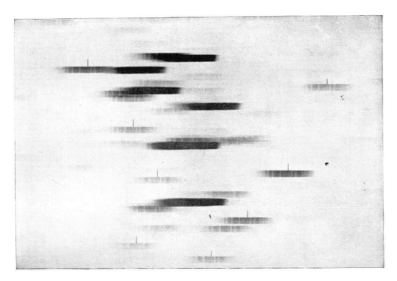

Figure 5.4. A Harvard objective prism plate of the Pleiades with the 8-inch Draper telescope, February 1910. The exposure was 54 minutes. Note the neodymium cell comparison absorption lines, shown marked, used for determining the radial velocity.

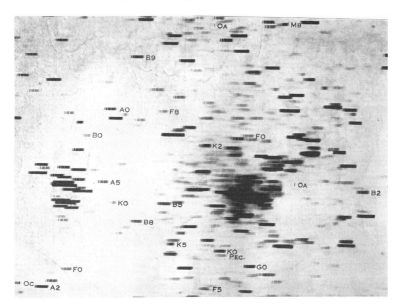

Figure 5.5. A Bache telescope objective prism photograph used for the Henry Draper Catalogue, showing spectra of stars in Carina, recorded May 1893. The exposure time was 140 minutes.

telescope with a 13° prism, giving the brighter limiting magnitude. Figure 5.4 shows an example of spectra recorded with the Draper telescope, while Fig. 5.5 shows an example of spectra from the Bache telescope in Peru.

The Henry Draper Catalogue was published between 1918 and 1924 in nine volumes of the *Harvard* *Annals* [27, 28, 29, 30, 31, 32, 33, 34, 35]. Edward Pickering died in 1919, by which time just three volumes had appeared in print, but Annie Cannon supervised the remainder of the work.

When Harlow Shapley (1885–1972) succeeded Pickering as Harvard director, the emphasis shifted considerably from the acquisition of new observational

Figure 5.6. The German astronomical station in La Paz, Bolivia, in about 1929, with its 30-cm Zeiss astrograph.

spectroscopic data for classification purposes, to the statistical analysis of the existing data (see, for example, [36, 37, 38]) in studies of Milky Way structure, and to the use of the higher dispersion spectra for astrophysical analysis using Saha's ionization theory, notably by Cecilia Payne (1900–79) [39]. However, Shapley did initiate the Henry Draper Extension (HDE) in 1923, partly to push the limiting magnitude to fainter stars, especially in the northern hemisphere. Once again, Annie Cannon classified nearly all of the 133 782 stars in the Henry Draper Extension [40, 41], the last publication appearing after her death in 1941. Many of the spectra for the Henry Draper Extension were recorded on a new astrograph, the 10-inch Metcalf telescope (made by the Revd Joel Metcalf (1866–1925), the noted Massachusetts amateur astronomer) with a triplet objective lens and a single objective prism, which gave a modest dispersion of 400 Å/mm at Hγ. This telescope was installed in Peru by 1923, but later was moved to the new Boyden Station in Bloemfontein, South Africa, in 1927. From Peru it could reach as far north as +45° declination [26]. Stars as faint as twelfth photographic magnitude are included in the HDE.

Several other large programmes of spectral classification with objective prism astrographs were undertaken in the first half of the twentieth century. Some are mentioned here. At the Hamburg-Bergedorf Observatory, Arnold Schwassmann (1870–1964), together

with his colleagues A. A. Wachmann (1902–90) and J. Stobbe (1900–43), undertook a large programme using the 30-cm f/5 Lippert astrograph between 1923 and 1933. The spectra were recorded at 400 Å/mm, and the aim was to reach to fainter limits than the Henry Draper Catalogue in the 115 northern areas of Kapteyn's *Selected Areas*. Indeed, the *Bergedorfer Spektraldurchmusterung* was complete to magnitude 12.0 and recorded some stars to magnitude 13.5. As many as 173 500 faint northern stars were classified on the Harvard system, a project comparable in size to the HD Catalogue itself [42] (see also [43]).

The Hamburg programme was complemented by the *Potsdamer Spektraldurchmusterung* undertaken by Friedrich Becker (1900–85), and with the later assistance of Hermann Brück. Stars in 91 southern *Selected Areas* were classified using a Zeiss f/5 30-cm objective prism astrograph installed near La Paz, Bolivia, in 1926 – see Fig. 5.6. The observations were completed in 1929, and the classifications comprised 68 000 stars to magnitude 12 [44, 45, 46] (see also [47] for a general review of the Potsdam programme in Bolivia).

At Mt Wilson Observatory the f/4.5 10-inch Cooke astrograph was also equipped with an objective prism of 15° angle. One extensive programme undertaken by Paul Merrill (1887–1961) and Milton Humason (1891–1972) from 1919 was the cataloguing of emission-line B stars (Be stars), based on the appearance of the red Hα line in low dispersion spectra (440 Å/mm at Hα). Merrill and his collaborators found many new Be stars, which often showed emission at Hα but not in higher members of the Balmer series [48, 49]. The blue photographically corrected Cooke triplet was not designed for work in the red, and this resulted in spectra that were out of focus, except for a small wavelength range near Hα, but this was adequate for the detection of the emission, and for cataloguing the properties of many of these stars. Special red-sensitized panchromatic plates were used, treated in pinacyanol or ammonia to increase their speed. However, three-hour long exposures were still required. The programme continued for two decades under Merrill's direction, with Humason being the main observer. A catalogue listing over 1000 Be stars in the northern Milky Way was the result [50, 51, 52, 53].

The Hα programme was continued after 1942 with a new objective lens by Frank Ross (1874–1960), corrected for red and yellow light, which gave good definition from the D lines to beyond Hα in the red. Figure 5.7 shows stars with Hα emission in Perseus. The same 15° prism now gave a dispersion of 385 Å/mm at Hα,[1] and with the more sensitive Kodak 103aE emulsion, exposure times could be halved to 90 minutes.

The Mt Wilson astrograph with its new Ross lens was installed at the Lamont–Hussey Observatory (owned by the University of Michigan) in Bloemfontein, South Africa, in 1949, and a survey of the southern sky was made between 1949 and 1951 by Karl Henize (1926–93). Catalogues of 459 southern planetary nebulae [54] and of other types of Hα emission-line stars (1929 objects) were the result [55]. The latter catalogue included some late-type emission-line stars (types M, S, C), some T Tauri stars, and many new Be stars.

A 10-inch astrograph similar to that at Mt Wilson was also used for objective prism spectroscopy by Alexander Vyssotsky (1888–1973) at the Leander McCormick Observatory of the University of Virginia. With a 7° prism, spectra with a dispersion of 330 Å/mm at Hγ were obtained. Vyssotsky used this instrument to identify M dwarfs in the Milky Way, based on spectrophotometric criteria

[56, 57]. Vyssotsky undertook an interesting comparison of some 5500 spectral types obtained from the McCormick Observatory with those obtained from Harvard, Mt Wilson, Hamburg and the Potsdam southern station in Bolivia [58]. All classification systems more or less agreed at type A0, but for K and M stars systematic differences amounting to two or even three decimal subclasses existed between observatories. Systematic effects were also quite large at F0 and G0, but at F5 they were generally less than one decimal subclass.

5.2.3 Spectrophotometry with objective prism astrographs

The objective prism instrument was widely used for spectrophotometry as well as for classification purposes in the 1920s to 1940s. Some of the spectrophotometry involved measuring line intensities to obtain more precise luminosity criteria in stellar spectra. Other work was on stellar continuum fluxes to measure temperatures.

In the former category, Donald Edwards (1894–1956) at the Norman Lockyer private observatory in Devon was using a 20° prism on a McLean 12-inch astrograph. He measured helium to hydrogen line intensity ratios of early B-type stars to obtain spectroscopic absolute magnitudes [59]. He extended this work to later B stars by using a line character parameter, based on the sharp or diffuse nature of the lines, and hence he calibrated his stars for absolute magnitude in terms of both spectral type and line character [60].

A similar programme for determining stellar luminosity was pursued by Bertil Lindblad (1895–1965) at Uppsala Observatory. The spectra were recorded on a 10-inch Cooke astrograph with a 6° prism, giving a dispersion of 340 Å/mm near the K line [61]. For A- and B-type stars, Lindblad measured the fluxes on each side of λ 390.7 nm, their ratio correlating well with absolute magnitude (which is largely a result of Stark broadening in the wings of the Balmer lines). For G and K stars he did spectrophotometry of the CN molecular bands at 388.3 and 421.6 nm, these features being stronger in the giant stars than the dwarfs. Lindblad also introduced a luminosity calibration based on hydrogen line width [62], which was developed further with Carl Schalén (1902–93) [63]. A Zeiss prism of 9. 7° was used on the new 15-cm Zeiss–Heyde astrograph at Uppsala for most of this work, giving a dispersion of about 270 Å/mm at Hγ.

---

[1] This figure is quoted by Merrill and Burwell [52]. Karl Henize gives 450 Å/mm at Hα for the dispersion [54].

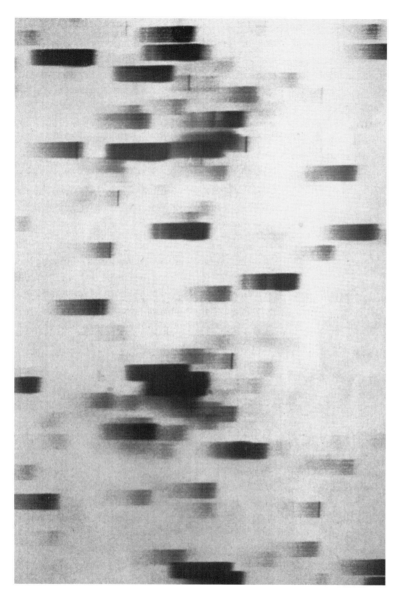

Figure 5.7. Mt Wilson 10-inch Cooke astrograph red objective prism spectra of the double cluster in Perseus, by Merrill and Burwell, 1949, showing a number of stars with Hα emission.

After Lindblad took up the directorship at Stockholm Observatory in 1927, he supervised the installation of several new instruments when the observatory was relocated to Saltsjöbaden in 1931. Included was a 40-cm Zeiss astrograph with a 4.8° prism (giving a dispersion of 220 Å/mm at Hγ) and the programme on spectroscopic luminosities, especially for late-type stars, was continued with this new larger instrument [64].

The objective prism also played an important role in early photographic spectrophotometry of stellar continuum fluxes. One of the early pioneers was Ralph Sampson (1866–1939) at the Royal Observatory Edinburgh. He used a 6-inch (150-mm) Cooke photovisual telescope of focal length 100 inches (2.5 m) onto which was mounted a Merz 12° dense flint glass prism [65, 66]. By tilting the plate holder through 20° to compensate for chromatic aberration, the spectra obtained were in sharp focus from 393 to 680 nm using Ilford Panchromatic plates. The dispersion was 280 Å/mm at Hα, and 51 Å/mm at the K line. A few strong lines in these low resolution (resolving power ~1000 in the yellow for average seeing) spectra were visible. However, the aim was to measure stellar continua, and the density was recorded using a photocell in a microdensitometer, using the spectrum of Polaris as a spectrophotometric standard. An attempt was made to obtain stellar temperatures from the continuous flux gradients, by applying Planck functions modified for the effects of limb darkening. These were not quite the first attempt at photographic spectrophotometry, but they represent the first use of an objective prism for this purpose, and an early use of a photoelectric recording microdensitometer.

Two other objective prism programmes in spectrophotometry are worthy of note. One was undertaken by Hans Kienle (1895–1975) and his colleagues at the Göttingen Observatory from about 1930. Two separate objective prism instruments were used. One was a 15-cm f/10 astrograph with an ultraviolet triplet objective lens and a 50° prism [67]. This gave blue spectra at a dispersion of about 76 Å/mm at Hγ. The other instrument was a 16-cm f/9 Newtonian reflector with 12° and 8° prisms, which, when used together, gave about 50 Å/mm dispersion at Hγ [68]. These instruments were used for photographic spectrophotometry of stellar energy distributions [69]. The refracting system was employed from the ultraviolet to 500 nm, the reflecting system to 685 nm. The spectra were traced on a photoelectric recording microdensitometer. Great care was taken with the photographic calibration, and for this purpose a coarse objective wire grating was placed in front of the objective prisms so as to give first order side spectra of lesser intensity. Kienle presented monochromatic magnitudes for 36 standard stars ranging in type from early B to M [69]. In a later paper, the absolute spectrum of several stars near or at spectral type A0 was calibrated relative to a tungsten lamp, which was in turn calibrated in the laboratory using a black-body spectrum [70] – see Fig. 5.8.

Finally the objective prism spectrophotometry of Daniel Barbier (1907–65), Daniel Chalonge (1895-1977) and their collaborators at the Institut d'Astrophysique in Paris is discussed. Theirs was the only programme to make extensive observations of stellar flux distributions below the Balmer discontinuity, after

preliminary surveys by William H. Wright (1871–1959) [71] and C. S. Yü [72] at Lick.

Barbier and Chalonge made their observations from the Jungfraujoch in Switzerland at an altitude of 3457 m. They took advantage of this high altitude site to make flux measurements down to 310 nm in the ultraviolet. Their quartz optics instrument had an aperture of just 90 mm, and a mean focal length of 60 cm, and it was equipped with two 30° quartz prisms (Fig. 5.9). The objective lens was tilted at 8°, so as to make the spectra wider by astigmatism, the width being 0.7 mm at 310 mm and 1.3 mm at Hβ. No trailing of the telescope was needed to produce additional widening. The length of the spectra from Hβ to 310 nm was 28 mm [73].

Between August 1934 and March 1939 the French astronomers made regular observing trips to the Jungfraujoch, and data were recorded on the ultraviolet fluxes of 204 stars. One paper explored how the size of the Balmer jump (defined as $D = \log_{10}(I_{370+}/I_{370-})$ varied with spectral type and luminosity, as well as the apparent mean position of the discontinuity itself, λ_1 (although theoretically it is at 364.7 nm, overlapping Balmer lines, especially in dwarf stars, shift the apparent wavelength to longer values) [74]. A second paper presented the Balmer jump (or discontinuity), D, apparent wavelength of the jump, λ_1, and gradients on each side of the jump, ϕ_1 and ϕ_2 [75] (see also [76] for an English review of this work). Figure 5.10 shows the Balmer jump for Vega in a spectrum recorded by Barbier and Chalonge.

The spectra were calibrated by observing the continuous spectrum from a hydrogen lamp which was used as an artificial star, erected some 600 m from the telescope. This lamp was in turn compared with a tungsten filament lamp, which they twice took to Göttingen (in 1936 and 1938) to calibrate against Kienle's black bodies [77, 78].

The flux gradients enabled black-body colour temperatures to be derived. The two temperatures disagreed markedly, which emphasized what is now well-known – that stars do not radiate like black bodies. Indeed, the Balmer jump itself was a manifestation of this departure from black-body radiation.

The objective prism programme on the Jungfraujoch was curtailed by World War II, and after the war it was continued with an ultraviolet Cassegrain slit spectrograph on a 25-cm telescope. The entire

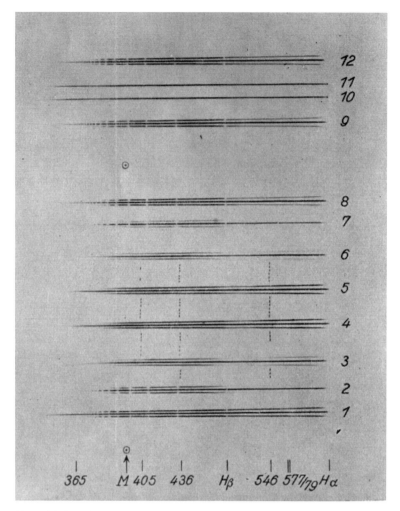

Figure 5.8. Objective prism spectra obtained by Hans Kienle, Göttingen, in 1940 for absolute stellar spectrophotometry. Spectra 1, 2, 7–9, 12 are of Vega, 10, 11 are of γ Cas, and 3–6 are of a white tungsten lamp. The triple spectra arise from an objective grating crossed with the objective prism for the purposes of photographic calibration.

history of the Jungfraujoch ultraviolet programme was described by Chalonge in 1951 [79].

5.2.4 Objective prism radial velocities

The 1950s finally saw come to fruition a method of measuring stellar radial velocities using the objective prism astrograph, after some six decades of largely unsuccessful experimenting. The idea dates back to 1887, when Edward Pickering proposed passing starlight through 'hyponitric fumes and other substances' to impose absorption features on the stellar spectra which would serve as a velocity calibration [80].

The advantages of objective prisms over slit spectrographs were clear – they could give high light throughput, thus allowing the observation of fainter stars, and they also permitted simultaneous spectra to be recorded. The basic problem, however, was to have a reference position on the plate against which Doppler-shifted stellar lines can be measured. Four solutions to this problem have been proposed. One is Pickering's absorption-line method, in which lines of an absorbing medium are superposed on the spectrum. The medium could be a gas or liquid solution held in the light path, or the terrestrial atmosphere or even the interstellar medium. Secondly there is the spectrum

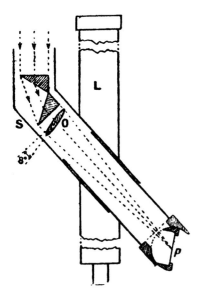

Figure 5.9. Objective prism spectrograph used by Barbier and Chalonge for ultraviolet spectrophotometry. The objective has an aperture of 90 mm and a focal length of 60 cm. Two 30° quartz prisms are used. The objective O is tilted by 8° to give astigmatic widening of the spectra perpendicular to the dispersion direction. L is a guide telescope.

length technique, which seeks to exploit the stretching of the spectrum of a receding star; that is, the differential shift between two lines is measured. Thirdly come attempts either to record laboratory comparison spectra alongside stellar spectra on objective prism plates, or to have the undispersed stellar images themselves act as reference marks from which to measure line shifts in stellar spectra. The last solution to be proposed was the reversion method, in which two spectra of opposing dispersions are recorded sequentially on the same plate by rotating the prism between exposures, and the distance between the same line in each spectrum is measured.

It is remarkable that the first three methods all originated with Pickering, who was also the first to experiment with the reversion technique using an objective prism. If he could have applied the objective prism to both spectral classification and radial velocity work, it would have been a considerable triumph. In this goal, Pickering was not successful, but the fact that from 1890 to 1920 such well-known names as Frost [81], Keeler [82], Deslandres [83], Maunder [84], Hale [85], Schwarzschild [86, 87] and Plaskett [88] all published papers on the technique for measuring objective

prism radial velocities, shows that the method was regarded as potentially rewarding.

In 1910, R. W. Wood proposed using the absorption feature at 427.3 nm resulting from the passage of light through a weak neodymium chloride ($NdCl_3$) solution [89] – see Fig. 5.11. Pickering claimed to achieve a random probable error of some 10 km/s using this technique [90]. A long series of tests was made at Harvard in the 1920s and 1930s on this method, notably by Bart Bok (1906–83) and Sidney McCuskey (1907–79) using the 16-inch Metcalf astrograph, but the probable errors still remained between ±10 and 13 km/s [91]. Only spectral types earlier than G0 could be measured, as the neodymium line became blended with a stellar iron line for the cooler stars. The truth is that stellar radial velocities using neodymium chloride were never very successful and in addition large systematic errors plagued the results.

Although Pickering was the first to suggest the spectrum length method of obtaining Doppler shifts, the method was taken up by Artemie Orbinsky at the University of Odessa [92]. John Plaskett in Ottawa also tried this technique [88]. However the shifts to be measured are small second order effects, and no useful results were obtained by either of these observers. Nor have proposals to include a laboratory comparison spectrum on objective prism plates ever come to fruition. Deslandres had suggested a slit and collimator in front of a small part of the prism for this purpose [83], but Keeler pointed out that this would result in lower resolution and vitiate the idea [82].

Soon after Pickering's suggestion of exposing undispersed stellar images on objective prism spectrograms to act as reference points for measuring Doppler shifts [14], Maunder at Greenwich followed up on this idea, by devising plans for a double astrograph with a common focal plane, one of the objectives being equipped with an objective prism [84]. Spectra and direct images were thus to be superposed on one plate, the latter serving as reference marks from which to measure Doppler shifts in the spectra. The method was too unwieldy to put into practice.

By far the most successful objective prism technique is the reversion method. The idea was first proposed by Hale and Wadsworth in 1896 [85] and independently by Pickering in the same year [93]. The method is really a photographic adaptation of Zöllner's reversion spectroscope of 1869 [94]. In this method,

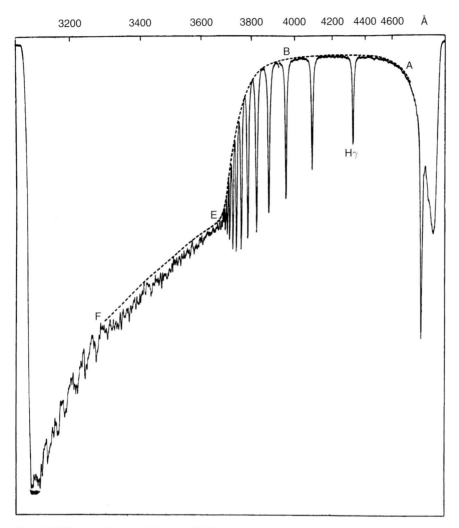

Figure 5.10. Tracing of an ultraviolet and visible light spectrum of α Lyrae, obtained by Barbier and Chalonge with the Arnulf objective prism spectrograph on the Jungfraujoch. Note the Balmer jump at E (365 nm), just on the short wavelength side of the Balmer lines.

two spectra are recorded side-by-side after rotating the prism through 180° between exposures. The displacements between a given line in the two spectra are thus twice those of a slit spectrograph.

Karl Schwarzschild in Potsdam used the reversion method in 1913 [3] and obtained results with probable errors of about ±7 km/s from one plate (see analysis by P. M. Millman (1906–90) [95]), which is better than other objective prism radial velocities, though still far short of the typical performance of conventional slit spectrographs. After World War II, Charles Fehrenbach (b. 1914) at the Observatoire de Haute-Provence

in France pioneered the reversion method [96, 97]. He used a special direct-vision compound prism, with crown and flint glass components, to eliminate field distortion introduced by the prism itself. His method is otherwise essentially that of Schwarzschild, with two reversed spectra side-by-side on the same plate – see Fig. 5.12. The technique was developed and reviewed over several decades.

By the early 1960s, two 40-cm astrographs were in use, one at Haute-Provence, the other at the Zeekoegaat Station in South Africa [98]. The dispersion was 110 Å/mm at 421 nm, and the field of view was 2°.

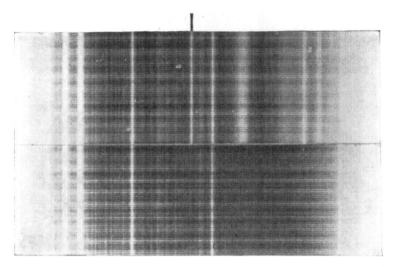

Figure 5.11. Stellar spectra of Procyon showing (upper) the neodymium chloride 427-nm absorption feature, and (lower) with no absorption. The spectra were recorded by Pickering in 1910.

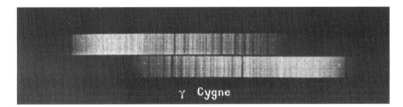

Figure 5.12. Reversion spectra of γ Cygni, recorded by Fehrenbach in 1947.

A catalogue of the radial velocities of stars in the direction of the Large Magellanic Cloud [99] was compiled from observations with the southern instrument. The high velocity of Cloud members allowed them to be readily distinguished from galactic foreground stars, with a precision of some 5 to 8 km/s, depending on spectral type. A review of the method was given by Fehrenbach in 1966 [7].

5.2.5 Objective prism surveys with Schmidt telescopes

The invention of a wide-field camera by Bernhard Schmidt (1879–1935) in Hamburg about 1930 [100], using an aspherized corrector plate to almost eliminate spherical aberration, and a spherical (hence achromatic) primary mirror, completely revolutionized wide-field imaging in astronomy. In a review in 1995, Jack MacConnell wrote:

> By 1930 Schmidt had developed the prototype of the telescope that bears his name at an unheard of

focal ratio of 1:1.75 and imaging a 15° field completely coma-free. There must have been a pent-up need for such an instrument because, by 1941, there were two dozen Schmidt telescopes with primary diameter of 25 cm or larger! The largest one operating in the world at the end of 1941, until it was surpassed by the Tonantzintla Schmidt a few months later, was the 61/91-cm f/3.5 Burrell Schmidt of Case Western Reserve Univ. ... equipped with a 4° prism ... Its first and primary user was Prof. J. J. Nassau who pioneered in the use of near-infrared emulsions with the prism in the late 1940s [101].

Jason Nassau (1892–1965) became one of astronomy's most active advocates for the objective prism technique in the mid twentieth century. His surveys of late-type stars (types M, S and C) in the Milky Way, using low dispersion spectra and Kodak I-N near infrared emulsion (the spectra, with a Wratten filter, were in the 680 to 880 nm range) were reviewed by

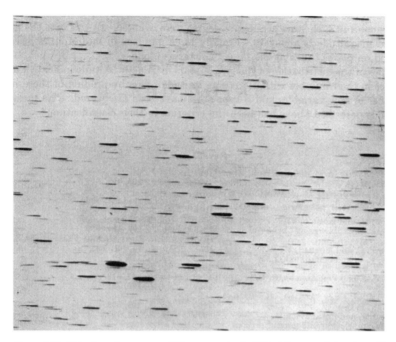

Figure 5.13. Objective prism spectra of M-type stars in the Milky Way, recorded by Jason Nassau at the Burrell Schmidt telescope with a 4° prism in 1956.

Nassau in 1956 [102], and again in 1963 [103]. The 4° prism gave a dispersion of just 1700 Å/mm at the A band, and this allowed M0 stars at infrared magnitude 13.0 to be reached; a 2° prism (3400 Å/mm) was also used, and here the limit was one magnitude fainter. Figure 5.13 shows objective prism spectra of M stars in the Milky Way recorded by Nassau in this programme.

Although the Schmidt telescope with an objective prism was essentially similar in principle to a refracting astrograph with such a prism, in practice the advantages were the achromatic performance (allowing for red spectra easily to be recorded), the generally faster focal ratio (resulting in shorter exposure times) and a wider field of view (giving more spectra per exposure). These advantages rendered the astrograph more or less obsolete for objective prism observations in the second half of the twentieth century.

A series of large Schmidt telescopes was built from the late 1940s through to the 1970s. One was the Curtis Schmidt at the University of Michigan in 1951 [104], an instrument with the same dimensions as the Burrell Schmidt (61 cm aperture, 91 cm mirror, 208 cm focal length) – see Fig. 5.14. The Curtis Schmidt was relocated to Cerro Tololo in Chile in 1967, where it was used for a reclassification of all the southern HD

stars on the MK system [105, 106]. The classification spectra were obtained with a 10° objective prism giving 108 Å/mm at Hγ. Not only were they used for the southern reclassification project, but William Bidelman and his colleagues surveyed the plates for luminous blue stars and stars with peculiar spectra (Ap stars, Be stars, weak-lined stars, C stars, Ba II stars, etc.) [107, 108, 109].

The Warner and Swasey Burrell Schmidt was also relocated to a better site, going to Kitt Peak in Arizona in 1979 (see [110]). This instrument is being used for reclassifying the northern HD stars on the MK system [111].

Several authors have used objective prism spectroscopy to catalogue OB stars in the Milky Way. One major survey, published in 1951 by Jason Nassau and William Morgan (1906–94), used the Burrell Schmidt to catalogue 918 early-type stars [112, 113]. The discovery of spiral structure in our Galaxy arose out of this work [114]. Other surveys in the southern hemisphere followed (for example [115, 116, 117]).

Another programme conducted by Morgan at Yerkes Observatory used a tiny 5-cm aperture Schmidt telescope with a 1° objective prism, giving spectra with a dispersion of 30 000 Å/mm. Such 'spectra' are really

Figure 5.14. Jason Nassau (right) and Carl Seyfert at the Burrell Schmidt telescope in Cleveland, Ohio.

no more than distorted star images, whose form is dependent on a star's energy distribution. Morgan, with Meinel (b. 1922) and Johnson (1921–80), showed that heavily reddened galactic OB stars were distinguishable from intrinsically red giants on such objective prism plates, the former having more ultraviolet light in their spectra [118].

Waltraut Seitter (1930–2007) used the Hoher List Schmidt telescope at Bonn to observe the emission line spectrum of Nova Herculis 1963. Three prisms were available, two of F3 flint glass (angles 7.3° and 2.7°) and also one of UBK7 ultraviolet-transmitting crown

glass (apex angle 3.1°) [119]. Flint glass prisms are preferred for most objective prism work, since they give greater dispersion for a given prism angle, or alternatively lighter and less absorbing prisms for a given dispersion. However, if observations into the ultraviolet below about 360 nm are required, then a crown glass such as UBK7 is mandatory.

At the Byurkan Observatory in Armenia, Benjamin Markarian (1913–85) used the 100-cm aperture (150-cm mirror diameter) Schmidt telescope to identify compact galaxies with a strong ultraviolet continuum [120]. Many of these were early-type

Table 5.1. *Table of selected Schmidt telescopes with objective prisms*

Observatory/telescope	d (cm) D (cm)	Year installed	Objective prisms dispersion	Notes
Warner & Swasey/Kitt Peak Burrell Schmidt	61 91	1941	4° 1700 Å/mm at A	to Kitt Peak in 1979
Tonantzintla Obs. Mexico	66 76.2	1945	3.96° 340 Å/mm (Hβ to Hγ)	
Mt Palomar Oschin Schmidt	125 183	1948		
Univ. Michigan/CTIO Curtis Schmidt	61 91	1951	1°50'; 10° 108 Å/mm (Hγ)	to CTIO, Chile in 1967
Hamburg-Bergedorf Calar Alto, Spain	80 120	1954		to Calar Alto, Spain in 1976
Southern Uppsala Mt Stromlo/Siding Spring	60 91	1957		to Siding Spring in 1982
Byurkan AO Armenia	102 132	1960	1.5°; 3°; 4° 1800; 900; 280 Å/mm (Hγ)	
Tautenburg, Germany K. Schwarzschild Observ.	134 200	1960	2600 Å/mm (Hγ)	
Uppsala Kvistaberg	100 135	1963	7° (80 cm) 199 Å/mm (He), 370 Å/mm (Hβ)	
Brorfelde Copenhagen	50 75	1966	15° 102 Å/mm (Hγ)	new optics in 1975 (45/77 cm)
Edinburgh ROE Monte Porzio, Italy	40 60	1967		
UK Schmidt Siding Spring, Australia	124 187	1973	44'; 2°15' 2400; 800 Å/mm	
ESO Schmidt La Silla, Chile	100 160	1974		decommissioned 1997
Kiso Schmidt Japan	105 150	1974	4°	

d: aperture of corrector plate; D diameter of spherical mirror.
CTIO: Cerro Tololo Interamerican Observatory, Chile; ROE: Roy. Observatory Edinburgh, UK; ESO: European Southern Observatory, La Silla, Chile.

galaxies, and some also had emission lines. The ultra-violet emission, at least for the sharp-lined objects, was presumed to have a non-stellar origin. A second survey for Markarian galaxies with strong ultraviolet emission was undertaken in the 1980s [121]. A southern survey for similar objects was made by Malcolm Smith (b. 1942) using the Curtis Schmidt in Chile [122]. The objects found included quasars (often showing Lyman α emission on Schmidt plates) and Seyfert galaxies and many galaxies with active nuclei. Both the Byurkan and Curtis Schmidt surveys used prisms with small angles (1. 5° at Byurkan; 1° 50′ at Cerro Tololo) and low dispersion. This enabled spectra as faint as magnitude 17 or 18 to be obtained.

Parker and Hartley reviewed the past and future use of objective prism surveys on Schmidt telescopes in 1997 [123]. Quasar surveys and surveys for emission line galaxies figured prominently in the 1980s and 1990s. Surveys for peculiar and emission-line stars have also continued. The use of high quantum efficiency and fine-grained Tech Pan film [124], with a blue quantum efficiency as high as 10 per cent, has given the Schmidt a new lease of life, although it is competing with multi-object fibre-fed spectrographs on conventional telescopes.

Finally, Tsvetkov and co-authors have reviewed the holdings of Schmidt plates in observatory archives [125]. The Warner and Swasey Observatory holds 14 000 objective prism plates in its archive, followed by Byurkan Observatory (4050 plates), Torun (2071), Asiago (2006), then Riga (1700) and the UK Schmidt at Siding Spring (1320). These are not necessarily the most active observatories using Schmidts with objective prisms, but the list gives some indication of where this type of observing was done in the second half of the twentieth century.

Table 5.1 gives a list of selected Schmidt telescopes that have engaged in objective prism spectroscopy.

REFERENCES

[1] Stephenson, C. B., *Vistas Astron.* **7**, 59 (1966)

[2] Bidelman, W. P., *ESO Conference on the Role of the Schmidt Telescope in Astronomy*, ed. U. Haug, ESO, SRC and the Hamburger Sternwarte, p. 53 (1972)

[3] Schwarzschild, K., *Publik. der astrophysik. Observ. Potsdam* **23** (Nr 69), 1 (1913)

[4] Haffner, H., *Astron. Nachrichten* **286**, 28 (1960)

[5] Bowen, I. S., in *Astronomical Techniques: Stars and Stellar Systems*, vol. 2, p. 34, ed. W. A. Hiltner, University of Chicago Press (1962)

[6] Geyer, E. H., *Veröff. der Landessternwarte Heidelberg-Königstuhl* **18**, 1 (1966)

[7] Fehrenbach, Ch., *Advances in Astron. & Astrophys.* **4**, 1 (1966)

[8] Stock, J. and Upgren, A. P., *Publ. Dept. de Astron., Univ. de Chile, Santiago* **2** (no. 1) 11 (1969)

[9] Fraunhofer, J., *Denk. der Münch. Akad. der Wiss.* **5**, 193 (1817). Also in *Gilberts Ann. der Phys.* **56**, 24 (1817)

[10] Fraunhofer, J., *Gilberts Ann. der Phys.* **74**, 337 (1823)

[11] Newall, H. F., *The Spectroscope and its Work*, London: The Society for Promoting Christian Knowledge (1910)

[12] Secchi, A., *Atti dell'Accademia Pontifica Romano de' Nuovi Lincei* (1872)

[13] Pickering, E. C., *Fortieth Annual Report of the Director of Harvard College Observatory* (1886) – report for the year 1885

[14] Pickering, E. C., *Ann. Harvard Coll. Observ.* **26** (part 1), i (1891)

[15] Pickering, E. C., *Proc. Amer. Phil. Soc.* **51** (no. 207), 564 (1912); also in *Pop. Astron.* **23**, 487 (1915)

[16] Pickering, E. C., *Ann. Harvard Coll. Observ.* **27**, 1 (1890)

[17] Maury, A. C. and Pickering, E. C., *Ann. Harvard Coll. Observ.* **28** (part 1), 1 (1897)

[18] Cannon, A. J., *Ann. Harvard Coll. Observ.* **56**, 65 (no. 4) (1912)

[19] Cannon, A. J. and Pickering, E. C., *Ann. Harvard Coll. Observ.* **28** (part 2), 131 (1901)

[20] Mouchez, E., *Rapport annuel de l'Observ. de Paris 1890*, p. 17 (1891)

[21] Lockyer, J. N., *Phil. Trans. R. Soc.* **184**, 675 (1893)

[22] McLean, F., *Phil. Trans. R. Soc.* **191**, 127 (1898)

[23] McLean, F., *Mon. Not. R. astron. Soc.* **56**, 428 (1896)

[24] Gill, D. F., *Mon. Not. R. astron. Soc.* **58**, 166 (1898)

[25] Gill, D. F., *Mon. Not. R. astron. Soc.* **61**, [66] (1901); also in *Proc. R. Soc.* **68**, 456 (1901)

[26] Shapley, H., *Harvard Coll. Observ. Circ.* 278 (1925)

[27] Cannon, A. J. and Pickering, E. C., *Ann. Harvard Coll. Observ.* **91**, 1 (1918)

[28] Cannon, A. J. and Pickering, E. C., *Ann. Harvard Coll. Observ.* **92**, 1 (1918)

[29] Cannon, A. J. and Pickering, E. C., *Ann. Harvard Coll. Observ.* **93**, 1 (1919)

[30] Cannon, A. J. and Pickering, E. C., *Ann. Harvard Coll. Observ.* **94**, 1 (1919)

[31] Cannon, A. J. and Pickering, E. C., *Ann. Harvard Coll. Observ.* **95**, 1 (1920)

[32] Cannon, A. J. and Pickering, E. C., *Ann. Harvard Coll. Observ.* **96**, 1 (1921)

[33] Cannon, A. J. and Pickering, E. C., *Ann. Harvard Coll. Observ.* **97**, 1 (1922)

[34] Cannon, A. J. and Pickering, E. C., *Ann. Harvard Coll. Observ.* **98**, 1 (1923)

[35] Cannon, A. J. and Pickering, E. C., *Ann. Harvard Coll. Observ.* **99**, 1 (1924)

[36] Shapley, H. and Cannon, A. J., *Harvard Coll. Observ. Circ.* 226 (1921)

[37] Shapley, H. and Cannon, A. J., *Harvard Coll. Observ. Circ.* 229 (1922)

[38] Shapley, H. and Cannon, A. J., *Harvard Coll. Observ. Circ.* 239 (1922)

[39] Payne, C. H., *Stellar Atmospheres*, Harvard Coll. Observ. Monographs, No. 1 (1925)

[40] Cannon, A. J., *Ann. Harvard Coll. Observ.* **100**, 1 (1925–36)

[41] Cannon, A. J. and Mayall, M. W., *Ann. Harvard Coll. Observ.* **112**, 1 (1949)

[42] Schwassmann, A., *Bergedorfer Spektraldurchmusterung*, vol. 1 (1935), 2 (1938), 3 (1947), 4 (1951), 5 (1953)

[43] Schwassmann, A., *Vierteljahresschrift der Astron. Ges.* **70**, 352 (1935)

[44] Becker, F., *Potsdamer Spektraldurchmusterung*, *Publik. der astrophysik. Observ. Potsdam* **27**, part 1 (1929); part 2 (1930); part 3 (1931)

[45] Brück, H. A., *Potsdamer Spektraldurchmusterung*, *Publik. der astrophysik. Observ. Potsdam* **28**, part 4 (1935)

[46] Becker, F., *Potsdamer Spektraldurchmusterung*, *Publik. der astrophysik. Observ. Potsdam* **28**, parts 5 and 6 (1938)

[47] Becker, F., *Himmelswelt* **46**, 41 (1936)

[48] Merrill, P. W., Humason, M. L. and Burwell, C. G., *Astrophys. J.* **61**, 389 (1925)

[49] Merrill, P. W., Humason, M. L. and Burwell, C. G., *Astrophys. J.* **76**, 156 (1932)

[50] Merrill, P. W. and Burwell, C. G., *Astrophys. J.* **78**, 87 (1933)

[51] Merrill, P. W. and Burwell, C. G., *Astrophys. J.* **98**, 153 (1943)

[52] Merrill, P. W. and Burwell, C. G., *Astrophys. J.* **110**, 387 (1949)

[53] Merrill, P. W. and Burwell, C. G., *Astrophys. J.* **112**, 72 (1950)

[54] Henize, K. G., *Astrophys. J. Suppl. ser.* 14, 125 (1967)

[55] Henize, K. G., *Astrophys. J. Suppl. ser.* 30, 491 (1976)

[56] Vyssotsky, A. N., *Astrophys. J.* **97**, 381 (1943)

[57] Vyssotsky, A. N., Janssen, E. M., Miller, W. J. and Walther, M. E., *Astrophys. J.* **104**, 234 (1946)

[58] Vyssotsky, A. N., *Astrophys. J.* **93**, 425 (1941)

[59] Edwards, D. L., *Mon. Not. R. astron. Soc.* **83**, 47 (1922)

[60] Edwards, D. L., *Mon. Not. R. astron. Soc.* **87**, 364 (1927)

[61] Lindblad, B., *Astrophys. J.* **55**, 85 (1922)

[62] Lindblad, B., *Nova Acta Reg. Soc. Scient. Upsala Ser. IV* **6**, No. 5 (1925)

[63] Schalén, C., *Arkiv för Mat. Astron. och Fysik* **19A**, No. 33 (1926)

[64] Lindblad, B. and Stenquist, E., *Astron. Iakttagelser och Undersökninger å Stockholms Observ.* **11** (no. 12) (1934)

[65] Sampson, R., *Mon. Not. R. astron. Soc.* **83**, 174 (1923)

[66] Sampson, R., *Mon. Not. R. astron. Soc.* **85**, 212 (1925)

[67] Wempe, J., *Zeitschr. für Astrophys.* **5**, 154 (1932)

[68] Strassl, H., *Zeitschr. für Astrophys.* **5**, 205 (1932)

[69] Kienle, H., Strassl, H. and Wempe, J., *Zeitschr. für Astrophys.* **16**, 201 (1938)

[70] Kienle, H., Wempe, J. and Beileke, F., *Zeitschr. für Astrophys.* **20**, 91 (1940)

[71] Wright, W. H., *Publ. Lick Observ.* **13**, 191 (1918)

[72] Yü, C. S., *Lick Observ. Bull.* 12 (no. 375), 104 (1926)

[73] Arnulf, A., Barbier, D., Chalonge, D. and Canavaggia, R., *J. des Observateurs* **19**, 149 (1936)

[74] Barbier, D. and Chalonge, D., *Ann. d'Astrophys.* **2**, 254 (1939)

[75] Barbier, D. and Chalonge, D., *Ann. d'Astrophys.* **4**, 30 (1941)

[76] Greenstein, J. L., *Astrophys. J.* **97**, 445 (1943)

[77] Barbier, D., Chalonge, D., Kienle, H. and Wempe, J., *Zeitschr. für Astrophys.* **12**, 178 (1936)

[78] Kienle, H., Chalonge, D. and Barbier, D., *Ann. d'Astrophys.* **1**, 396 (1938)

[79] Chalonge, D., *Contrib. de l'Inst. d'Astrophys., Sér. A*, No. 97, 1 (1951)

[80] Pickering, E. C., *Henry Draper Memorial, First Annual Report*, p. 9 (1887)

[81] Frost, E. B., *Astrophys. J.* **2**, 235 (1895)

[82] Keeler, J. E., *Astrophys. J.* **3**, 311 (1896)

[83] Deslandres, H. A., *Astron. Nachrichten* **139**, 241 (1896)

[84] Maunder, E. W., *Observ.* **19**, 84 (1896)

[85] Hale, G. E. and Wadsworth, F. L. O., *Astrophys. J.* **4**, 54 (1896)

[86] Schwarzschild, K., *Astron. Nachrichten* **194**, 241 (1913)

[87] Schwarzschild, K., *Publik. der astrophysik. Observ. Potsdam* **23**, 1 (no. 69) (1913)

[88] Plaskett, J. S., *Publ. Dominion Observ. Ottawa* **1**, 171 (1914)

[89] Wood, R. W., *Astrophys. J.* **31**, 460 (1910)

[90] Pickering, E. C., *Astrophys. J.* **31**, 372 (1910)

[91] Bok, B. J. and McCuskey, S. W., *Ann. Harvard Coll. Observ.* **105**, 327 (1937)

[92] Orbinsky, A., *Astron. Nachrichten* **138**, 9 (1895)

[93] Pickering, E. C., *Harvard Coll. Observ. Circ.* 13 (1896)

[94] Zöllner, J. C. F., *Astron. Nachrichten* **74**, 305 (1869)

[95] Millman, P. M., *Harvard Coll. Observ. Circ.* 357 (1931)

[96] Fehrenbach, C., *Ann. d'Astrophys.* **10**, 306 (1947)

[97] Fehrenbach, C., *Ann. d'Astrophys.* **11**, 35 (1948)

[98] Duflot, M., Duflot, A. and Fehrenbach, C., *J. des Observateurs* **46**, 109 (1963)

[99] Fehrenbach, C. and Duflot, M., *Astron. & Astrophys. Suppl.* **13**, 173 (1974)

[100] Schmidt, B. V., *Mitt. der Hamburger Sternwarte* **7**, no. 36 (1932)

[101] MacConnell, D. J., in 'The future utilisation of Schmidt telescopes', ed. J. Chapman, R. Cannon, S. Harrison and B. Hidayat, *Astron. Soc. Pacific Conf. Ser.* **84**, 323 (1995)

[102] Nassau, J. J., *Vistas Astron.* **2**, 1361 (1956)

[103] Nassau, J. J. and Velghe, A. G., *Astrophys. J.* **139**, 190 (1963)

[104] Stebbins, J., *Publ. Observ. Univ. Michigan* **10**, 1 (1951)

[105] Houk, N. and Cowley, A., *Intl. Astron. Union Symp.* **50**, 70 (1973)

[106] Houk, N. and Cowley, A., *University of Michigan Catalogue of Two-dimensional Spectral Types for HD Stars*, University of Michigan Press, vol. 1 (1975), vol. 2 (1978), vol. 3 (1982); Houk, N. and Smith-Moore, M., *ibid.*, vol. 4 (1988)

[107] Bidelman, W. P., MacConnell, D. J. and Frye, R. L., *Intl. Astron. Union Symp.* **50**, 77 (1973)

[108] Stephenson, C. B. and Sanduleak, N., *Publ. Warner & Swasey Observ.* **1** (No. 1), 1 (1971)

[109] Sanduleak, N. and Stephenson, C. B., *Astrophys. J.* **185**, 899 (1973)

[110] Pesch, P., In *Objective Prism and Other Surveys*, ed. A. G. D. Philip and A. R. Upgren, p. 3 (1991)

[111] Houk, N. and von Hippel, T., *Astron. Soc. Pacific Conf. Ser.* **84**, 292 (1995)

[112] Nassau, J. J. and Morgan, W. W., *Astrophys. J.* **113**, 141 (1951)

[113] Nassau, J. J. and Morgan, W. W., *Publ. Observ. Univ. Michigan* **10**, 43 (1951)

[114] Morgan, W. W., Osterbrock, D. and Sharpless, S., *Astron. J.* **57**, 3 (1952)

[115] Klare, G. and Szeidl, B., *Veröff. der Landessternwarte Heidelberg-Königstuhl* **18**, 9 (1966)

[116] Geyer, E. H., *Intl. Astron. Union Symp.* **50**, 82 (1973)

[117] Nordström, B. and Sundman, A., *Intl. Astron. Union Symp.* **50**, 85 (1973)

[118] Morgan, W. W., Meinel, A. B. and Johnson, H. M., *Astrophys. J.* **120**, 506 (1954)

[119] Seitter, W. C., *Veröff. der Univ.-Sternwarte Bonn* no. 67 (1963)

[120] Markarian, B. E., *Astrofizika* **3**, 55 (1967)

[121] Markarian, B. E. and Stepanian, J. A., *Astrofizika* **19**, 639 (1983)

[122] Smith, M. G., *Astrophys. J.* **202**, 591 (1975)

[123] Parker, Q. A. and Hartley, M., in *Wide-field Spectroscopy*, ed. E. Kontizas, M. Kontizas, D. H. Morgan and G. P. Vettolani. Kluwer Academic Publishers, *Astrophys. & Space Sci.* Library **212**, 17 (1997)

[124] Parker, Q. A., Phillipps, S. and Morgan, D., in 'The future utilisation of Schmidt telescopes', ed. J. M. Chapman, R. D. Cannon, S. J. Harrison and B. Hidayat, *Astron. Soc. Pacific Conf. Ser.* **84**, 96 (1995)

[125] Tsvetkov, M., Stavrev, K., Tsvetkova, K., Mutafov, A., Michailov, M.-E., in 'The future utilisation of Schmidt telescopes', ed. J. M. Chapman, R. D. Cannon, S. J. Harrison and B. Hidayat, *Astron. Soc. Pacific Conf. Ser.* **84**, 148 (1995)

6 · Ultraviolet and nebular spectroscopy

6.1 ULTRAVIOLET AND NEBULAR SPECTROSCOPY

6.1.1 Henry Draper and William Huggins, pioneers in ultraviolet stellar spectroscopy

Observational studies of the near ultraviolet region of stellar spectra have a long history, which goes back to the early days of stellar spectrum photography. The very first spectrum ever recorded by photography was by Henry Draper in 1872. He used his 28-inch reflector and a spectrograph with a quartz prism, and the then relatively new innovation of a dry emulsion glass plate. He noted:

> In the photographs of the spectrum of Vega there are eleven lines, only two of which are certainly accounted for, two more may be calcium, the remaining seven, though bearing a most suspicious resemblance to the hydrogen lines in their general characters, are as yet not identified [1].

The key to Draper's success was in part his use of the new dry plates, which were so much more convenient than the wet collodion plates used previously in astronomical photography. But also his use of a silvered-glass reflecting telescope and a spectrograph with a quartz prism allowed him not only to go below the approximately 400 nm wavelength limit of the human eye, but below the approximately 380 nm limit for the transmission of flint glass used in the lenses of achromatic refractors. The near ultraviolet spectrum of a star was thus recorded for the first time, enabling spectral lines in the region below 400 nm to be studied.

William Huggins in 1876 had similar success to that of Draper. He also used a reflecting telescope and dry photographic plates. However his reflector used speculum for its primary mirror, and not silver-on-glass. His spectrograph, in which the prisms were from

Iceland spar (crystalline calcite, mainly calcium carbonate) and lenses from quartz, gave the necessary transmission to ultraviolet light to record the spectrum of Vega to well below the Balmer limit [2]. Of the seven lines recorded, only the Hγ and Hδ lines were previously known to Huggins, and the others were ultraviolet members of the Balmer series to H$_{11}$ (377 nm).

Huggins described his spectrograph in more detail in 1880 [3]. The Iceland spar prism of 60° apex angle was made by the firm of Adam Hilger and chosen for its good ultraviolet transparency and higher dispersion than quartz. The whole spectrograph was mounted inside his 18-inch Cassegrain reflector after removing the secondary mirror, so as to act as a prime focus spectrograph – see Fig. 6.1. Spectra were widened by moving the star along the slit in discrete displacements. The slit was observed by means of a small Galilean telescope looking through the hole in the Cassegrain primary, thereby allowing the telescope to be guided by keeping a star in view. Comparison spectra of Jupiter and Venus were recorded alongside stellar spectra, and by removing the spectrograph from the telescope, spark spectra could also be photographed. By this time H. W. Vogel (1834–98) in Berlin had identified the new ultraviolet lines as originating from hydrogen [4], and Huggins' spectra of Sirius, Vega and other early-type stars showed the hydrogen series as far as H$_{16}$ at 370 nm [3].

From about 1880 Huggins had designed and built a more convenient ultraviolet spectrograph for his reflecting telescope. The new instrument was used at the Cassegrain focus, and had two 60° Iceland spar prisms, as well as quartz lenses, and a convenient means of slit viewing for guiding [5] – see Fig. 6.2.

By 1890, Huggins had obtained a spectrum of Sirius showing six absorption lines in the far ultraviolet between 333.8 and 319.9 nm [6]. His ability to go to such short wavelengths resulted from his use of a

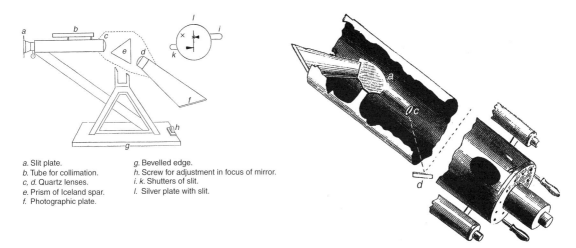

a. Slit plate.
b. Tube for collimation.
c, d. Quartz lenses.
e. Prism of Iceland spar.
f. Photographic plate.

g. Bevelled edge.
h. Screw for adjustment in focus of mirror.
i, k. Shutters of slit.
l. Silver plate with slit.

Figure 6.1. The first ultraviolet spectrograph of William Huggins, 1876. The instrument is shown (right) mounted inside the tube of the 18-inch reflector, at prime focus.

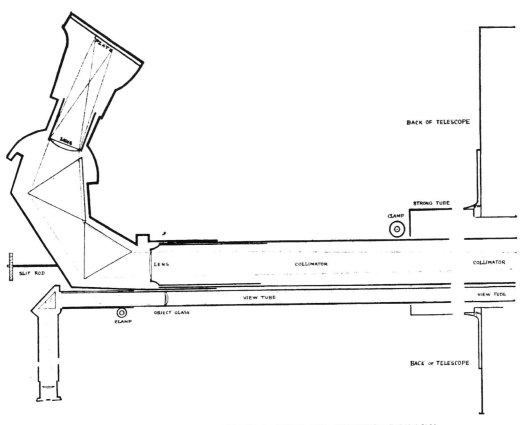

THE TULSE HILL ULTRA-VIOLET SPECTROSCOPE

Figure 6.2. Two-prism ultraviolet Cassegrain spectrograph of Huggins, 1880.

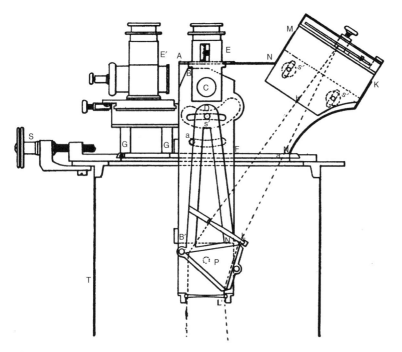

Figure 6.3. Prime focus ultraviolet spectrograph designed by Keeler and used by Palmer and Campbell on Lick's Crossley telescope, 1903. P is the quartz prism, E the focussing eyepiece, E′ is an offset guiding eyepiece, L is the concave quartz collimator lens, L′ is the camera lens and KK is the box housing the plateholder. The whole instrument is mounted on the side of the Crossley by means of the tube, T.

speculum mirror, as this metal has a useful reflectance of about 50 per cent down to the atmospheric limit (~300 nm), whereas silver loses its reflectance abruptly for $\lambda < 340$ nm. The lines discovered by Huggins were later shown to be telluric ozone (O_3) features by Alfred Fowler (1868–1940) and Lord Rayleigh (1842–1919) in 1917 [7].

Although many observers copied the success of Draper and Huggins and embarked on stellar spectrum photography (notably H. C. Vogel in Potsdam), the widespread use of reflecting telescopes and quartz or Iceland spar optical components followed only somewhat later. Johannes Hartmann (1865–1936) in Potsdam commissioned an ultraviolet spectrograph for the 80-cm refractor from the firm of Toepfer and Son (also in Potsdam) [8]. Although 285 nm could be reached in the laboratory, on the 80-cm refractor the limit was 360 nm. Two 30° quartz prisms were the dispersing elements for this instrument. As with Huggins' spectrographs, the plates were tilted to compensate for chromatic aberration in the camera.

6.1.2   Ultraviolet spectrographs at Lick Observatory

At the beginning of the twentieth century, the Crossley 36-inch silver-on-glass reflector at Lick Observatory, which had been installed in California in 1896, was one of the world's major new reflectors. It played an important role in ultraviolet spectroscopy after a slitless quartz prism spectrograph was installed at the prime focus in 1901 [9] – see also [10]. Harold Palmer (1878–1960) at Lick noted: 'such an instrument, to preserve and utilize the enormous advantages of the silver-on-glass reflecting telescope for work in the violet and ultra-violet regions, called for a design radically different from those of conventional spectrographs' [9]. The instrument had been designed by James Keeler, but was only completed two weeks before his early death. William Campbell and Palmer were therefore the first observers to commission it on the Crossley. It had a 50° quartz prism, placed 15 cm inside the prime focus. The collimator was a concave quartz lens. Figure 6.3 shows the optical layout. The spectra obtained were recorded

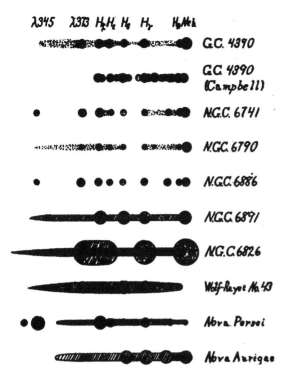

Figure 6.4. Drawings of ultraviolet spectra recorded on the Keeler spectrograph on the Crossley telescope by Harold Palmer, 1903.

from about 345 nm to Hβ, with a mean dispersion of about 50 Å/mm. Some drawings of ultraviolet spectra by Harold Palmer, based on those photographed on the Crossley telescope, are shown in Fig. 6.4.

A new two-prism quartz spectrograph was brought into operation on the Crossley telescope at Lick by William Hammond Wright with the support of Campbell in about 1917 [11]. The new instrument was also mounted at prime focus. It could be used either with a slit or in slitless mode. In the latter case, a concave quartz lens served as collimator. The instrument is shown in Fig. 6.5. Spectra to 330 nm could be recorded. Wright used this spectrograph to observe strong ultraviolet lines (e.g. 3313, 3342, 3444 Å) in the spectra of planetary nebulae. In this sense, it was an early nebular spectrograph as well as being an ultraviolet instrument. A spectrum of the planetary nebula NGC 7662, recorded by Wright in 1917, is shown in Fig. 6.6.

Nebulae are extended objects of low surface brightness. To record their spectra, wide slits or slitless operation is desirable and fast (short focal length) cameras are therefore needed to concentrate the spectrum onto a small region of photographic plate. As it happens, the Campbell and Wright spectrograph on the Crossley had an f/5.5 camera comprising a single quartz lens of focal length 11 inches (279 mm).

Another ultraviolet quartz spectrograph, quite similar in design to Campbell and Wright's instrument at Lick, was commissioned by J. S. Plaskett for the 72-inch telescope in Victoria, BC in 1922. It had two prisms and the lenses were of UV crown glass and the dispersion was about 50 Å/mm at 360 nm. The focal plane was severely curved, but by using film, the entire region from 320 nm to 700 nm could be recorded [12]. Presumably below 340 nm the spectra became rapidly weaker owing to the falling reflectivity of silvered mirrors. The f/5 camera was of 200 mm focal length.

6.1.3 Cameras for nebular and ultraviolet spectrographs

The problems of making fast camera objectives with a wide field of view for spectrographs were discussed by Hartmann in 1904. He listed the necessary requirements for a spectrographic camera. These were a flat focal plane, the complete elimination of spherical aberration over the whole image, and the production to be from as few lenses as possible. The lenses should be thin, not glued and they should be as transparent as possible. Chromatic aberration was only of secondary importance, given that tilting the plate could compensate for it [13]. Hartmann described how spherical aberration could be removed by aspherizing the surfaces of compound lenses, so that a field of view as great as 15° could result, and hence a much greater length of spectrum could be recorded. However, such lenses still were slow (40 mm aperture, 320 mm focal length, f/8 in Hartmann's example).

Very fast refracting camera lenses were only devised in the 1930s, and these had an immediate and beneficial effect on nebular spectrograph design. William Rayton of the Bausch and Lomb Company described in 1930 a seven-element compound camera objective of 50 mm aperture and 32 mm effective focal length operating at f/0.59, and similar in speed to a microscope objective [14]. This objective was used in a new spectrograph at Mt Wilson by Milton Humason

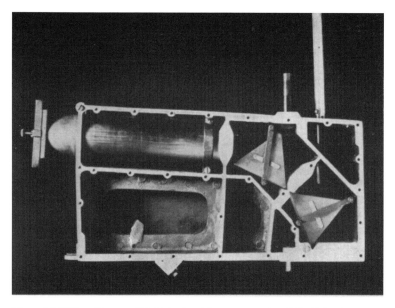

Figure 6.5. William Hammond Wright's ultraviolet spectrograph at Lick, 1917, shown in slitless mode.

Figure 6.6. Slitless ultraviolet spectrum of the planetary nebula, NGC 7662, recorded by Wright on the Crossley telescope, 1917.

in 1930, and it resulted in the exposure times of faint extragalactic 'nebulae' being reduced by more than a factor of two [15].

In 1936 R. J. Bracey proposed increasing the speed of the Rayton lens yet further to f/0.36 by immersing the air gap between the last plane lens surface and the photographic plate with a suitable liquid, thereby increasing the speed by the refractive index of the fluid [16].

An example of a spectrograph designed to record the night sky spectrum in the violet to red spectral domain and with a fast refracting f/0.65 camera was built by Jean Cojan in France in 1947 [17]. The camera had five lenses, two of them in a glued doublet.

Yet another prime focus faint object nebular ultraviolet spectrograph was built for the Crossley telescope prime focus by Nicholas Mayall (1906–93) in 1936 [18]. Note that the Crossley had been aluminized (instead

of silvered) for the first time in December 1933, giving the mirror a reflectivity below the 340 nm limit for silver. This new instrument was a two-prism spectrograph, and Zeiss UV glass was used. The camera was an f/1.3 triplet of focal length 60 mm, and thus was much faster than the Campbell–Wright camera of two decades earlier. Nebular spectra to 335 nm (the transmission limit for UV glass) could be recorded, and a study of the [OII] 372.7 nm doublet in extragalactic objects was made. Thirteenth magnitude galaxies could be recorded in eight hours or less.

An average gain in the speed of the Crossley nebular spectrograph of about 120 per cent was achieved by coating all 14 air–glass surfaces (collimator, prisms and camera lenses) with a thin antireflection layer of lithium fluoride [19]. Exposures were reduced by factors of two to three. This represents an impressive gain, and was an important technical development in spectrograph optics.

6.1.4 Nebular slit spectrographs at Yerkes and McDonald observatories

In 1937 Otto Struve (1897–1963) described a novel design of ultraviolet nebular spectrograph mounted on the outside of the tube of the Yerkes 40-inch refractor [20] – see Fig. 6.7. This instrument comprised a large slit (about 25 × 635 mm) mounted at the top end of the tube and a prism box and fast f/1 Schmidt camera at the bottom end, 17.7 m from the slit. Two 62° quartz prisms were in front of the Schmidt plate. There was no collimator and the telescope optics were not used. The large distance of the slit from the prisms and corrector plate assured nearly collimated rays, and the relatively large aperture of the corrector plate (94 mm) assured that a large angular extent on the sky was seen (about 16' × 2°).

Whereas a conventional spectrograph employs a large entrance pupil (defined by the telescope's objective or primary mirror) and a narrow slit (so as to limit rays to a small angular range $\theta_s \sim 1''$ on the sky), the Struve instrument used a small entrance pupil (defined by the slit), but accepted rays from a large angular range (defined by the camera aperture and slit-to-camera separation). The Schmidt camera was focussed on the slit, as for a normal spectrograph. For objects of angular extent at least 16', this arrangement was the fastest possible. The elimination of light losses in telescope objective and collimator optics gave large speed gains, enabling the spectra of faint extended nebulosity to be recorded. The dispersion was 250 Å/mm at Hδ, and the slit image on the plate was just 0.14 mm with a 25 mm slit, giving a wavelength resolution of 35 Å, and the resolving power was only about 120. However, the ultraviolet spectra of faint emission nebulae, such as the North America nebula (NGC 7000), could be recorded in three hours with this instrument [21]. Neither conventional slit spectrographs nor objective prism instruments could match this performance, because of the low surface brightness and large angular extent of nebulae, and because of the high sky brightness.

The following year Struve and his colleagues built a much larger nebular slit spectrograph at McDonald Observatory, with now 46 m between the slit and the f/1 Schmidt camera. The whole instrument was folded using a fixed plane mirror, and it was pointed by making the slit from a steerable plane mirror which subtended 6' by 1.5° on the sky – see Fig. 6.8. It was erected in the open on the side of Mt Locke (Fig. 6.9). Two quartz prisms and an f/1 Schmidt camera were once again employed [22]. This so-called 150-foot nebular slit spectrograph was used to record the spectra of faint emission-line nebulosity in the Milky Way as well as night sky spectra.

6.1.5 Ultraviolet spectroscopy with aluminized reflectors and Schmidt cameras

Although the Crossley telescope played a dominant role in the development of ultraviolet stellar spectroscopy in the first three decades of the twentieth century, the fact that until 1933 it used a silvered mirror effectively prevented any observations below about 340 nm (or a little less if exposures were long). Early experiments by Robley Williams (1908–95) and John Strong (1926–92) at Cornell University into chromium and aluminium reflecting surfaces deposited in a vacuum led to Williams and his colleagues obtaining spectra

Figure 6.7. Otto Struve's ultraviolet nebular spectrograph mounted on the side of the tube of the 40-inch refractor at Yerkes Observatory, 1937.

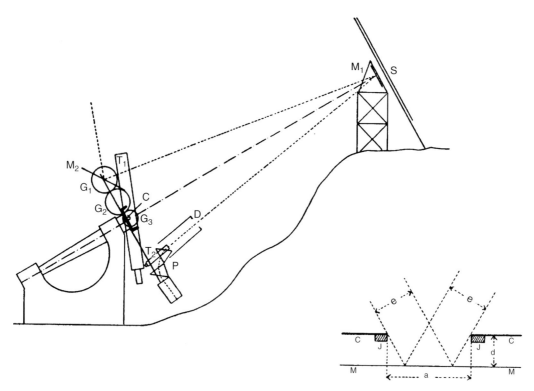

Figure 6.8. Diagram of Otto Struve's 150-foot nebular spectrograph at McDonald Observatory, 1938. M_1 is a fixed plane fold mirror, M_2 is the steerable plane slit mirror, G_1, G_2, G_3 are gears, T_1 is a guide telescope, T_2 an alignment telescope, P are quartz prisms, S is an opaque shield, D a diaphragm, C a mounting bracket.

of Vega to 310 nm with a chromium-on-glass mirror in 1932 [23]. By the following year, S. L. Boothroyd at Cornell had aluminized a 15-inch mirror, and with this he obtained spectra at the Lowell Observatory to 310 nm, using a two-prism quartz spectrograph [24]. Williams undertook photographic spectrophotometry of 14 bright early-type stars with this instrument from 435 down to 300 nm [25]. This is the practical atmospheric limit, which Huggins had almost reached nearly half a century earlier, with his speculum mirror telescope.

After 1930, the use of fast Schmidt cameras in ultraviolet and nebular spectrographs immediately circumvented many problems of fast refracting cameras, not least the light losses from multiple lens components. In France, Albert Arnulf (1898–1984) designed quartz prism spectrographs with fast f/1 and f/2 spherical camera mirrors [26] but with no Schmidt corrector plate. Instead, the correction for spherical aberration was provided through the two quartz lenses

of the collimator, with plane or spherical surfaces – see Fig. 6.10. Such an instrument with f/15 collimator and f/2 camera was tested by Daniel Barbier and Daniel Chalonge at Haute-Provence Observatory for observations of gaseous nebulae [26] – see Fig. 6.11.

An adaptation of the Arnulf design was a night sky spectrograph designed by Joseph Bigay (1910–82) at the Observatoire de Lyon. Here the collimator was of the two-mirror Cassegrain design, but the primary mirror of the Cassegrain collimator was over-corrected for spherical aberration. This therefore played the same role as the classical Schmidt plate, and allowed the camera, as with the Arnulf instrument, to consist of a simple spherical mirror operating at f/0.9 [27].

Another proposal has been to correct the spherical aberration of a fast spherical camera mirror by aspherizing the surface of a plane reflection grating. Such an instrument for fast ultraviolet nebular spectroscopy was built for the prime focus of the Canada–France–Hawaii telescope by G. Lemaître at Marseille [28] and

Figure 6.9. McDonald Observatory 150-foot nebular spectrograph on Mt Locke, 1938.

was known as UV-PRIM – see Fig. 6.12. It was based on a proposal by Harvey Richardson [29] to take advantage of higher throughput by eliminating the corrector plate of a conventional Schmidt camera.

6.2 ULTRAVIOLET SPECTROSCOPY FROM ABOVE THE EARTH'S ATMOSPHERE

6.2.1 Some technical aspects of ultraviolet spectroscopy

The absorption bands first seen by Huggins in the near ultraviolet spectrum of Sirius (320–334 nm) are the longest wavelength manifestation of stratospheric ozone in stellar spectra. Below 320 nm, the absorption by ozone rapidly increases, such that the observation of stellar spectra in the 200–300 nm wavelength domain necessitates observations from an altitude of at least 40 to 50 km. Between 100 and 200 nm, atmospheric molecular oxygen (O_2) is even more opaque than ozone, and observations from above 100 km are essential. Below 91.2 nm, interstellar neutral hydrogen renders the interstellar medium essentially opaque, but from 100 nm down to 91.2 nm the terrestrial atmosphere is steadily increasing in opacity. To be completely free of the effects of the Earth's atmosphere, observations in ultraviolet astronomy ($91.2 < \lambda < 300$ nm) must be made from above 200 km (see [30] for details).

For observations in the mid ultraviolet domain (200–300 nm), it follows that balloons (altitude $\sim$ 40 km), rockets and satellites are possible telescope platforms, but below 200 nm (the far or rocket ultraviolet) only rockets (altitude greater than 100 km) and satellites are usable. Although balloon ultraviolet spectroscopy has been used, the early advances in this field came from rockets in the mid 1960s and continuing into the 1970s. Satellite ultraviolet spectroscopy has had some notable milestones since 1968, with satellites such as OAO-2 (launched December 1968), TD-1 (March 1972), Copernicus (August 1972) and IUE (January 1978). Some observations were also made from manned space vehicles (Gemini XI and XII, Skylab, Salyut, Soyuz-13). This review does not cover the recent ultraviolet spectroscopy from the Hubble Space Telescope (launched April 1990), nor the extreme ultraviolet observations from the EUVE satellite (launched June 1992).

The optical technology required for ultraviolet spectroscopy involves the use of optical components with crystalline or glassy materials that have high transmittance for wavelengths greater than some short wavelength cutoff, which is usually fairly sharply defined. Common materials and their cutoffs are given in Table 6.1. Various high-silica ultraviolet glasses may be usable in the mid ultraviolet domain ($\lambda > 200$ nm).

For reflecting surfaces, pure aluminium has some useful reflectivity right down to the Lyman limit (91.2 nm), provided there is no oxide layer. To prevent such a layer from forming, thin films of lithium fluoride or magnesium fluoride can be applied which also serve to enhance the reflectance by interference in the thin layer. The lithium fluoride coating gives a usable reflectance to about 100 nm, or even to the Lyman limit, whereas more durable magnesium fluoride is limited to wavelengths greater than about 120 nm.

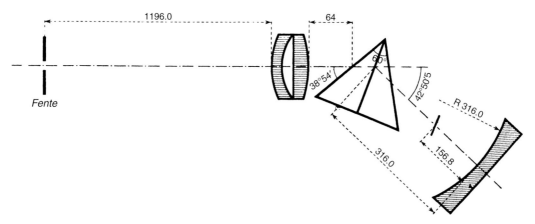

Fente

Figure 6.10. Albert Arnulf's f/2 nebular spectrograph with spherical camera mirror, 1943.

Table 6.1. *Ultraviolet cutoff wavelengths of some materials used in optics*

Material	Cutoff (nm)
lithium fluoride (LiF)	105.0
magnesium fluoride (MgF$_2$)	113.0
calcium fluoride (CaF$_2$)	124.0
barium fluoride (BaF$_2$)	137.0
aluminium oxide (Al$_2$O$_3$)	145.0
fused silica (SiO$_2$)	161.0

The data are from [30]

Detectors for ultraviolet spectroscopy in the 1960s and 1970s fell into three categories: photographic emulsion, photomultiplier tubes and various photoelectronic imaging devices, which include the SEC vidicon (a type of television tube). For photography, a special ultraviolet emulsion is required below 220 nm, such as Kodak Short Wave Radiation (SWR) emulsion [31].

Spectrographs and spectrometers used for ultraviolet spectroscopy from space have been designed to cover a variety of resolution domains from low resolution observations ($\Delta\lambda \sim 1.0$ nm), which are useful for determining stellar fluxes and their distribution with wavelength, to medium resolution ($\Delta\lambda \sim 0.1$ nm), which is useful for a survey of the stronger lines in stellar spectra, as well as studies of mass loss from stars with P Cygni line profiles and of strong interstellar lines, and finally to high resolution observations

($\Delta\lambda \sim 0.01$ nm), which allow detailed studies of many weaker lines and of line profiles.

Of course spectroscopy in space has required many other major technological challenges to be overcome. Not only is the efficiency or transparency of the optics critical, but spectrometers have to be compact and lightweight, and there are issues of pointing and stability, of launch and (for rockets) recovery or (for satellites) of telemetry, the effects of the Earth's radiation belts on electronics, and many others. It is perhaps surprising that so much was accomplished in ultraviolet astronomy in the two decades from about 1965, given the huge technical problems that had to be surmounted.

6.2.2 Ultraviolet spectroscopy from rockets and balloons

Ultraviolet astronomical spectroscopy at wavelengths below the atmospheric limit was born in 1946, when a team from the US Naval Research Laboratory launched a German V-2 rocket from Palestine, Texas. The rocket carried an ultraviolet solar spectrograph in its tail fin, which enabled the first solar ultraviolet lines in the 210 to 300 nm region to be observed, including the Mg II doublet lines near 280 nm [32]. The spectrograph comprised a 15 000 gr/in concave grating, with the spectrum (resolution ~ 1.5 Å) being recorded on film placed on the Rowland circle – see Fig. 6.13. Three flights were undertaken in these pioneering experiments between October 1946 and October 1947,

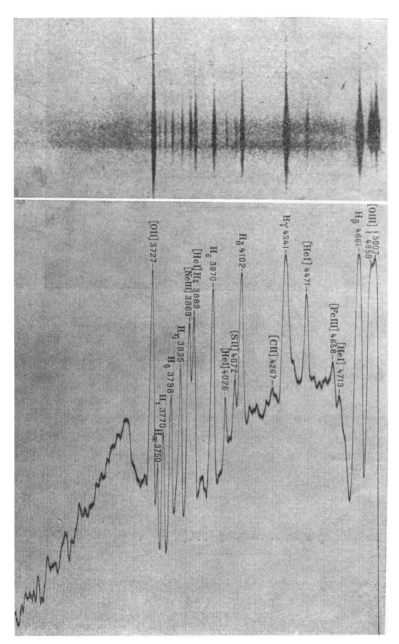

Figure 6.11. Orion nebula spectrum taken by Barbier and Chalonge in a 90-minute exposure using the Arnulf f/2 spectrograph at Haute-Provence. A 25-cm aluminized-mirror telescope was used to collect the light.

and these enabled a rich absorption-line spectrum from 220 to 300 nm to be identified.

As it happens, another group from the Johns Hopkins University undertook their experiments in ultraviolet solar spectroscopy at this same time. Their first flight was also in October 1946, but ended in failure when the film cassette was damaged on impact after descent, hence fogging the film. Later flights in April and July 1947, using a new concave grating spectrograph, were, however, successful [33]. The spectra were

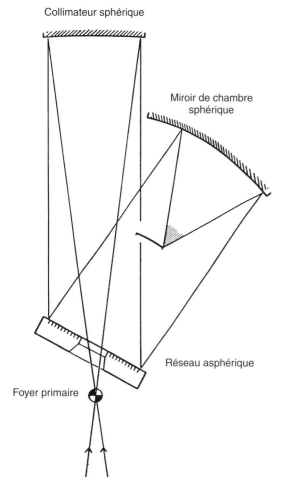

Collimateur sphérique

Miroir de chambre
sphérique

Réseau asphérique

Foyer primaire

Figure 6.12. Optical scheme of Lemaître's UV-PRIM aspherized plane grating spectrograph at the Canada–France–Hawaii telescope, 1980.

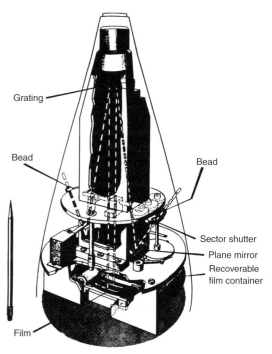

Grating

Bead

Bead

Sector shutter
Plane mirror
Recoverable
film container

Film

Figure 6.13. Schematic diagram of a rocket spectrograph for ultraviolet solar spectroscopy from a V–2 rocket, flown in 1946 and 1947 from Texas by E. Durand *et al.* from the US Naval Research Laboratory.

at a dispersion of 34 Å/mm and had about 1 Å of resolution over the 230–300 nm range. Both flights rose to well over 100 km altitude, allowing excellent solar spectra to be obtained. The analysis and line identifications were similar to those by the Naval Research team.

The first detection of ultraviolet radiation from below the atmospheric limit coming from a celestial source other than the Sun was made by E. T. Byram and his colleagues at the Naval Research Laboratory in November 1955 during an Aerobee rocket flight [34]. A collimated photon counter in the 122 to 134 nm band detected discrete celestial sources in the Puppis–Vela region.

Successful low resolution stellar spectrophotometry followed a few years later, when in November 1960 Theodore Stecher (b. 1930) and James Milligan at NASA's Goddard Space Flight Center launched an Aerobee rocket carrying four photoelectric grating spectrometers, which scanned with the rocket's rotation [35]. Two of these instruments operated in the 122.5–300 nm range, two from 170 to 400 nm. The spectrometers comprised a baffle collimator, two plane gratings and a parabolic f/1 camera imaging onto a slit in front of a photomultiplier tube – see Fig. 6.14. Slow rocket rotation gave rise to the wavelength scan, with the wavelength resolution in the spectrometers being 5.0 or 10.0 nm. Absolute fluxes of some early-type stars were recorded, and the first indications were found of an excess interstellar extinction around 220 nm, later ascribed to graphite grains [36, 37, 38].

The next major advance came in 1965 when Don Morton (b. 1933) and Lyman Spitzer (1914–97) at Princeton recorded the first rocket ultraviolet stellar spectra to show lines [39]. The use of a star-pointing

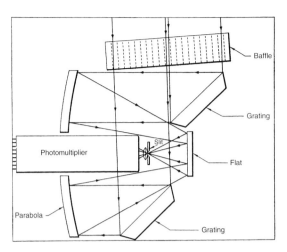

Figure 6.14. Dual grating spectrometer used by Stecher and Milligan for the first rocket ultraviolet spectrophotometry of stars in 1960.

Figure 6.15. Two objective grating Schmidt cameras used on an Aerobee rocket for the first ultraviolet stellar spectroscopy by Morton and Spitzer, June 1965.

gyroscopic rocket control system known as IACS (Inertial Attitude Control System) made such observations possible. Their Aerobee rocket carried two f/2 Schmidt cameras with reflection objective gratings and a 10° field of view. One camera had a calcium fluoride corrector plate, enabling wavelengths greater than 125 nm to be recorded (hence eliminating night sky Lyman α emission); the other had a quartz corrector plate for $\lambda > 170$ nm. Spectra, with a wavelength resolution of $\Delta\lambda \sim 0.1$ nm, were recorded on film. Figure 6.15 shows the two objective grating cameras and the film cassette.

After two earlier unsuccessful flights in late 1964 and early 1965, the third flight (June 1965) also had technical problems involving the parachute that returned the spectrometers to the ground. On flight number three, the instrument package crashed and was destroyed, but the film from one spectrograph camera (that with the calcium fluoride corrector) was nevertheless recovered. The spectra of just two early-type B stars were recorded on film not damaged in the crash landing. They were π and δ Scorpii (see Fig. 6.16). Numerous ultraviolet absorption lines were recorded. In particular, the strong ultraviolet resonance lines of Si IV (139.8 and 140.3 nm) and C IV (155.0 nm) were seen in stellar spectra for the first time.

Further Aerobee flights followed. In October 1965 Morton launched a pair of similar objective grating cameras, except that the quartz corrector plate used

earlier was replaced by lithium fluoride, so as to include the Ly α line, with the hope that the Ly α night sky emission would not obliterate all stellar details [40]. The Ly α line in absorption was recorded in δ and ζ Orionis (both are luminous O9.5 stars). The discovery of P Cygni line profiles for the Si IV and C IV resonance lines in ϵ and ζ Orionis, with the blue-shifted absorption components indicating a mass loss rate of about 1900 km/s, was a surprise discovery from this flight, suggesting mass loss rates as high as 10^{-6} solar masses per year.

The year 1965 can be regarded as when stellar ultraviolet spectroscopy from rockets came of age, and several other groups joined in the rocket race. These included George Carruthers (b. 1939) at the Naval Research Laboratory and Andrew Smith at Goddard. Carruthers used an objective reflection grating spectrograph with an electronographic camera [41]. This detector converts the ultraviolet image into an electron image which is focussed onto nuclear track emulsion. The absence of transmissive optics allowed observations of 16 stars down to 95.0 nm. The reflection

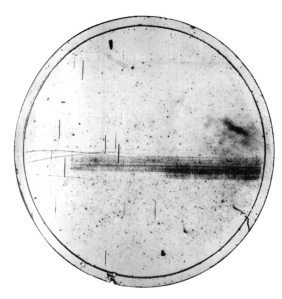

Figure 6.16. Ultraviolet rocket spectra of δ Sco and π Sco by Morton and Spitzer in 1965. The π Sco spectrum extends over three-quarters of the diameter from 126 to 218 nm; that of δ Sco over the right-hand third of the diameter, from 126 to 172 nm. The dark vertical lines are undispersed zero-order spectra of stars in Libra.

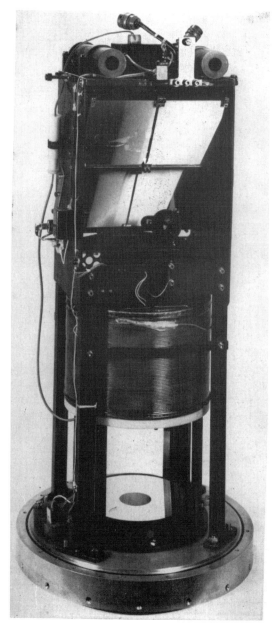

Figure 6.17. Objective grating spectrograph with electronographic camera used by G. Carruthers for ultraviolet stellar spectroscopy in 1965.

gratings were aluminium coated with lithium fluoride. The instrument first flew in March 1967. The Carruthers objective reflection grating instrument with electronographic camera is shown in Fig. 6.17.

In 1970 Carruthers detected the Lyman resonance molecular absorption bands of molecular hydrogen in the spectrum of the highly reddened O8 star, ξ Persei with this instrument. Eight strong bands in the 101.3 to 110.8 nm range were detected, indicating a high column density of H_2 in the direction towards this star [42].

Andrew Smith also used an all-reflecting spectrograph, in which a concave grating focussed light directly onto a curved film plane [43]. The collimator was a baffle. The resolution was 0.8 Å, and spectra of α Virginis [43] and ζ Puppis [44] were recorded to 92.8 nm in respectively June 1967 and March 1968.

Meanwhile an improved all-reflecting spectrograph with a far ultraviolet capability was also used by the Princeton group. This had an objective grating and two-mirror f/2 camera [45]. This instrument flew in September 1966 and the spectra of ten OB stars in Orion were recorded. The strength of the interstellar Ly α line at 121.6 nm was demonstrated in these results.

Meanwhile several groups pursued stellar ultraviolet spectrophotometry from rockets. Theodore Stecher at Goddard continued his work, using now a stabilized pointed instrument giving 1.0 nm resolution [46]. His instrument was a concave grating spectrometer with photomultiplier detectors, mounted on a 13-inch Cassegrain telescope. Absolute fluxes of four bright stars were obtained in November 1966 (α CMa, ϵ CMa, ζ Pup, γ^2 Vel), with enough resolution to show the stronger lines present. Other low resolution spectrophotometry on bright stars was undertaken by Frank Stuart at Kitt Peak Observatory [47] in February 1968, and by Albert Boggess (b. 1929) and Yoji Kondo (b. 1933) at Goddard in November 1965 [48].

This very active early phase in ultraviolet spectroscopy from rockets continued into the 1970s, by which time the first data from satellites were also becoming available. Rockets that spend little more than five minutes in the 100–150 km altitude range are not much competition for a satellite at over 200 km working for several years. Yet the early foundations in this new branch of astronomy certainly came from the rocket pioneers.

Ultraviolet astronomy from balloons is possible in the mid ultraviolet (200–300 nm) but not the far ultraviolet ($\lambda < 200$ nm). Thus a balloon at 40 km altitude receives only about 10 per cent of the radiation at 250 nm that is incident on the top of the atmosphere, as a result of stratospheric ozone. Thus observations of bright stars are still possible. Yoji Kondo at NASA developed a helium balloon-borne spectroscopy programme known as BUSS (balloon-borne ultraviolet stellar spectrometer) in the 1970s. The first generation in the BUSS programme used a scanning Ebert–Fastie spectrometer with a 2160 gr/mm grating that gave spectra in the second order, and the detector was an image dissector tube [49] – see Fig. 6.18. The spectrometer gave 0.5 Å resolution, and the goal was to observe the Mg II lines at 280 nm in a variety of different bright stars. Successful flights to 40 km altitude were made in June and October 1971. Further stars were observed in March 1975, and the strong Mg II h and k chromospheric emission lines were recorded in several G and K giant stars [50]. Kondo noted that the data were monitored in real time on the ground, enabling an integration to be terminated when the signal-to-noise ratio was sufficient.

A second generation of the BUSS programme used an échelle spectrograph and SEC vidicon detector, a combination similar to that which had been successfully adopted for the IUE satellite [51]. The BUSS

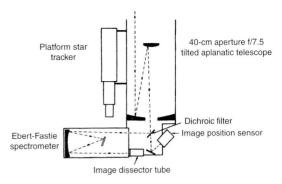

Figure 6.18. First balloon-borne ultraviolet spectrometer (BUSS-I) for studying stellar Mg II lines at 280 nm, 1971.

payload of 800 kg comprised a 40-cm f/7.5 Cassegrain telescope linked to a compact R2 79 gr/mm échelle spectrograph with a 15° fused silica cross-dispersing prism, of which one face had a reflecting coating. A catadioptric camera sent the light to an SEC vidicon television camera, for which the quantum efficiency was over 10 per cent in the mid ultraviolet range. The camera delivered 25-μm resolution in the image plane. The camera could record all of the spectrum from about 200 to 340 nm in a single exposure with 0.1 Å resolution, and the acquisition of the spectral image could be viewed in real time from the ground during an exposure. Figure 6.19 shows the BUSS-II instrument.

This BUSS-II payload flew from Palestine, Texas in May and September 1976, and 53 spectra of 33 stars brighter than magnitude 5.0 were acquired. Observations of the Mg II doublet, and of many interstellar, circumstellar or chromospheric lines were made.

6.2.3 The Orbiting Astronomical Observatories

One of the first projects of the National Aeronautics and Space Administration (NASA) after its foundation in 1957 was to plan for a series of Orbiting Astronomical Observatories for the pursuit of ultraviolet stellar astronomy. Although this was the early era of the so-called 'space race', it was also a time before any ultraviolet stellar spectra (other than that of the Sun) had been observed at all from space. To plan for three or four ambitious observatories at that time therefore required the considerable vision of Arthur Code (b. 1923) at Wisconsin, Lyman Spitzer at Princeton and others who were behind this proposal.

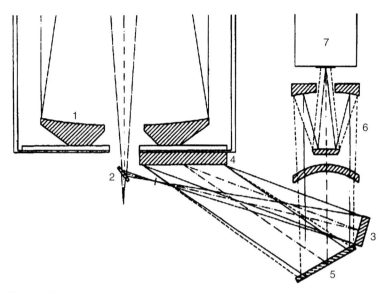

Figure 6.19. Second generation balloon-borne spectrometer (BUSS-II), comprising an échelle spectrograph with vidicon detector. 1: primary mirror of telescope; 2: dichroic mirror; 3: collimator; 4: échelle; 5: cross-dispersing prism; 6: camera; 7: vidicon detector. This instrument flew in 1976.

The first observatory (OAO-1), which was designed to do broad-band photometry, failed in 1966 soon after reaching its orbit. OAO-2, launched in December 1968, took on some of the photometric programmes of its predecessor. It also had two scanning grating spectrometers for low resolution spectroscopy of the ultraviolet fluxes from hot stars, as part of the Wisconsin instrument package.

OAO-2 carried two ultraviolet spectrometers [52, 53]. They were objective grating instruments with a slat collimator passing light with a 3° field of view to the 15 × 20-cm gratings. These scanned the spectrum by grating rotation in discrete steps. Light diffracted from the gratings passed to a parabolic camera mirror, which focussed the beam back through a hole in the grating centre and onto a slit, behind which was placed a photomultiplier tube – see Fig. 6.20. The first of these spectrometers scanned fluxes in the 180 to 380 nm range, the second from 105 to 200 nm. Both instruments were pointed at the same star, although simultaneous scanning was not possible. Resolutions were 2.0 or 20 nm for the first spectrometer, 1.0 or 10 nm for the second. The exit slits in front of the photomultipliers corresponded to two arc seconds on the sky. Thus, for extended objects of greater angular extent than this, the wavelength resolution was somewhat degraded.

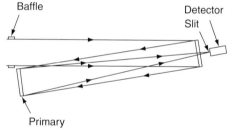

Figure 6.20. Arthur Code's Wisconsin scanning ultraviolet spectrometer on OAO-2, placed in orbit December 1968.

Stars were, however, the main objects studied, and OAO-2 recorded the ultraviolet fluxes of 330 stars. For the scans with 1-nm resolution, several of the stronger ultraviolet lines in stellar spectra, such as $Ly\,\alpha$ 121.6 nm, Si IV and C IV were visible. Figure 6.21 shows a sample of OAO-2 scans, with these strongest ultraviolet lines clearly seen.

The next mission in the OAO series carried a 0.9-m Cassegrain telescope and an Ebert–Fastie scanning spectrometer for medium or low resolution spectral scans. It too used photomultipliers for detectors. This instrument failed on launch in November 1970.

The final mission in the series was OAO-3, known as the Copernicus satellite. It was launched

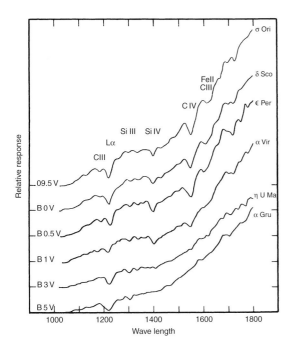

Figure 6.21. Spectral scans of early-type main-sequence stars showing the variation of line strengths as a function of spectral type. From data obtained with the OAO-2 spectrometer no. 2 and a resolution of 1.0 nm.

in August 1972 and was dedicated to high resolution spectroscopy. It operated for over eight years, until February 1981. The instrument on board was a high resolution scanning spectrometer designed by the Princeton group and using a concave grating in a Paschen–Runge mount [54] – see Fig. 6.22. In this mounting, the scanning is achieved by moving the detectors, in this case four photomultiplier tubes, on the Rowland circle. Two of the photomultipliers worked in the far ultraviolet, two in the mid ultraviolet, as shown in Table 6.2.

The far ultraviolet was eliminated from tube V1 by means of a fused silica cutoff filter, and from V2 by a sapphire prism. Pulse counting electronics counted the photons received in 14-s integrations undertaken between movements of the carriages on which the photomultipliers were mounted.

A major goal for the Copernicus satellite was the investigation of interstellar absorption lines resulting from gas in the galactic disk scattering the light from distant OB stars. Many of the resonance lines from neutral or ionized light elements lie in the far ultraviolet,

and indeed Copernicus was able to observe lines of H I, C I, C II, N I, N II, O I, Mg I, Mg II, Si II–IV, P II, S I–IV, Cl II, Ar I, Mn II and Fe II in the spectra of various reddened stars, notably ξ Per, α Cam, λ Ori, ζ Oph and γ Ara [55]. These were the first observations to show that the composition of the interstellar gas was often depleted in these elements relative to the solar composition. For the light elements considered here, deficiencies of factors of ten were common where the line of sight passed through dust clouds.

6.2.4 Ultraviolet spectroscopy with Europe's TD-1 satellite

In 1972, the European Space Research Organisation (ESRO) launched the TD-1 satellite for ultraviolet and high energy astronomy on a Thor-Delta rocket (hence the satellite's name). TD-1 carried two ultraviolet astronomy experiments. One, known as the S2/68 package, was a UK–Belgian collaboration for low resolution spectral scans in the 135–255 nm range. The other experiment was called S59, from the Netherlands, and was designed for medium resolution studies of strong lines in the mid ultraviolet region.

The S2/68 instrument scanned with a plane grating in the 135–255 nm range. The telescope was an off-axis paraboloid of 27.5 cm diameter. The orbital motion of the satellite, in a nearly polar orbit that precessed by $1°$ per day, gave the wavelength scan that was detected by three photomultiplier tubes behind exit slits. The resolution was 3.5–4.0 nm. This is not enough to see lines, but the absolutely calibrated fluxes were invaluable, especially for studies of interstellar extinction [56].

A catalogue of low resolution ultraviolet data from S2/68 on TD-1 was published in 1976 [57]. It contained data for 1356 stars brighter than ninth magnitude, with absolute fluxes in $\mathrm{erg\,cm^{-2}\,s^{-1}\,\mathring{A}^{-1}}$.

The other ultraviolet spectroscopy package on TD-1 undertook medium resolution spectroscopy. The proposal originated with Cornelis de Jager (b. 1921) at Utrecht. A 26-cm telescope observed early-type stars and the spectra were obtained by a 1200 gr/mm concave grating after collimating the light with an off-axis paraboloid. Three photomultiplier tubes were mounted behind 30-μm slits, so as to give 1.8 Å wavelength resolution. Spectra in three regions some 10.0 nm wide and centred on about 210, 250 and

Table 6.2. *Table of photomultipliers on the Copernicus OAO-3 satellite*

Photomultipliers	Grating order	Photocathode	Slit (μm)	Disp. (Å/mm)	λ range (nm)	Res. (Å)
U1	2	KBr	23	2.1	71–150	0.05
V1	1	bialkali + MgF$_2$ window	24	4.2	164–318.5	0.10
U2	2	KBr	98	2.1	75–164.5	0.20
V2	1	bialkali + MgF$_2$ window	96	4.2	148–327.5	0.40

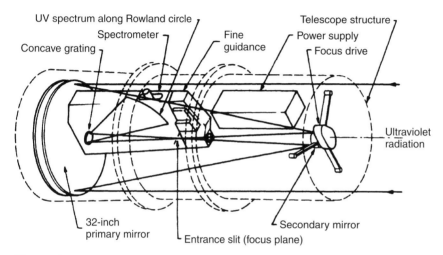

Figure 6.22. The Princeton telescope and high resolution ultraviolet spectrometer on the OAO-3 Copernicus satellite, launched August 1972.

280 nm were recorded. The last of these three was able to include the strong ionized magnesium h and k absorption lines at 279.6 and 280.3 nm (analogues of the longer wavelength Ca II H and K lines), and the associated magnesium emission lines, which are indicative of stellar chromospheric activity [58]. The instrument is seen in Fig. 6.23.

6.2.5   The International Ultraviolet Explorer

The International Ultraviolet Explorer was one of the most successful space astronomy missions ever flown. IUE was originally devised in the 1960s as a UK and ESRO project, largely at the initiative of Robert Wilson (1927–2002) at University College, London. In 1971, NASA formally became the major partner for the project. High and low resolution ultraviolet spectroscopy was the goal, but two key aspects of the design distinguished IUE from the OAO and TD-1 projects.

Firstly, échelle gratings were used, giving a compact two-dimensional format which imaged onto vidicon cameras, giving a huge multiplexing advantage over single-channel photomultiplier scanning. Secondly, a geosynchronous orbit was selected facilitating continuous interactive real-time control of the observatory and assessment of the data by the observer on the ground.

IUE had a 45-cm Cassegrain telescope with Ritchey–Chrétien optics (see Fig. 6.24). Two entirely separate échelle spectrographs were installed, one for the short wavelength range (115–195 nm), the other for the long wavelength range (190–320 nm) [59]. Collimation was by means of an off-axis paraboloid mirror. These were all-reflecting spectrographs, and hence a concave grating after the échelle was chosen to give order separation (using the grating in first order) as well as to image the two-dimensional échelle format

Figure 6.23. The S59 spectrometer on the TD-1 satellite was devised by C. de Jager at Utrecht. The 26-cm telescope primary mirror is hidden from view, but the spherical secondary is mounted on the tripod. The concave grating is in the assembly at upper left.

Table 6.3. *Table of IUE spectrograph characteristics*

	SWR spectrograph	LWR spectrograph
échelle (gr/mm)	101.9	63.2
échelle blaze angle	45.5°	48.1°
concave grating (gr/mm)	313	200
radius of curvature (m)	1.37	1.37
resolving power (high res.)	12 000	13 000
resolving power (low res.)	260	320
wavelength range (nm)	115–195	190–320

onto the detector. For the low resolution mode, the simple device of inserting a plane mirror in front of the échelle was adopted. Then the concave grating was the sole dispersing element. For the long wavelength spectrograph, the second order from the cross-dispersion grating was suppressed by using silica coatings on the collimator and 45° mirrors (the latter directed the beam into the spectrograph). Table 6.3 gives some of the main characteristics of the IUE spectrograph. Fig. 6.25 shows the IUE short wavelength spectrograph.

IUE was of course devised over a decade before CCD detectors became available, at a time that the SEC (secondary electron conduction) vidicon was being used in ground-based astronomy as a two-dimensional television-type of detector with a linear response. It was therefore the detector of choice, given the technology available in the early 1970s. To give the cameras ultraviolet sensitivity, image converters from ultraviolet to the visible were however necessary, and these were coupled to the bialkali vidicon photocathode through a fibre optics array. The readout system of the vidicon scans

the image stored on the potassium chloride target, giving in effect 768 pixels across the image diameter with 37-μm resolution.

The IUE satellite was launched in January 1978 with the goal of a five-year lifetime. In the event, it operated for an amazing 18 years, with operations being finally terminated in September 1996 as a result of gyroscope failure. By that time 104 000 spectra had been recorded. The achievements of IUE in ultraviolet spectroscopy are too numerous to cover in this historical summary. Needless to say, it revolutionized ultraviolet spectroscopy. Blair Savage has noted some of the many highlights as: studies of stellar mass loss phenomena across the Hertzsprung–Russell diagram, studies of the chromospheres of cool stars, studies of active galactic nuclei and their variability, studies of the spectral evolution of novae and supernovae, studies of the interstellar medium, and the discovery of a hot component of the interstellar gas from the interstellar C IV (155 nm) line, and studies of the composition of comets [60].

6.2.6 Ultraviolet spectroscopy from manned spacecraft

Soon after the first rocket spectra were recorded in the ultraviolet, several programmes for ultraviolet spectroscopy from manned spacecraft were undertaken using objective prism or objective grating cameras recording onto film.

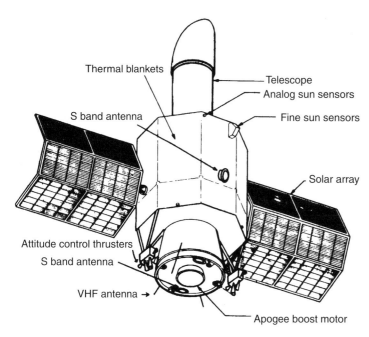

Figure 6.24. The International Ultraviolet Explorer satellite, launched January 1978.

The first such experiment was in July 1966 on board Gemini X. An objective reflection grating camera with an ultraviolet-transmitting f/3.3 lens was used by astronauts John Young and Michael Collins [61]. The camera was mounted on a bracket on the outside of the spacecraft and operated by the astronauts during extravehicular activity (i.e. a spacewalk). However, no lines were seen in the spectra of the bright stars recorded, as a result of image motion in the dispersion direction.

The experiments were repeated with objective grating (183 Å/mm) and objective prism (1400 Å/mm at 250 nm) cameras on board Gemini XI in September 1966 and Gemini XII, with Karl Henize from Northwestern University as the principal investigator. The camera had a quartz–lithium fluoride lens able to transmit to 210 nm. The 10° objective prism was quartz [62, 63]. An objective grating spectrum of Canopus taken by astronauts Charles Conrad and Richard Gordon in September 1966 with a resolution of about 15 Å recorded the Mg II h and k doublet near 280 nm for the first time in any astronomical spectrum other than that of the Sun.

The objective grating spectra of nine bright O and B stars in Orion obtained from Gemini XI

were used for spectrophotometric measurements in the 260–300 nm range by T. H. Morgan et al. [64]. The continuous fluxes in the ultraviolet were in agreement with other observations and with fluxes computed from model atmospheres.

The Soviets also undertook ultraviolet photographic spectroscopy on manned space stations. The first such experiments were in June 1971 on the Salyut station, and a slitless grating camera was used to record spectra in the 200–380 nm range. The 28-cm telescope was of the Mersenne type, in which a convex secondary produced a collimated beam. This was sent to a concave reflection grating and 32 Å/mm spectra of 5-Å resolution were then recorded on film [65]. Figure 6.26 shows the instrument used. The space crew were responsible for pointing the telescope and commencing an exposure, as well as for bringing exposed film cassettes back to Earth. The continuous spectra of bright stars were produced from subsequent microdensitometer tracings. Some strong lines were also visible in these spectra, including the Mg II doublet near 280 nm in the spectrum of Vega [66].

The ultraviolet experiments by Grigor Gurzadyan (b. 1922) and J. B. Ohanesyan were continued in

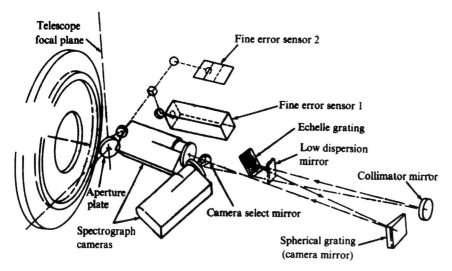

Figure 6.25. The IUE short wavelength spectrograph.

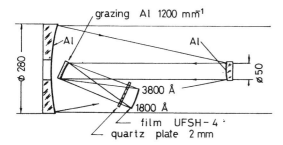

Figure 6.26. Slitless ultraviolet grating spectrograph 'Orion' on the Salyut space station, 1971.

Figure 6.27. Ultraviolet objective prism spectra photographed on the Soyuz-13 spacecraft near the star Capella in 1973. The spectra extend from 500 down to 200 nm and are brighter at the longer wavelength end. The limiting magnitude is about 13.

December 1973, using cosmonauts on board Soyuz-13. A new version of the camera known as Orion-2 was employed. This was a 4° quartz objective prism spectrograph on a 24-cm Cassegrain telescope and was used for obtaining spectra in the 200–500 nm range. It allowed spectra of a large number of faint stars (to a magnitude of about 12–13) to be recorded with a resolution of about 8 Å at 200 nm (see [67] for a summary and list of publications arising from this programme). Several lines between 200 and 300 nm were visible in the tracings of the brighter stars, as discussed by Gurzadyan in 1974 [68]. Figure 6.27 shows a sample of the photographic recording of a large number of low resolution ultraviolet objective prism spectra from Soyuz-13.

An objective prism spectrograph (known as the S019 spectrograph) was also flown on all three missions on the international Skylab space station. In this case, the wavelength extended down to 130 nm by using a 4° calcium fluoride prism [69]. The prism was mounted on a 15-cm Ritchey–Chrétien Cassegrain telescope. Magnesium fluoride coatings were applied to the telescope mirrors, while a two-lens field flattener was from calcium fluoride and lithium fluoride. The wavelength resolution of this system was 2 Å at 140 nm, and 12 Å at 200 nm.

REFERENCES

[1] Draper, H., *Amer. J. Sci. Arts* (3) **18**, 419 (1879)

[2] Huggins, W., *Proc. R. Soc.* **25**, 445 (1876)

[3] Huggins, W., *Phil. Trans. R. Soc.* **171**, 669 (1880)

[4] Vogel, H. W., *Monatsber. der k. Preuss. Akad. der Wissenschaften zu Berlin*, p. 586 (1879) and p. 192 (1880). See also *ibid. Astron. Nachrichten* **96**, 327 (1880)

[5] Huggins, W., *Astrophys. J.* **1**, 359 (1895)

[6] Huggins, W., *Proc. R. Soc.* **48**, 216 (1890)

[7] Fowler, A. and Strutt, R. J., *Proc. R. Soc.* **93**, 577 (1917)

[8] Hartmann, J., *Zeitschr. für Instr. Kunde* **25**, 161 (1905)

[9] Palmer, H. K., *Astrophys. J.* **18**, 218 (1903)

[10] Palmer, H. K., *Lick Observ. Bull.* **2**, 46 (1902)

[11] Wright, W. H., *Lick Observ. Bull.* **9**, 52 (no. 291) (1917)

[12] Plaskett, J. S., *Pop. Astron.* **31**, 20 (1923)

[13] Hartmann, J., *Zeitschr. für Instr. Kunde* **24**, 257 (1904)

[14] Rayton, W. B., *Astrophys. J.* **72**, 59 (1930)

[15] Humason, M. L., *Astrophys. J.* **71**, 351 (1930)

[16] Bracey, R. J., *Astrophys. J.* **83**, 179 (1936)

[17] Cojan, J., *Ann. d'Astrophys.* **10**, 33 (1947)

[18] Mayall, N. U., *Publ. astron. Soc. Pacific* **48**, 14 (1936)

[19] Mayall, N. U. and Wyse, A. B., *Publ. astron. Soc. Pacific* **53**, 120 (1941)

[20] Struve, O., *Astrophys. J.* **86**, 613 (1937)

[21] Greenstein, J. L. and Henyey, L. G., *Astrophys. J.* **86**, 620 (1937)

[22] Struve, O., van Biesbroeck, G. and Elvey, C. T., *Astrophys. J.* **87**, 559 (1938)

[23] Williams, R. C. and Sabine, G. B., *Astrophys. J.* **77**, 316 (1933)

[24] Boothroyd, S. L., *Astrophys. J.* **80**, 1 (1934)

[25] Williams, R. C., *Astrophys. J.* **80**, 7 (1934)

[26] Arnulf, A., *Ann. d'Astrophys.* **6**, 21 (1943)

[27] Bigay, J., *Ann. d'Astrophys.* **13**, 72 (1950)

[28] Lemaître, G., *Astron. & Astrophys.* **103**, L14 (1981). See also *Astrophys. & Space Sci. Library* **92**, 137 (1980). Intl. Astron. Union Coll. 67: Instrumentation for Astronomy with large optical telescopes, ed. C. M. Humphries, Dordrecht: Reidel

[29] Richardson, E. H., *Astrophys. & Space Sci. Library* **92**, 129 (1980). Intl. Astron. Union Coll. 67: Instrumentation for Astronomy with large optical telescopes, ed. C. M. Humphries, Dordrecht: Reidel

[30] Wilson, R. and Boksenberg, A., *Ann. Rev. Astron. Astrophys.* **7**, 421 (1969)

[31] Schoen, A. L. and Hodges, E. S., *J. Opt. Soc. Amer.* **40**, 23 (1950)

[32] Durand, E., Oberly, J. J. and Tousey, R., *Astrophys. J.* **109**, 1 (1949)

[33] Hopfield, J. J. and Clearman, H. E., *Phys. Rev.* **73**, 877 (1948)

[34] Byram, E. T., Chubb, T. A., Friedman, H. and Kupperian, J. E., *Astron. J.* **62**, 9 (1957)

[35] Stecher, T. P. and Milligan, J. E., *Astrophys. J.* **136**, 1 (1962)

[36] Hoyle, F. and Wickramasinghe, N. C., *Mon. Not. R. astron. Soc.* **124**, 417 (1962)

[37] Stecher, T. P. and Donn, B., *Astrophys. J.* **142**, 1681 (1965)

[38] Stecher, T. P., *Astrophys. J.* **142**, 1683 (1965)

[39] Morton, D. C. and Spitzer, L., *Astrophys. J.* **144**, 1 (1966)

[40] Morton, D. C., *Astrophys. J.* **147**, 1017 (1967)

[41] Carruthers, G. R., *Astrophys. J.* **151**, 269 (1968)

[42] Carruthers, G. R., *Astrophys. J.* **161**, L81 (1970)

[43] Smith, A. M., *Astrophys. J.* **156**, 93 (1969)

[44] Smith, A. M., *Astrophys. J.* **160**, 595 (1970)

[45] Morton, D. C., Jenkins, E. B. and Bohlin, R. C., *Astrophys. J.* **154**, 661 (1968)

[46] Stecher, T. P., *Astrophys. J.* **159**, 543 (1970)

[47] Stuart, F. E., *Astrophys. J.* **157**, 1255 (1969)

[48] Boggess, A. and Kondo, Y., *Astrophys. J.* **151**, L5 (1968)

[49] Kondo, Y., Guili, R. T., Modisette, J. L. and Rydgren, A. E., *Astrophys. J.* **176**, 153 (1972)

[50] Kondo, Y., Morgan, T. H. and Modisette, J. L., *Astrophys. J.* **207**, 167 (1976)

[51] Hoekstra, R., Kamperman, T. M., Wells, C. W. and Werner, W., *Appl. Optics* **17**, 604 (1978)

[52] Code, A. D., *Publ. astron. Soc. Pacific* **81**, 475 (1969)

[53] Code, A. D., Houck, T. E., McNall, J. F., Bless, R. C. Lillie, C. F., *Astrophys. J.* **161**, 377 (1970)

[54] Rogerson, J. B., Spitzer, L., Drake, J. F. *et al.*, *Astrophys. J.* **181**, L97 (1973)

[55] Morton, D. C., Drake, J. F., Jenkins, E. B. *et al.*, *Astrophys. J.* **181**, L103 (1973)

[56] Boksenberg, A., Evans, R.G., Fowler, R.G. *et al.*, *Mon. Not. R. astron. Soc.* **163**, 291 (1973)

[57] Jamer, C., Macau-Hercot, D., Monfils, A. *et al.*, *UV Bright-star Spectrophotometric Catalogue*, ESA-SR-27, Paris (1976)

[58] Kamperman, T. M., van der Hucht, K. A., Lamers, H. J. and Hoekstra, R., *Sky and Tel.* **45**, 85 (Feb. 1973)

[59] Boggess, A., Carr, F. A., Evans, D. C. *et al.*, *Nature* **275**, 372 (1978)

[60] Savage, B. D., in *The Century of Space Science*, ed. J. Bleeker, J. Geiss and M. Huber, vol. 1, chap. 13, p. 287, Dordrecht: Kluwer Academic Publishers (2001)

[61] Henize, K. G. and Wackerling, L. R., *Sky and Tel.* **32**, 204 (Oct. 1966)

[62] Henize, K. G., Wray, J. D. and Wackerling, L. R., *Bull. Astron. Inst. Czechoslovakia* **19**, 279 (1968)

[63] Henize, K. G., Wackerling, L. R. and O'Callaghan, F. G., *Science* **155**, 1407 (1967)

[64] Morgan, T. H., Spear, G. G., Kondo, Y. and Henize, K. G., *Astrophys. J.* **197**, 371 (1975)

[65] Gurzadyan, G. A. and Ohanesyan, J. B., *Space Sci. Rev.* **13**, 647 (1972)

[66] Gurzadyan, G. A. and Ohanesyan, J. B., *Astron. & Astrophys.* **20**, 321 (1972)

[67] Gurzadyan, G. A., *Trans. Int. Astron. Union* **16A** (Part 3), 217 (1976)

[68] Gurzadyan, G. A., *Sky and Tel.* **48**, 213 (Oct. 1974)

[69] O'Callaghan, F. G., Henize, K. G. and Wray, J. D., *Appl. Optics* **16**, 973 (1977)

7 · Multi-object spectrographs

7.1 LOW RESOLUTION MULTI-OBJECT SPECTROSCOPY[1]

For almost a century since the objective prism spectrograph was first used at Harvard in 1885, this was the only solution available to astronomers for multi-object spectroscopy – the simultaneous recording of the spectra of more than one object. All other astronomical spectroscopy was undertaken sequentially, one object at a time. The limitations of the objective prism caused by overlapping spectra and sky brightness, as well as variable resolving power (dependent on the seeing) have already been discussed.

From about 1980, two new solutions for multi-object spectroscopy, which overcame some of the problems of the objective prism, were introduced. One is the aperture plate combined with a low dispersion element, which is often a combined prism and transmission grating (or 'grism' – see Section 2.8). The other is the use of optical fibres. The early history (roughly the first decade) of both these techniques is described here.

7.2 APERTURE PLATE MULTI-OBJECT SPECTROSCOPY

The aperture plate was introduced by Harvey Butcher (b. 1947) at Kitt Peak in 1980 [3] as a way of undertaking multi-object spectroscopy of faint objects, all of which lay within the 5 arc minute field of view at the Cassegrain focus of the 4-m Mayall telescope. The aperture plate is a black anodized aluminium sheet in the focal plane, with holes drilled in it, typically with a diameter corresponding to 2.5 arc seconds on the sky.

The plate eliminates the sky background except at the apertures, whose locations correspond to the precise positions of galaxies in a cluster. The Kitt Peak aperture plate system used an achromatic doublet lens as collimator, a grism and the so-called cryogenic camera, which was an 800×800-pixel CCD with a fast f/1 Schmidt camera – see Fig. 7.1. Spectra as faint as $m_V \sim 22$ were possible with this system, at resolving power of 200 to 400. The apertures were variously for objects or for adjacent sky spectra, and rotation of the whole instrument about its main optical axis allowed selection of a position angle that minimized spectrum overlaps. Up to 50 apertures per plate were possible. Figure 7.2 shows a sample aperture plate used on this system.

The disadvantage of the aperture plate technique is the need to prepare plates in advance, based on the precisely known coordinates of objects in the field of view. Also, the spectra are not all in register on the detector in relation to their wavelengths. On the other hand, the efficiency of the system is high, allowing faint objects to be reached. A key optical element of the system is the grism, which is an optical transmission grating cemented to one face of a thin prism so as to give a 'direct vision' device in which both components of the grism contribute to the dispersion. The concept of a grism was first proposed by Bowen and Vaughan in 1973 [4].

As an example of the use of the multi-aperture plate system, the spectrophotometry of the Crab nebula filaments in 1981 by Richard Henry and others using the aperture plate instrument at Kitt Peak can be cited [5]. Here an aperture plate with 14 slits, each of 4 arc seconds, was made for recording simultaneous spectra of gaseous filaments in the Crab nebula at a resolution of 2.5 nm, in order to study physical properties and structure within the supernova remnant. Another example showing spectra of a number of faint galaxies in a cluster, recorded by Butcher, is shown in Fig. 7.3.

---

[1] This section made considerable use of review articles by Fred Watson on 'Multi-object fibre spectroscopy' [1] and by John Hill on 'The history of multi-object fibre spectroscopy' [2].

night as direct imaging (used for finding the coordinates of the objects to be observed). This limitation was overcome at the CFHT (Canada–France–Hawaii telescope) Cassegrain focus grism spectrograph and camera [8]. Here a direct CCD image was processed immediately after readout to provide data for the automated aperture plate fabrication, based on the positions of the objects in the $5' \times 7'$ field selected for spectroscopy. The apertures were punched in the plate within two minutes and multi-object spectroscopy with some 50 spectra using a grism could commence within ten minutes of the direct image readout. No a-priori knowledge of the objects' coordinates was required. Early tests with this system produced over 300 spectra of faint objects (to $m_R \sim 21$) in $3\frac{1}{2}$ nights. The same spectra would have taken 60 nights if observed sequentially, which represents an impressive 18-fold gain in efficiency.

A similar aperture plate system was implemented at the European Southern Observatory with EFOSC (the ESO Faint Object Spectrograph and Camera) – [9] see Section 8.2. In spectrographic mode, this is a Cassegrain grism instrument with an f/2.5 camera designed for low resolution spectroscopy on the 3.6-m telescope. EFOSC was adapted to multi-aperture spectroscopy with rapid mask fabrication in 1986 [10]. Galaxies in clusters to about $m_B \sim 22$, with up to 28 spectra per exposure, were obtained with this system in its initial trials in 1986.

7.3 MULTI-OBJECT SPECTROSCOPY USING OPTICAL FIBRES AND APERTURE PLATES

The concept of using an optical fibre to transmit starlight from the focal plane of a telescope to a spectrograph goes back at least to the 1950s. The paper by Narinder Kapany (b. 1927), written as an appendix to John Strong's 1958 textbook on classical optics, proposed using fibres as a way of reformatting a stellar image into a linear slit, so as to act as an image slicer [11]. At this time Kapany was a graduate student at Imperial College, London. He is widely recognized as the father of fibre optics technology. He was born in India, and after obtaining a doctorate in London in 1955, he then worked in California for most of his professional career. Early experiments in glass

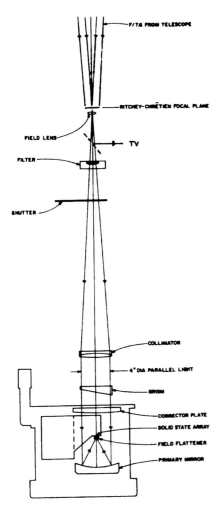

Figure 7.1. Optical diagram of Harvey Butcher's grism and aperture plate system at the Cassegrain focus of the Mayall 4-m telescope at Kitt Peak, 1982.

Other aperture plate systems followed that at Kitt Peak. A prime focus spectrograph and camera known as PFUEI (Prime Focus Universal Extragalactic Instrument) was built for the Hale 5-m telescope by James Gunn and James Westphal [6]. It used a transmission grating in spectrographic mode. The f/3.52 prime focus beam passed through a focal reducer so as to be at a fast focal ratio of f/1.4 at the CCD detector. The PFUEI was adapted for multi-object spectroscopy in 1982 [7].

One of the limitations of aperture plates is the time taken for plate fabrication for each new field. Initially this prevented spectroscopic observations on the same

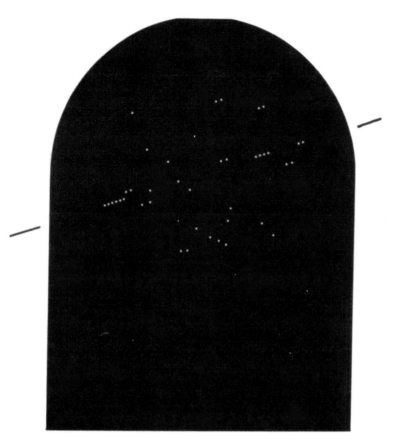

Figure 7.2. Shape and hole pattern of a typical aperture plate used at Kitt Peak. The direction of dispersion is perpendicular to a line indicated by the two tick marks.

fibre technology had been carried out by Hopkins and Kapany at Imperial College [12] and by van Heel in Delft [13].

However, it was not until low-loss fused-silica fibres became available in the early 1970s that their efficient use in astronomy became practicable. One suggestion by Roger Angel (b. 1941) for building large telescopes was to link a number of small telescopes by optical fibres [14]. In 1978, Hubbard, Angel and Gresham undertook the first experiments in astronomical spectroscopy using a fibre-fed spectrograph [15]. An image tube spectrograph with a fast solid Schmidt camera was linked to the f/5 prime focus of the 2.3-m Steward Observatory telescope using 20 metres of fused-silica fibre – see Fig. 7.4. Spectra of the 15th magnitude Seyfert galaxy, NGC 4151, were obtained in five minutes' exposure in the initial tests, which is about a magnitude slower than when no fibre is used.

The Steward fibre-coupled spectrograph was followed a year later by the fibre-coupled spectrograph at the Lunar and Planetary Laboratory's Mt Lemmon Observatory in Arizona. This instrument was operated by William Heacox (b. 1942) for radial-velocity measurements [16].

These early experiments in fibre-fed spectroscopy at the University of Arizona led to the first instrument for multi-object spectroscopy. This was the Medusa spectrograph, implemented by John Hill and his colleagues on the 2.3-m (90-inch) telescope at Steward Observatory [17] – see Fig. 7.5. The aim was the simultaneous acquisition of the spectra of galaxies in a cluster, for studies of the dynamics within the cluster. For this early form of multi-object spectroscopy, an aperture plate in the Cassegrain focal plane was still used, with up to 32 holes drilled in it in the locations of the objects to be observed within a 30 arc minute

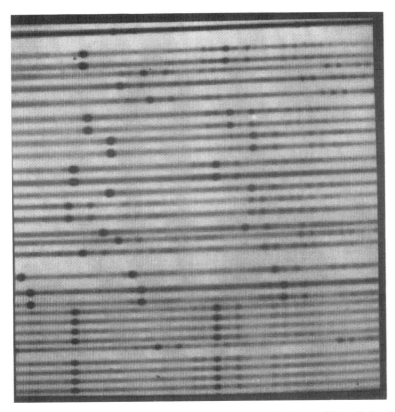

Figure 7.3. Portion of the spectra from aperture plate exposure of a cluster of faint galaxies observed by Harvey Butcher at the 4-m Mayall telescope at Kitt Peak in 1982. At the top, the bright spectrum is that of a plate alignment star. Short wavelengths are to the left. The wavelength range is about 500 or 600 nm. The shortest wavelength emission feature is [OII] 372.7 nm at a redshift of $z \sim 0.4$.

field [18]. The optical fibres used were of 300-μm core diameter, corresponding to 3 arc seconds on the sky. Once the aperture plate had been fabricated, fibres were inserted into the holes and held there with an epoxy glue. The fibres were short, and transmitted light to a linear slit, just 25 cm behind the f/9 Cassegrain focal plane, from where the light entered a normal spectrograph. Forty fibres were arranged in a linear format of length 25 mm at the spectrograph entrance, some of them being used for the helium-argon comparison lamp (and hence not coming from the aperture plate).

In the first test of Medusa in December 1979, Hill and his colleagues obtained low resolution image-tube spectra at 250 Å/mm in the galaxy cluster Abell 754, with eight measurable spectra. By March 1980 as many as 26 simultaneous galaxy spectra had been obtained in Abell 1904 – see Fig. 7.6. Further details of the Medusa spectrograph were presented by Hill and his

colleagues in 1982 [19], and results of its performance in observing galaxy clusters were published in the *Astronomical Journal* [20]. As discussed, Medusa was not as efficient as multi-object systems without fibres, because of focal ratio degradation in the fibres, other fibre losses such as reflection and absorption, and small positioning errors in the aperture plate. However, redshifts measured with Medusa were often better than with slit spectrographs, because guiding errors in the telescope still result in constant ray paths within the spectrograph, whereas slit spectrographs generate significant radial-velocity systematic errors if they are poorly guided.

The Medusa spectrograph can be considered as a prototype system that was soon followed by instruments that positioned the fibres in the focal plane automatically for each new field. The MX spectrometer at Steward Observatory was one such device [21]

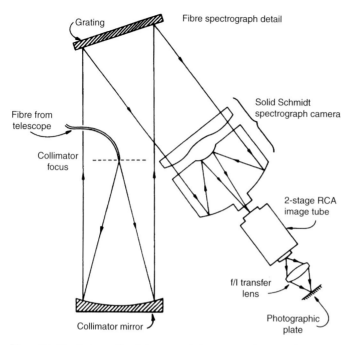

Figure 7.4. The first ever fibre-fed astronomical spectrograph was used by Hubbard, Angel and Gresham at the Steward Observatory 2.3-m telescope in 1978. A 20-m length of fused-silica fibre linked the prime focus of the telescope to the spectrograph on the observatory floor.

that was in operation from 1985 – see Section 7.4. Before then, however, other observatories were quick to follow with aperture-plate optical-fibre systems for multi-object spectroscopy. In particular, the Optopus system at the European Southern Observatory, which allowed for 50 independent 200-μm fibres, was developed for the 3.6-m telescope at La Silla [22]. Here the fibres were held in the aperture plate holes by magnetic connectors. Another instrument was developed for the Anglo–Australian 3.9-m telescope by Peter Gray, and known as FOCAP (Fibre optic coupled aperture plate) [23]. FOCAP operated at the f/8 Cassegrain focus and had first light in December 1981. Initially FOCAP had 25 fibres each 2 metres long and mounted in brass ferrules. An improved version, FOCAP2, was built early in 1983 with 50 fibres which had better performance with respect to focal ratio degradation, and a later development introduced 400-μm fibres (2.7 arc seconds on the sky) so as to admit more light, especially in poor seeing conditions.

Observations with FOCAP were recorded using the Image Photon Counting System (IPCS), an image-tube detector coupled to a television camera,

which enabled individual photon events to be counted and the image to be built up and displayed in real time during the exposure. During its six-year lifetime, FOCAP ensured that the productivity of the Anglo-Australian telescope was high, with the result that more faint galaxy and quasar spectra were recorded with this system during these years than with any other telescope–spectrograph combination [2]. An example of the data coming from FOCAP, in this case for the galaxy cluster IC 2082, is given in a paper by R. S. Ellis *et al.* [24].

One interesting development in multi-object spectroscopy with optical fibres linked to aperture plates was on the UK Schmidt telescope, also on Siding Spring (as part of the Anglo-Australian Observatory). Here Fred Watson developed a fibre system in 1986 known as FLAIR (Fibre-linked array image reformatter), with a 6.6-degree field on the Schmidt telescope for multi-object spectroscopy with up to 40 fibres [25, 26]. The FLAIR aperture plate was simply a photographic positive plate on 1-mm thick glass; the galaxies were therefore transparent images on a dark background. The fibres were glued to the emulsion side,

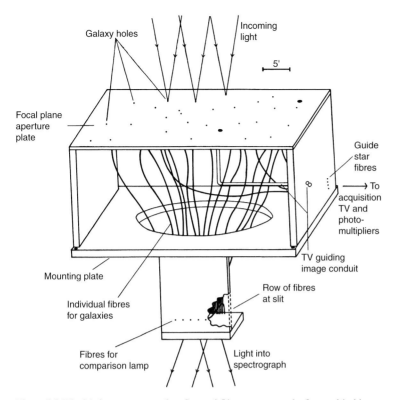

Figure 7.5. The Medusa spectrograph at Steward Observatory was the first multi-object spectrograph in 1979. The diagram shows the 32 optical fibres that link the aperture plate with the slit plane of the spectrograph.

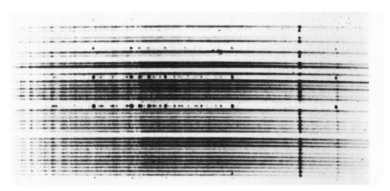

Figure 7.6. Twenty-six measurable spectra of galaxies in the cluster Abell 1904 recorded in March 1980 at the Steward Observatory by John Hill and his colleagues. This is the first successful multi-object spectrograph observation using optical fibres. The H and K lines and the G band are plainly visible.

and the plate was held in a normal Schmidt photographic plate holder. This technique ensured a good registration between the fibre inputs and the images in the curved focal plane of the telescope. The fibres in this system were only 40 μm in diameter (equivalent to 2.7 arc seconds). The bench-mounted spectrograph for this system was somewhat unusual, having a very

fast f/1.7 collimator lens ($f_{coll} = 50$ mm) and a slightly slower refracting camera ($f_{cam} = 85$ mm). A rotatable plane grating gave a dispersion of either 95 or 190 Å/mm, and a resolution of $\Delta\lambda \geq 5.8$ Å.

Another multi-object system on a Schmidt telescope was built for the 1-m Kvistaberg telescope in Uppsala, Sweden. Here each fibre input was terminated by a button carrying a small 45° mirror. The buttons were mounted on a glass plate which in turn was at the telescope's focal plane [27]. The 48 fibres were mounted on the glass plate after matching the object positions with those from a direct photographic exposure.

The problems of matching fibre entrances to telescope focal planes and fibre exits to spectrographs have been discussed by John Hill, Roger Angel and Harvey Richardson, and by many others since [28]. Focal ratio degradation represents a loss of efficiency (measured by the resolving power–throughput product) of any fibre-fed spectrograph. It can be minimized by using fast fibre inputs, at about f/3, which corresponds to the prime focus of many telescopes. For Cassegrain fibre feeds at say f/9, a microlens can be cemented to the fibre input to achieve a faster effective focal ratio.

A beam of fast focal ratio emerges from the fibre in an even larger cone as a result of focal ratio degradation. If an existing Cassegrain spectrograph is to be matched to the light emerging from the fibre, then a microlens can be used here too to adjust the focal ratio to the collimator available [28]. An example of a fibre system for multi-object spectroscopy using microlenses was designed for the Isaac Newton telescope on La Palma [29]. This had an f/3.3 prime focus input without microlenses, and the output went to an f/15 spectrograph after passing through a microlens on each fibre. A similar system with microlenses was implemented for the William Herschel telescope. The microlens of course magnifies the fibre exit as seen by the spectrograph, and this in turn affects the resolving power if the fibre diameter acts as a slit.

7.4 MULTI-OBJECT SPECTROSCOPY WITH AUTOMATED FIBRE POSITIONING

The problems of aperture plate systems with fibres for multi-object spectroscopy were well known, even in the early 1980s, when aperture plate systems were first being used. They included the time-consuming preparation of the plates, the lack of flexibility while observing, the imperfect positioning of the fibres on a plate and the inability to tweak fibre positions on a plate in real time to maximize throughput. Moreover,

the operational costs of an aperture plate system were not negligible (see for example [30]).

In order to overcome these drawbacks, multi-object spectroscopy systems using fibres were developed in which the fibres were positioned automatically by computer-controlled robotic actions in real time immediately prior to each observation of a new field. As early as February 1981, a prototype automated fibre positioner was tested at the prime focus of the Hale 5-metre telescope [31]. This only had two fibres, in order to test the principle of the system. Soon thereafter, John Hill at Steward Observatory began the development of an automated fibre system, which was ultimately to be a successor to Medusa. This new system was known as the MX spectrometer (because of its multiplexing capability) [21], which is shown in Fig. 7.7.

The first discussion of the design concept for MX was in early 1982 [19]. Its first use on the 2.3-m telescope at Steward was in 1985. The basic feature of MX was 32 fibre probes arranged in a 'fishermen around the pond' configuration – see Fig. 7.8. Each probe could be positioned within a sector of the 45-arc minute field to within a few tenths of an arc second. Each probe was positioned by two microprocessor-controlled stepper motors, one controlling the fibre's radial position, the other controlling its angle about a pivot point just outside the field of view. Each fibre probe carried two 200-μm diameter fibres, one for an object and the other for the neighbouring sky. When observing a new field, the fibre probes could be accurately positioned on the 32 objects in the field within 90 seconds, thus greatly enhancing observing efficiency relative to aperture plate systems, such as Medusa.

The MX spectrometer used fibres about 30 cm in length, and these carried the light to a linear slit at the entrance to the spectrograph mounted behind the Cassegrain focus of the 2.3-m Steward telescope. One clear benefit of an automated system such as MX is its ability to tweak individual fibre positions for maximum light throughput, thereby overcoming any small positioning errors arising from either the coordinates supplied, or from the setting precision.

Hill and Lesser [21] discuss the minimum and maximum number of fibres useful or feasible in multi-object spectrographs. For a marginal gain in the observing efficiency from the multiplexing advantage, at least

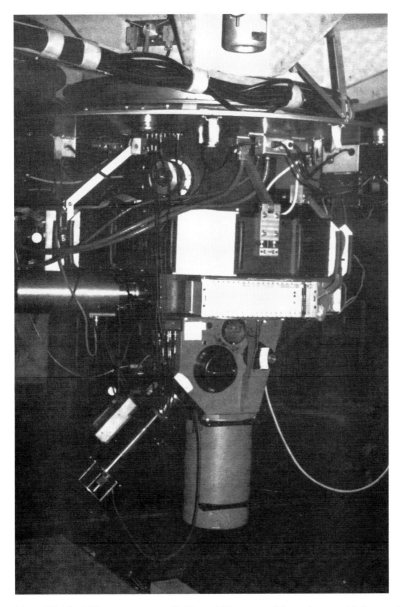

Figure 7.7. The MX spectrometer on the Steward Observatory 2.3-m telescope with Cassegrain spectrograph and CCD detector.

six objects need to be observed. The maximum number of fibres is set by the limited space in the telescope's focal plane, the space available at the entrance slit of the spectrograph, or the area available on the detector to accept spectra. It may also be limited by the density of faint galaxies on the sky (of the order of 10^3/square degree), though instrumental factors are more likely to dominate. Fibre-optic multi-object

spectroscopy systems can accept more spectra on the available CCD area, as the spectra are aligned for maximum packing efficiency on the area available. Aperture plate systems using slits but no fibres do not make such efficient use of the CCD area, as the spectra are not aligned. These therefore have fewer objects in a given field, and are generally limited to a smaller field of view, but they have the advantage of a somewhat fainter

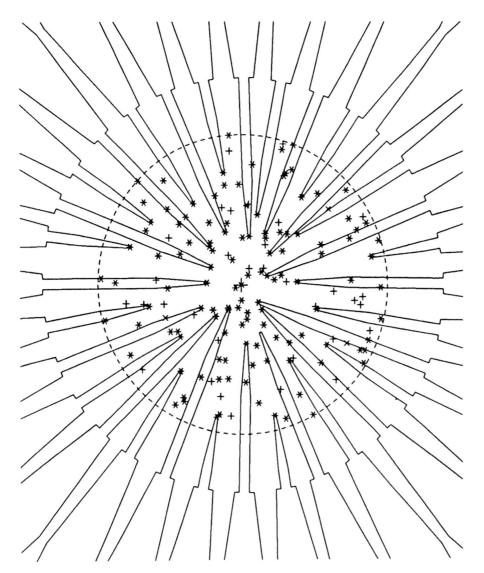

Figure 7.8. MX spectrometer probes in the focal plane of the Steward Observatory 2.3-m telescope arranged on a target pattern in the cluster Abell 2634. The field of view is 45 arc minutes, which corresponds to a diameter of 28 cm.

limiting magnitude. T. E. Ingerson also considered the optimal number of fibres for multi-object spectroscopy, and noted that the benefit of more fibres is often not proportional to their number, especially if the additional objects are for purely statistical purposes [32].

Another solution to automated fibre positioning for multi-object spectroscopy was devised in a joint project of the Anglo-Australian Observatory and Durham University (UK) by Ian Parry and Peter Gray.

The instrument was Autofib, and designed as a replacement for FOCAP on the Anglo-Australian telescope [30]. The philosophy behind Autofib made an interesting contrast with that of MX. Autofib was a single robotic fibre-probe positioner mounted on an x-y carriage. It picked up fibre probes sequentially (one at a time) from just outside the field of view, using a magnetic clamp, and placed them down on a flat steel plate in the Cassegrain focal plane of the telescope, where they were held in position by a tiny magnet on each probe. A small prism cemented to each fibre entrance reflected light from the telescope through 90° into the fibre.

Sixty-four 300-μm (2.0″) fibres were available for objects or sky. Fibre placement was to a precision of about 20 μm ($\simeq 0.15''$), which was about twice the precision achieved with FOCAP, and the fibres were also better aligned with the telescope axis. Because Autofib is a sequential robotic fibre positioner, it took just under ten minutes to position all the fibres, which is longer than MX, a disadvantage offset by the greater number of object fibres available.

Autofib was commissioned on the Anglo–Australian telescope in February 1987. A second generation of the instrument was built for the 4.2-m William Herschel telescope on La Palma by late 1988 [33].

It is not proposed to review other robotic multi-object spectroscopy fibre systems here in any detail. They included Argus on the 4-m Cerro Tololo telescope in Chile [32], Medisis on the 3.6-m Canada–France–Hawaii telescope [34], Decaspec on the 2.4-m McGraw-Hill telescope [35], the Lick Observatory 3-m telescope multi-object spectroscopy system [36] and Spider, a system for the 60-cm and 5-degree field of view Schmidt telescope at the Beijing Astronomical Observatory [37]. The first three of these instruments were all based on the MX principle of individually and simultaneously positioned fibres. On the other hand, the Lick Observatory instrument and Beijing's Spider used the Autofib principle of sequential positioning by a single robot.

A useful review of fibre-fed multi-object spectrographs has been written by Ian Parry in 1997 [38]. Some notable instruments built or in the planning stages after 1990 are referenced there (such as the Anglo–Australian 2dF instrument (see Section 8.6), and the Sloan Digital Sky Survey instrument (see Section 8.9), which have respectively 400 and 600 fibres. It is noted that by the end of 1990, Parry lists 18 multi-object spectroscopy instruments that had been put into operation at major observatories, seven in the United States, four in Australia, three in Chile, three in western Europe (including two at La Palma) and one in China. Of these 18, eight used aperture plates, four used robotic fibre positioners of the MX type, while six were based on the Autofib concept in which a single robot positioned all fibres sequentially. At least

one system (Medisis on the Canada–France–Hawaii telescope, noted above) is not in Parry's table.

7.5 SPECTROSCOPY WITH AN INTEGRAL FIELD UNIT

A somewhat different concept in multi-object spectroscopy using optical fibres is that of the integral field unit. Here the fibres are not positioned on individual objects in the field of view, but are instead packed together so that multiple spectra are obtained from a closely packed array of points in an extended object. When fibres are covering a field, the device is termed an integral field unit, and such an array was built by C. Vanderriest at the Meudon Observatory near Paris in 1980 [39]. It comprised 169 fibres of 100-μm core diameter in a hexagonal array, while an additional 36 fibres were used for obtaining spectra of the nearby sky. The fibre exits were placed into a linear slit 23 mm long.

In an integral field unit of this type, about 25 per cent of the light is lost in the interstitial space between the circular fibres. One way to overcome this is to use an array of square and contiguous microlenses, each one imaging the entrance pupil of the telescope and accepting about 1″ × 1″ of the field [40]. If the array is limited to 4 × 4 lenslets, then 16 spectra over a limited field of view can be recorded without overlaps on the detector. No fibres are used. However, if each lens feeds light into an optical fibre, then the array can be larger; the fibres can then be placed into a linear slit to feed a spectrograph [40].

REFERENCES

[1] Watson, F. G., *J. Brit. Astron. Assoc.* **93**, 193 (1983)

[2] Hill, J., *Astron. Soc. Pacific Conf. Ser.* **3**, 77 (1988) ed. S. C. Barden

[3] Butcher, H., *Proc. Soc. Photo-instrumentation Engineers (SPIE)* **331**, 296 (1982)

[4] Bowen, I. S. and Vaughan, A. H., *Publ. astron. Soc. Pacific* **85**, 174 (1973)

[5] Henry, R. B. C., MacAlpine, G. M. and Kirshner, R. P., *Astrophys. J.* **278**, 619 (1984)

[6] Gunn, J. E. and Westphal, J. A., in 'Solid state imagers for astronomy', ed. J. C. Geary and D. W. Latham, *Proc. Soc. Photo-instrumentation Engineers (SPIE)* **290**, 16 (1981)

[7] Dresler, A. and Gunn, J. E., *Astrophys. J.* **270**, 7 (1983)

[8] Fort, B., Mellier, Y., Picet, J. P., Rio, Y. and Lelievre, G., *Proc. Soc. Photo-instrumentation Engineers (SPIE)* **627**, 321 (1986)

[9] Buzzoni, B., Delabre, B., Dekker, H. *et al.*, *European South. Observ. Messenger* **38**, 9 (1984)

[10] Dupin, J. P., Fort, B., Mellier, Y. *et al.*, *European South. Observ. Messenger* **47**, 55 (1987)

[11] Kapany, N. S., Appendix N, p. 553 in J. Strong, *Concepts of Classical Optics*, San Francisco: W. H. Freeman & Co. (1958)

[12] Hopkins, H. H. and Kapany, N. S., *Nature* **173**, 39 (1954)

[13] van Heel, A. C. S., *Nature* **173**, 39 (1954)

[14] Angel, J. R. P., Adams, M. T., Boroson, T. A. and Moore, R. L., *Astrophys. J.* **218**, 776 (1977)

[15] Hubbard, E. N., Angel, J. R. P. and Gresham, M. S., *Astrophys. J.* **229**, 1074 (1979)

[16] Heacox, W. D., in *Optical and Infrared Telescopes for the 1990s*, ed. A. Hewitt, vol. II, p. 702, Kitt Peak Nat. Observ., Tucson (1980)

[17] Hill, J. M., Angel, J. R. P., Lindley, D., Scott, J. and Hintzen, P., in *Optical and Infrared Telescopes for the 1990s*, ed. A. Hewitt, p. 370, Kitt Peak Nat. Observ., Tucson (1980)

[18] Hill, J. M., Angel, J. R. P., Lindley, D., Scott, J. and Hintzen, P., *Astrophys. J.* **242**, L69 (1980)

[19] Hill, J. M., Angel, J. R. P., Lindley, D., Scott, J. and Hintzen, P., in 'Instrumentation in astronomy IV', ed. D. L. Crawford, *Proc. Soc. Photo-instrumentation Engineers (SPIE)* **331**, 279 (1982)

[20] Hintzen, P., Hill, J. M., Lindley, D., Scott, J. S. and Angel, J. R. P., *Astron. J.* **87**, 1656 (1982)

[21] Hill, J. M. and Lesser, M. P., *Proc. Soc. Photo-instrumentation Engineers (SPIE)* **627**, 303 (1986)

[22] Lund, G. and Enard, D., in 'Instrumentation in astronomy V', ed. A. Boksenberg and D. L. Crawford, *Proc. Soc. Photo-instrumentation Engineers (SPIE)* **445**, 65 (1984)

[23] Gray, P. M., in 'Instrumentation in astronomy V', ed. A. Boksenberg and D. L. Crawford, *Proc. Soc. Photo-instrumentation Engineers (SPIE)* **445**, 57 (1984)

[24] Ellis, R. S., Gray, P. M., Carter, D. and Godwin, J., *Mon. Not. R. astron. Soc.* **206**, 285 (1984)

[25] Watson, F. G., in 'Instrumentation in astronomy VI', *Proc. Soc. Photo-instrumentation Engineers (SPIE)* **627**, 787 (1986)

[26] Watson, F. G., *Astron. Soc. Pacific Conf. Ser.* **3**, 125 (1988)

[27] Pettersson, B., *Astron. Soc. Pacific Conf. Ser.* **3**, 133 (1988)

[28] Hill, J. M., Angel, J. R. P. and Richardson, E. H., in 'Instrumentation in astronomy V', ed. A. Boksenberg and D. L. Crawford, *Proc. Soc. Photo-instrumentation Engineers (SPIE)* **445**, 85 (1984)

[29] Powell, J. R., *Proc. Soc. Photo-instrumentation Engineers (SPIE)* **627**, 125 (1986)

[30] Parry, I. R. and Gray, P. M., *Proc. Soc. Photo-instrumentation Engineers (SPIE)* **627**, 118 (1986)

[31] Tubbs, E. F., Goss, W. C. and Cohen, J. G., in 'Instrumentation in astronomy IV', ed. D. L. Crawford, *Proc. Soc. Photo-instrumentation Engineers (SPIE)* **331**, 289 (1982)

[32] Ingerson, T. E., *Astron. Soc. Pacific Conf. Ser.* **3**, 99 (1988)

[33] Parry, I. R., *Astron. Soc. Pacific Conf. Ser.* **3**, 93 (1988)

[34] Felenbok, P., Guerin, J., Fernandez, A., Tournassoud, P. and Vaillant, R., *Astron. Soc. Pacific Conf. Ser.* **3**, 174 (1988)

[35] Fabricant, D. F., *Astron. Soc. Pacific Conf. Ser.* **3**, 170 (1988)

[36] Craig, W. W., Halley, C. J., Stewart, R. E., Blaedel, K. L. and Brodie, J. P., *Astron. Soc. Pacific Conf. Ser.* **3**, 153 (1988)

[37] Shunde, W., *Astron. Soc. Pacific Conf. Ser.* **3**, 183 (1988)

[38] Parry, I. R., in *Wide-field Spectroscopy*, ed. E. Kontizas, M. Kontizas, D. H. Morgan and G. P. Vettolani, p. 3, Dordrecht: Kluwer Academic Publishers (1997)

[39] Vanderriest, C., *Publ. astron. Soc. Pacific* **92**, 858 (1980)

[40] Courtès, G., *Astrophys. & Space Sci. Lib.* **92**, 123 (1982)

8 · Ten pioneering spectrographs of the late twentieth century

In this chapter, ten outstanding spectrographs which had their first light in the late twentieth or early twenty-first centuries are presented. The section covers five high resolution instruments, four at low resolution and one spectrograph with multiple capabilities that includes low, medium and high resolution modes. All were commissioned between 1984 and 2003 and they are presented in chronological order of their entry into service.

The choice of ten representative spectrographs from the several dozen commissioned over this two-decade period was a matter of some difficulty. The final choice was my own personal one. I wanted to show that spectrograph design has become a pursuit of great innovation, ingenuity and also complexity, at times verging towards a creative art-form involving cutting edge optical technology. In this period, huge developments were made in the use of optical fibres, larger and improved efficiency CCD detectors, the production of large mosaic gratings, the development of grisms and volume phase holographic gratings, the design and manufacture of multi-element dioptric cameras with specialized antireflection coatings, high reflection coatings on mirrors, and the extension of the wavelength range both down into the near ultraviolet at the atmospheric limit and also to the far red near one micrometre. Issues such as high wavelength stability and night sky subtraction have been addressed by designers of high resolution spectrographs, while at low resolution, an ever greater multi-object capability with rapid and automated fibre placements on a field plate have been the themes developed for the instruments presented here.

I would have liked to have included spectrographs not only covering a wide range of types and capabilities, but also with a wide geographical distribution of where they were developed and deployed and by whom. Unfortunately, some major instruments just missed out in my short list of ten, such as the University College London échelle spectrograph, UCLES, on the Anglo-Australian telescope, to cite one notable example.

I make no apology for including my own instrument, Hercules, in the list, even though it is mounted on a mere 1-metre telescope. It has accordingly made much less of an impact than those on large telescopes used by the international community. But in terms of performance, innovation and originality, I believe that Hercules matches the best of high resolution instruments developed elsewhere.

This chapter does no more than give the reader the flavour of the extraordinary innovation of the optical designers working in astronomy today. The foundation of much of observational astronomy of the current epoch rests on the success of their work.

8.2 THE ESO FAINT OBJECT SPECTROGRAPH AND CAMERA: EFOSC

The European Southern Observatory's faint object spectrograph and camera, EFOSC, is the first of the spectrographs described here to be constructed. The design is quite simple in concept, and comprises dioptric (all-refracting) collimator and camera lenses and a grism as the dispersing element. The optical train is linear, which makes it a simple matter to withdraw the grism, and to convert the instrument to a focal-reducing system for direct imaging. The optical light train in spectrographic mode is shown in Fig. 8.1.

EFOSC was designed from 1981, and had its first test run on the 3.6-m telescope (at the f/8 Cassegrain focus) at La Silla, Chile, in 1984, before entering regular service the following year [1].

The collimator consists of two doublet lenses, while the f/2.5 camera has seven lenses in three groups.

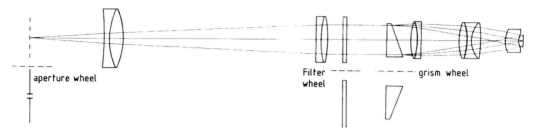

Figure 8.1. The optical design of the European faint object spectrograph and camera (EFOSC), in spectrographic mode. EFOSC was commissioned in 1984 at La Silla in Chile.

A theoretical image quality from spot diagrams of 0.3 to 0.5 arc seconds is achieved over the field of view, in the wide wavelength range of 350 nm to 1 μm.

At the time of the EFOSC construction, fast dioptric cameras were quite unusual in spectrograph design. The advent of new glasses with special properties (EFOSC uses Schott FK54 in the camera), the use of ray-tracing programs (such as Zemax) able to optimize lens parameters, and the application of efficient antireflection coatings on air–glass surfaces, were the factors that made such cameras possible from the early 1980s.

EFOSC was designed around an RCA 512 × 320-pixel CCD camera with 30-μm pixels. By today's standards, that was a small detector with large pixels, but representative of CCD technology of that time. One pixel corresponded to $0.67''$ on the sky in the direct imaging mode.

EFOSC has eight different grisms for different dispersions and resolving powers and also different blaze wavelengths. The grism rulings range from 1000 gr/mm to 150 gr/mm, and the corresponding resolving powers vary from about 120 to 1400. In addition, there is also an échelle grism mode with prism cross-dispersion, which gives a resolving power of 2000.

The first observations with EFOSC [2] showed that a 20th magnitude quasar spectrum could be recorded in 30 minutes at medium dispersion (23 nm/mm) using a 300 gr/mm grism.

EFOSC also has a multi-object spectroscopy (MOS) capability. In this mode, the slit plate in the Cassegrain focal plane is replaced with an aperture plate with punched holes prepared the previous day [3, 4]. Observations of galaxies to $V = 22.5$ were reported.

Imaging polarimetry and spectropolarimetry are also possible with EFOSC. These modes are achieved by inserting a Wollaston prism into the beam at the position of the filter wheel, which is in the collimated beam between collimator and camera, close to the grism wheel. The overall mechanical layout of EFOSC is shown in Fig. 8.2.

By 1987, EFOSC was accounting for about one-third of all time allocated on the 3.6-m ESO telescope. To cater for this demand, a further instrument based on the same design concept, EFOSC-2, was built for the NTT and 2.2-m telescopes on La Silla [5]. Copies of EFOSC have also been built for other observatories (for example IFOSC is at the IUCAA 2-m telescope in India). Further technical details of EFOSC are in the operator's manual [6].

The success of EFOSC is based on its high efficiency and its versatility. The average optical throughput of EFOSC in imaging mode over the 350 nm to 1 μm range in wavelength is 75 per cent [1]; this quantity peaks at 82 per cent at 550 nm (the efficiency does not include telescope or detector). For EFOSC spectroscopy these figures are a little lower as a result of the blaze function of the grism, but typically they peak at 60 per cent for the blaze wavelength of each grism.

As for versatility, EFOSC can convert rapidly between imaging and spectroscopic modes, as all functions are fully automated, with a series of wheels in the optical path carrying slits, filters and dispersing elements. Changing the instrument mode therefore only takes a few seconds.

These properties of EFOSC have made it a pioneering instrument for the 1980s through into the twenty-first century.

8.3 THE KECK HIGH RESOLUTION ECHELLE SPECTROMETER: HIRES

The Keck High Resolution Echelle Spectrometer, HIRES, was designed by Steven Vogt (b. 1949) (University of California, Santa Cruz) to be a

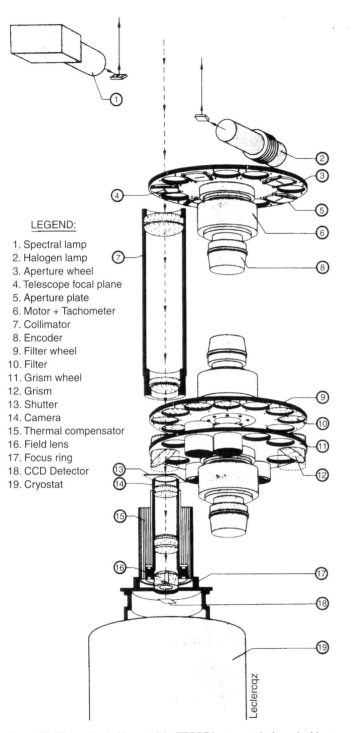

LEGEND:

1. Spectral lamp
2. Halogen lamp
3. Aperture wheel
4. Telescope focal plane
5. Aperture plate
6. Motor + Tachometer
7. Collimator
8. Encoder
9. Filter wheel
10. Filter
11. Grism wheel
12. Grism
13. Shutter
14. Camera
15. Thermal compensator
16. Field lens
17. Focus ring
18. CCD Detector
19. Cryostat

Figure 8.2. The mechanical layout of the EFOSC instrument is shown in this cutaway diagram. Note the three wheels, of which the lowest holds the grisms.

state-of-the-art high resolution instrument on the Keck I 10-m telescope, mounted at the f/15 Nasmyth focus. Vogt described the design philosophy thus: 'The Keck, as the world's largest telescope at the world's best site, is first and foremost a threshold instrument. At any given resolution and S/N, it must be able to go significantly fainter than any other telescope in the world' [7]. The thresholds that Vogt had in mind were pushing high resolution ($R \sim 30-50 \times 10^3$) spectroscopy to faint magnitude limits, on, for example, quasars to $V = 19.5$, and to be able to reach stars with very high signal-to-noise ($\sim$200 per pixel) at a resolving power of $R \sim 10^5$ and to magnitude $V = 11$ in one hour.

The resolving power times slit width product $R.\theta_s$ ($= \frac{2L\sin\theta_B}{D}$) is a measure of a spectrograph's figure of merit (see Section 2.4), and equals 39 000 arc seconds for HIRES. To achieve this on a large telescope ($D = 10$ m), a very large échelle grating is required. In the case of HIRES, the échelle is a three-grating mosaic. Each component measures 30 $\times$ 40 cm. The three are mounted facing down on a large granite slab and are aligned to high precision. The beam size is $A = 30$ cm and the length of the échelle mosaic is $L = 1.20$ m, making it the largest mosaic grating at present in use.

The parameters for the échelle are 52.68 gr/mm and the blaze angle is 70.5°. The quasi-Littrow angle is $\theta = 5.0°$, which is a fairly large value in order to separate incident and diffracted beams from the échelle within the space available.

A decision was taken early on not to use prism cross-dispersion, popular in some other high resolution instruments (such as the Hamilton échelle at Lick [8]), but large plane mosaic gratings in first or second order, placed after the échelle in the light path. These give more cross-dispersion, thereby allowing for sky subtraction, for spectra with long slits on extended objects or eventual image slicing, albeit at the expense of wavelength coverage (this arises both from the cross-grating blaze function and from the fact that large order separation limits the number of orders on a given CCD detector). The overall layout of HIRES is shown in Fig. 8.3.

Apart from the huge échelle mosaic, the other key component in HIRES is the fast f/1 catadioptric camera designed by Harland Epps. The requirements for the camera were to provide excellent spot diagrams over a large wavelength range (310 nm to 1 μm or more). The root mean square image size from a point source in the slit plane should ideally be less than the pixel size of 15 μm. The field of view needed is 6 to 8°. The need for a fast camera is apparent from Equation 2.126, where it is seen that the speed of the camera goes as $1/D$ for a CCD with a given pixel size (where D is the telescope aperture), provided the Nyquist sampling theorem is to be satisfied.

Epps and Vogt have described the stages through which the HIRES camera design progressed [9]. Conventional Schmidt cameras did not perform adequately, as the white pupil on the échelle is too far from the Schmidt corrector plate (at some three camera focal lengths).

Eventually a new f/1 camera design was found with two corrector plates placed in front of a large spherical mirror. The plates are fused silica and all optical surfaces are spherical. The mean image quality over the wavelength range 0.3 to 1.1 μm is 12.6 μm over all field angles up to 6.7° diameter. Figure 8.4 shows details of the HIRES camera, with its two large corrector lenses and field-flattener.

The camera has a small field-flattening lens just ahead of the CCD, and this also serves as a dewar window. The CCD is mounted on the central optical axis, so this results in some light loss, as it obstructs part of the beam at central wavelengths. Earlier versions of the so-called HIRES 'super camera' were described by Epps in 1990 [10]. These earlier designs had aspheric surfaces on the large corrector lenses.

Full technical details of the HIRES instrument are given by Vogt *et al.* in 1994 [11].

HIRES had its first light in 1993. The scientific achievements of the instrument over the first six years of operation have been reviewed by Vogt [12]. The research cited covered the beryllium abundances in stars (using the ultraviolet capability of HIRES to reach 313 nm), the cosmic deuterium-to-hydrogen ratio from high redshift quasar spectra, studies of CI fine structure lines in quasars (these lines are sensitive to the temperature of the cosmic microwave background at early epochs), Lyman α forest studies of quasar spectra, studies of the velocity dispersion of faint stars in dwarf spheroidal galaxies in the Local Group, Doppler imaging of spotted stars, and the search for extrasolar planets. References for all this research are given in Vogt's review.

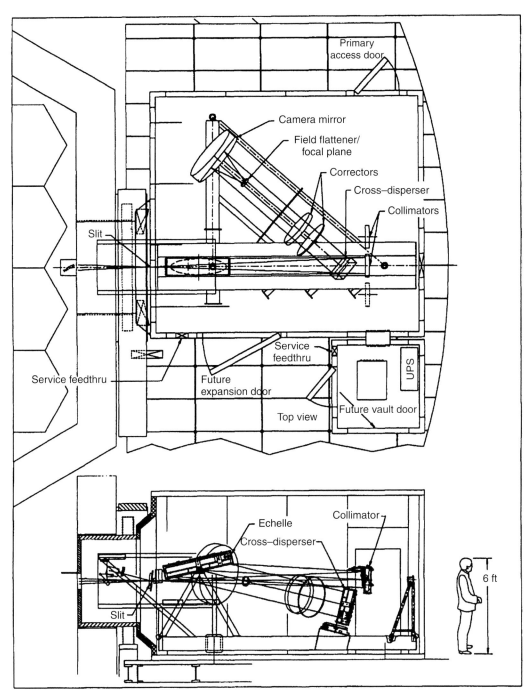

Figure 8.3. The mechanical layout of the Keck telescope HIRES, in plan view (above) and side elevation (below). Note the échelle and cross-dispersion gratings, and the large optical elements of the Epps catadioptric camera.

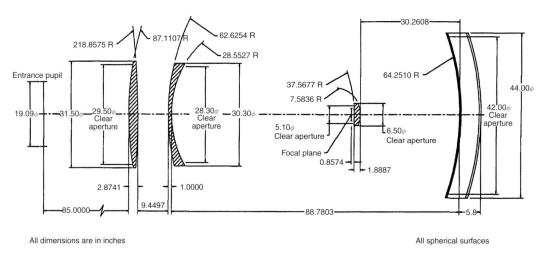

All dimensions are in inches   All spherical surfaces

EPPS design 7465 (11/15/90)

Figure 8.4. Details of the Epps design for the HIRES camera. The camera features a large spherical concave mirror and two large aberration correctors with spherical surfaces. A small field-flattener lens is placed in front of the on-axis CCD detector.

8.4 THE KECK LOW RESOLUTION IMAGING SPECTROMETER: LRIS

The Keck Low Resolution Imaging Spectrometer is a highly efficient low resolution instrument mounted at the Cassegrain f/15 focus of the Keck 10-m telescope, which is therefore able to reach the faintest objects. It was designed by Beverley Oke (1928–2004) and his colleagues at Caltech [13]. The optical design was by Harland Epps at Santa Cruz and Lick Observatory. LRIS has been in operation since 1995, mainly on Keck I at the Mauna Kea observatory in Hawaii. A high level of automation characterizes the instrument's functions, which enables remote observing on LRIS to be undertaken by observers on the US mainland.

LRIS has two arms, blue and red, with the beam being separated by a dichroic filter, which reflects blue light and transmits in the red. The blue arm records 310 to 550 nm, while the red covers 550 nm to about 1.10 μm. The overall optical layout is shown in Fig. 8.5.

LRIS can be used either for direct imaging, recording blue and red images simultaneously, with a 6 × 8-arc minute field of view, and on separate CCD cameras, or for spectroscopy. Initially, three different gratings (300, 600 or 1200 gr/mm) were provided for the red side, and these gave resolutions of 4.99, 2.55 or 1.31 Å, corresponding to resolving powers of about 1.3×10^3, 2.6×10^3 or 5.0×10^3 at Hα.

Mechanical slit masks can be inserted into the telescope's focal plane, allowing about 40 objects to be recorded with the 300 gr/mm grating. The collimator is an off-axis paraboloid, and this gives a collimated beam size of 141 mm.

The blue side of LRIS uses a grism as the dispersing element. For direct imaging, the dispersing elements can be removed, and filters can be inserted directly in front of the cameras. The f/2.2 cameras ($f_{cam} = 305$ mm) are all-refracting dioptric designs by Harland Epps [10] at Lick Observatory, with seven lens elements. The two independent cameras for blue and red are essentially the same in their design, though the antireflection coatings are optimized for the different wavelengths of each side of the instrument.

The first lens element, a calcium fluoride singlet (diameter 228 mm) provides most of the power. It is followed by a low power triplet that corrects for chromatic aberrations, and then there is a doublet to provide the remaining positive power, and this is followed by a field-flattening lens just in front of the CCD window.

The entrance window for the whole instrument is a large field lens which images the telescope pupil close to the dispersing elements, as well as keeping out dust from the optics.

LRIS has been used for redshift surveys of very faint galaxies (for example, for recording spectra of a very faint 27th magnitude Hubble Deep Field galaxy

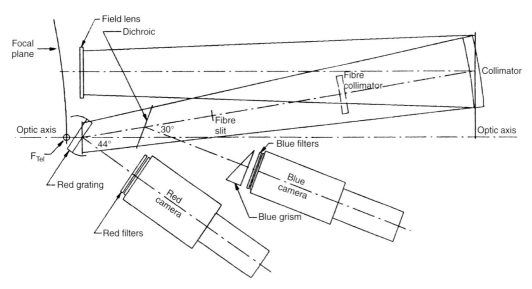

Figure 8.5. The mechanical layout of the Keck Low Resolution Imaging Spectrometer (LRIS). Note the blue and red arms, which have respectively grism and plane grating dispersing elements.

with redshift $z \sim 5.6$ by Ray Weymann (b. 1934) *et al.* in 1998 [14]) and for a study of the kinematics and metallicity of faint halo stars in the Andromeda galaxy, M31, in 2002 [15].

8.5 ELODIE AT HAUTE-PROVENCE

ELODIE is a high resolution fibre-fed échelle spectrograph explicitly designed for the determination of precise stellar radial velocities, in order to search for planets by the Doppler method. ELODIE is installed on the 1.93-m telescope at the Observatoire de Haute-Provence, and its construction was a joint project of Haute-Provence, Marseille and Geneva observatories. First light was in 1993.

In many respects, ELODIE is a pioneering instrument with many new or innovative features being used. These include an R4 ($\theta_B = 76.0°$, dimensions 102×408 mm) échelle grating, which results in a compact instrument with a 102 mm beam size. Nearly all échelle spectrographs before ELODIE used a smaller blaze angle. In addition, ELODIE uses André Baranne's white pupil concept (see Section 2.9), in which the white pupil on the échelle is reimaged and demagnified to a 75-mm white pupil at the camera.

Another innovation is the use of both a crown glass grism in first order (8.63°; 150 gr/mm) and a

flint glass prism (40° apex angle) for cross-dispersion, thereby giving nearly constant order separation in the échelle format. The optical and mechanical layouts of ELODIE are shown in Fig. 8.6.

ELODIE is designed to operate in the wavelength range 390.6 to 681.1 nm over 67 diffraction orders. This therefore avoids the far-red spectral orders, which have few stellar lines suitable for radial-velocity determination, and many telluric lines, especially the strong A and B bands of molecular oxygen. Below about 390 nm absorption in the fibre and the low quantum efficiency of the CCD makes this ultraviolet part of the spectrum less useful for radial-velocity work.

The camera of ELODIE is a dioptric f/3 camera with an effective focal length of 300 mm, giving a dispersion of 2 Å/mm in the blue. The resolving power is $R = 42\,000$. ELODIE is fed from the f/15 Cassegrain focus of the 1.93-m telescope using 20 metres of 100 µm core diameter optical fibre, thus enabling the spectrograph to be in a stable environment off the telescope.

Another feature of ELODIE is the use of a double scrambler in the fibre [16], thereby eliminating the effects of guiding errors of a star imaged on the fibre entrance. The scrambler comprises two microlenses inserted in a small gap in the fibre. These scramble the fibre output in both space and angle, thus giving a stable

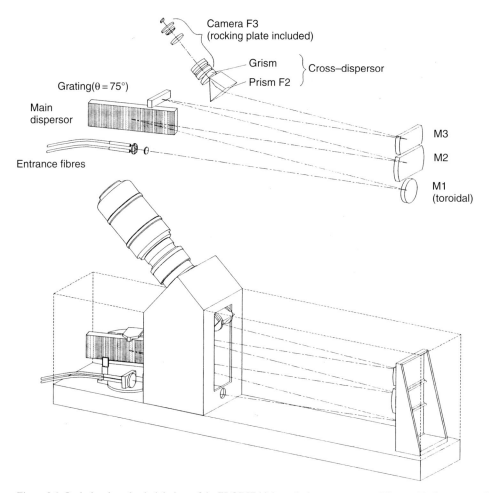

Figure 8.6. Optical and mechanical designs of the ELODIE high resolution spectrometer. The notable features are the high angle of incidence of the light on the R4 échelle grating, the mirrors M2 and M3 which reimage the white pupil on the échelle to the camera, and the use of a prism and a grism for the cross-dispersion.

illumination within the spectrograph, which is vital for precise velocity measurements. Without a double scrambler ELODIE typically gave a velocity precision of 15 m/s on late-type stars with sharp lines. This precision improved to some 4 m/s with the scrambler [17]. Ninth magnitude stars can be observed in one hour with a signal-to-noise ratio of about 100 on the 1024 × 1024-pixel CCD camera.

To achieve a good wavelength calibration, a second fibre is used in ELODIE for the simultaneous exposure of the spectrum of a thorium–argon hollow-cathode lamp. This helps reduce the error from wavelength shifts arising from changes in atmospheric temperature and pressure.

A full description of the ELODIE spectrograph is given by André Baranne, Didier Queloz (b. 1966), Michel Mayor (b. 1942) and their colleagues [18]. A second instrument of the same design, and known as CORALIE, was later built for the 1.2-m telescope for the European Southern Observatory at La Silla in Chile.

ELODIE has automated on-line data analysis, whereby Doppler shifts are measured by cross-correlation with a template spectrum. The position of the cross-correlation peak is then a measure of the radial velocity.

ELODIE has been used to detect extrasolar planets by means of the periodic Doppler shifts induced in the spectra of the parent stars by a massive orbiting planet. By 1998, 324 northern solar-type dwarf stars with $m_V < 7.65$ were being monitored by ELODIE for such Doppler-shift variations. The first detection of an extrasolar planet orbiting a solar-type star was announced in 1995, just 18 months after ELODIE was first commissioned [19]. The star was the nearby G2IV star, 51 Pegasi. The discovery was one of the most significant astronomical discoveries of the 1990s decade, and the first of some 200 such extrasolar planets found by all observers over the following decade.

8.6 THE 2DF SPECTROGRAPHS AT THE ANGLO-AUSTRALIAN TELESCOPE

The 2dF (two-degree field) system was conceived at the Anglo-Australian Observatory in the late 1980s. The design was driven by the scientific requirements for the redshifts of hundreds of thousands of faint galaxies to be recorded in order to map the spatial distribution of visible matter in the Universe, as well as by the practical realization that a wide-field imaging telescope could also act as an excellent platform for multi-object spectroscopy using optical fibres.

In the case of the Anglo-Australian telescope (AAT), this goal was realized by providing an aberration-corrected two-degree diameter field at the prime focus, which in turn enabled up to 400 fibres to be precisely positioned so as to accept light from a corresponding number of objects, be they faint galaxies, quasi-stellar objects or (on occasions) stars. Certainly the 2dF system represents more than just a spectrograph. In fact it comprises a four-lens aberration corrector, an atmospheric dispersion compensator, an automated fibre positioner with gripper arm, a tumbler mechanism (explained shortly), 800 optical fibres, a focal-plane imager and two low-resolution fibre-fed spectrographs and associated CCD detectors. All these are mounted at or near the prime focus on a new 5-m diameter top end for the 3.9-m AAT. The 2dF system not only represents the highest level of development of fibre MOS techniques at the end of the twentieth century; it also highlights the fact that a spectrograph is not any more just an instrument that can be bolted onto a telescope at will, as was once the case. At this level of instrumentation, the spectrograph is integrated

into a highly automated opto-mechanical system comprising telescope, fibres, spectrographs and detectors, as 2dF perfectly demonstrates. A full description of the 2dF instrument has been given by Ian Lewis *et al.* [20]. Numerous earlier reports described progress during the construction and commissioning stages in the early 1990s, such as [21, 22].

The heart of 2dF is the ability to produce sub-arc-second images over a two-degree field (for the AAT this field has a diameter of 481 mm). Four corrector lenses are used; the first two are large crown (BK7) plus flint (F2) glass doublets, almost one metre in diameter. The design work was undertaken by Charles Wynne in Cambridge, Richard Bingham at the Royal Greenwich Observatory, and later by Damien Jones at CSIRO in Australia. The final design was described by Jones in 1994 [23]. The first two lens elements of the corrector are in fact wedge-shaped prism lenses which can be rotated about their central axis, thus providing for atmospheric dispersion compensation (the ADC) for zenith angles up to 60 degrees.

The 2dF facility uses two stainless-steel field plates for fibre placements by means of magnetic buttons at the entrance end of each fibre. One of these plates is in the focal plane of the telescope and receiving light from 400 objects, while the other is being configured for the next field. The two plates are interchanged with the tumbler mechanism, which rotates through 180 degrees. Complete reconfiguration takes an hour or less, so there is little loss of time between fields, assuming exposures of about one hour. Thus each fibre can be placed on a field plate in five or six seconds. The typical precision of a placement is $\pm 11\,\mu m$ (about $\pm 0.16''$). The fibres themselves are 8 metres in length and 140 μm core diameter fibre with high ultraviolet transmission.

The 2dF spectrographs are mounted on the periphery of the top-end ring. Each accepts 200 fibres arranged in a linear slit. The use of two spectrographs was dictated by the 1k × 1k CCD detectors at that time available, and which can accommodate 200 spectra on each chip. The spectrographs have off-axis f/3.15 Maksutov collimators and a fast f/1.2 Schmidt camera, and they use conventional plane gratings. Each spectrograph has six different gratings available, from 300 to 1200 gr/mm. The choice of grating determines the blaze wavelength and resolving power obtained. Data from the spectrographs are reduced on-line in real

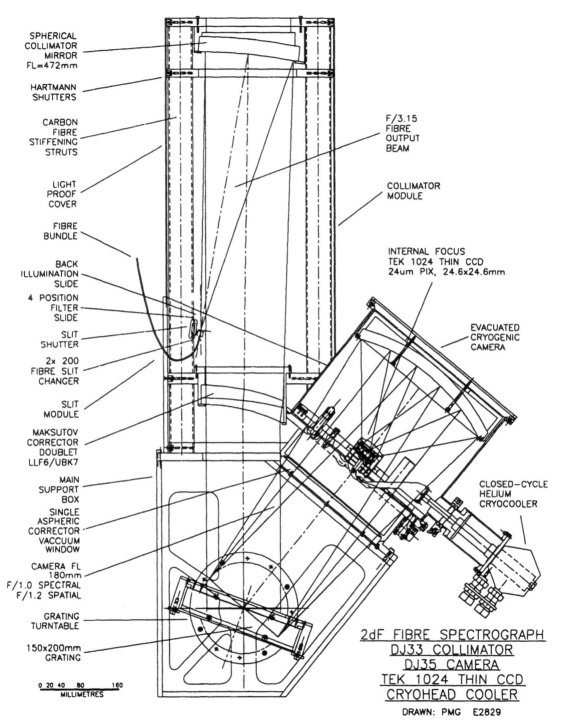

SPHERICAL
COLLIMATOR
MIRROR
FL=472mm

HARTMANN
SHUTTERS

CARBON
FIBRE
STIFFENING
STRUTS

LIGHT
PROOF
COVER

FIBRE
BUNDLE

BACK
ILLUMINATION
SLIDE

4 POSITION
FILTER
SLIDE

SLIT
SHUTTER

2x 200
FIBRE SLIT
CHANGER

SLIT
MODULE

MAKSUTOV
CORRECTOR
DOUBLET
LLF6/UBK7

MAIN
SUPPORT
BOX

SINGLE
ASPHERIC
CORRECTOR
VACCUUM
WINDOW

CAMERA FL
180mm
F/1.0 SPECTRAL
F/1.2 SPATIAL

GRATING
TURNTABLE

150x200mm
GRATING

F/3.15
FIBRE
OUTPUT
BEAM

COLLIMATOR
MODULE

INTERNAL FOCUS
TEK 1024 THIN CCD
24um PIX, 24.6x24.6mm

EVACUATED
CRYOGENIC
CAMERA

CLOSED−CYCLE
HELIUM
CRYOCOOLER

0 20 40 80   160
MILLIMETRES

2dF FIBRE SPECTROGRAPH
DJ33 COLLIMATOR
DJ35 CAMERA
TEK 1024 THIN CCD
CRYOHEAD COOLER

DRAWN: PMG   E2829

Figure 8.7. Diagram of one of the 2dF spectrographs, showing the 200-fibre optical fibre feed, Maksutov collimator, plane grating and f/1.2 Schmidt camera. Two identical spectrographs were commissioned for the 2dF facility in 1996.

time [24]. Figure 8.7 shows the optical layout of a 2dF spectrograph.

The 2dF instrument has been used for two major surveys. The 2df galaxy redshift survey comprises about 250 000 galaxies in selected areas of the sky covering about 2000 square degrees and to magnitude 19.5. The routine observing was commenced in 1997 and took less than five years to complete [25, 26]. Numerous papers have discussed the results and their implications for structure in the Universe and cosmology. The 2dF quasi-stellar object redshift survey was undertaken at the same time as the galaxy survey. The data resulted in about 25 000 quasar redshifts from 750 square degrees of sky [27, 28]. Several studies using 2dF have also been made of stars, in the galactic thick disk and halo, in the globular cluster 47 Tucanae and in the Magellanic Clouds.

The success of the 2dF facility has led to a second generation 2dF spectrograph to be commissioned, using VPH gratings as the main dispersing elements. This is the AAOmega multi-object spectrograph [29]. The AAOmega instrument is mounted off the telescope and linked with 30 metres of optical fibre. Significantly higher efficiency and resolving power are achieved than with the original grating 2dF spectrographs.

8.7 THE SPACE TELESCOPE IMAGING SPECTROGRAPH: STIS

STIS is the Space Telescope Imaging Spectrograph, and was installed on the Hubble Space Telescope in February 1997 to replace the two first generation spectrographs (FOS and GHRS). STIS is an exceptionally versatile and complex instrument, designed to undertake a wide range of functions from very low resolution spectroscopy ($R \sim 50$ using a prism), through to high resolution échelle spectroscopy ($R > 10^5$). There is also a non-dispersed imaging mode. It is arguably the most complex and versatile astronomical spectrograph of the end of the twentieth century.

The wavelength range of STIS covers 115 nm in the far ultraviolet to 1.0 μm in the near infrared, and three detectors are used; two of them are MAMA (multi-anode microchannel arrays) detectors (2048×2048) [30, 31] for the ultraviolet (115–310 nm), while

a single 1024×1024-pixel CCD is used for the visible and near-infrared regions (305 nm – 1.0 μm).

A brief outline of the optical system of STIS is given here; a full account has been given by Bruce Woodgate (b. 1939) *et al.* [32]. STIS has five spectroscopic modes. The low ($R = 500-1000$) and medium ($R = 5000-10^4$) resolution spectral imaging modes use concave parabolic gratings in first order and long slits (50 arc seconds). The gratings are in the Wadsworth configuration, in which the grating focusses the spectrum directly onto the detector, which is placed normal to the grating (see Section 1.7). The full wavelength range of 115 nm to 1 μm is available using the three detectors, though in restricted intervals for each exposure. On the other hand, medium ($R = 23 000-35 000$) and high resolution ($R \sim 105 000$) échelle modes use plane cross-dispersion gratings ahead of the échelle and a camera mirror to focus the spectrum on the appropriate MAMA detector. The optical arrangement is a Czerny–Turner mounting (see Section 1.7). The échelle modes only operate in the ultraviolet (as this type of work cannot be done from the ground). Finally, the objective spectroscopy mode ($R \sim 50$) is for very low resolution ultraviolet spectroscopy using a prism as the dispersion element. After the prism, a plane fold mirror directs the light to the camera. The optical layout of STIS is shown in Fig. 8.8.

STIS has a total of 16 gratings, made up of four échelle gratings ($\theta_B = 32$ to $70°$), two concave gratings ($\theta_B = 0.67°$ and $14.7°$) and ten plane gratings for cross-dispersion. The gratings are surprisingly small, 25 mm in diameter, partly because they are located near a system pupil. High resolving powers can still be achieved with such small dispersing elements, because HST delivers close to diffraction-limited images for which very narrow slits can be used (for example, in échelle mode, there are 16 slits from $0.025''$ to $0.5''$).

The science undertaken by STIS is very wide-ranging. At lower resolving power it includes studies of active galactic nuclei, evolution of galactic halos, protogalaxies and the intergalactic medium. At high resolving power STIS has been used for interstellar medium studies in the Galaxy, observations of stellar atmospheres, magnetic phenomena in late-type stars and the chemistry of the outer Solar System planets, and studies of extrasolar protoplanetary systems.

Woodgate and Kimble have presented more on STIS capabilities [33] and both Kimble [34]

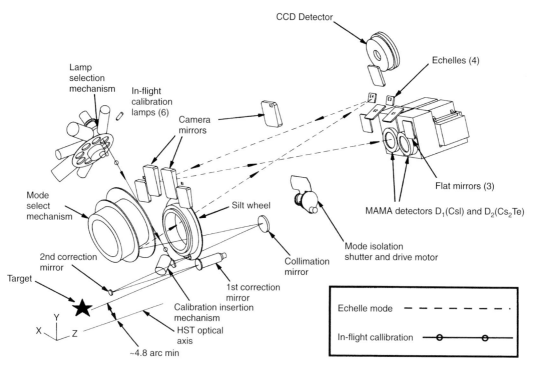

Figure 8.8. The optical layout of the Space Telescope Imaging Spectrograph (STIS).

and Baum *et al.* [35] discussed early in–orbit performance. The Hubble Space Telescope website (www.stsci.edu/hst/stis) gives more details of instrument design and performance.

Unfortunately STIS suffered from a power supply failure in August 2004, and has been unused in a stored safe mode since then. There are prospects of restoring it into service at the proposed fourth servicing mission to the telescope.

8.8 THE ULTRAVIOLET ECHELLE SPECTROGRAPH AT ESO PARANAL: UVES

UVES is the European Southern Observatory's ultraviolet échelle spectrograph, mounted on the unit telescope no. 2 (UT2) Kueyen, one of the four 8-m telescopes that comprise the Very Large Telescope (VLT) at Paranal in Chile. It receives the light at the f/15 Nasmyth focus, but preslit optics converts the beam to f/10.

The scientific goals for UVES were very similar to those of Keck HIRES, namely, on the one hand,

the study of absorption lines in very faint high redshift quasars, and on the other, the study of stellar spectra at high resolving power and signal-to-noise. Like HIRES, an ultraviolet capability close to the atmospheric limit was stipulated in the design goals, and an overall wavelength range of 320 nm to 1.10 μm was specified. Not surprisingly, some of the solutions of UVES are similar to those adopted for HIRES – in particular a large mosaic échelle grating, and grating cross-dispersion for maximum order separation. The principal dissimilarities are that UVES uses a dichroic filter before the slit, to divide the spectrograph into two arms (essentially two independent spectrographs) for blue and red light, the design employs Baranne's white pupil concept, and the instrument uses dioptric (refracting) cameras, instead of the catadioptric solution at HIRES.

The design process for UVES was commenced in 1992 by Hans Dekker and Sandro D'Odorico at ESO as an in-house project. At first two identical instruments were to be constructed for telescopes UT2 and UT3; in 1994 a single but more versatile instrument was adopted. Construction took five years and cost about

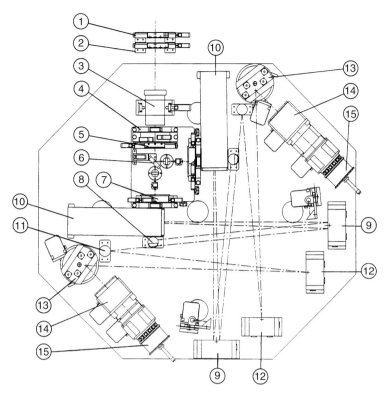

Figure 8.9. The optical layout of the Ultraviolet Echelle Spectrograph (UVES) at ESO Paranal, Chile. The components are: 1: calibration system, 2: image slicer, 3: image derotator, 4: filter wheel, 5: pupil stop, 6: mode selector, 7: slit, 8: fold mirror, 9: main collimator, 10: R4 échelle, 11: intermediate spectrum mirror, 12: pupil transfer collimator, 13: cross-dispersion grating, 14: camera, 15: CCD detector.

7.1 million German marks; commissioning took place in 1999 and routine observations followed in 2000.

The full design parameters of UVES were described by Dekker *et al.* in 2000 [36]. Earlier reports were published in 1992 [37] and in 1995 [38]. The 1995 report discussed the design changes of 1994, which included the addition of an atmospheric dispersion compensator, an iodine cell for precision radial velocities, a depolarizer and an exposure meter.

The échelle gratings of UVES are mosaics of two échelles for each of the blue and red arms. The dimensions of each component are 204×408 mm and the composite mosaic grating is about 21 by 84 cm. The dimension perpendicular to the grooves is $L = 84$ cm, there being a 14 mm gap between the two components of the mosaic. The échelles are both R4 (blaze angle $\theta_B = 76°$), but with 41.59 gr/mm for the blue arm and 31.6 gr/mm for the red (the coarser red ruling gives

slightly shorter red orders so as better to match the detector dimensions).

The beam size of UVES is just 20 cm, and the use of a pupil transfer parabolic mirror images a white pupil at the cross-disperser and this enables a very small quasi-Littrow angle of $\theta = 0.8°$ to be used, thereby leading to high efficiency in the order centres and fewer problems of line tilt which are inherent in non-Littrow R4 designs. The optical design in which the parabolic collimator is used in double pass is due to Bernard Delabre. Figure 8.9 shows the UVES optical layout.

Four different cross gratings have been provided for different order separations and blaze angles, two for each arm of the spectrograph. As for the cameras, these are multi-component fast dioptric designs. The blue camera (300–500 nm) has a focal ratio of f/1.8 and seven components (silica and calcium fluoride), while the red one (420–1.10 μm) operates at f/2.5 and has eight components.

The design philosophy for large high resolution échelle spectrographs on 8–10-m class telescopes has been discussed by Cathy Pilachowski (b. 1949) *et al.* [39]. There it is pointed out that the white pupil used in UVES (but not in HIRES) results in a camera of lesser aperture and hence less extreme overall focal ratio; in addition, the separate blue and red arms simplify the design of dioptric cameras.

Spectrographs for large telescopes able to observe very faint objects require sky subtraction, and this in turn necessitates a large order separation, so that the sky spectra are adequately recorded adjacent to the object spectra in the échelle format. For this reason, the largest telescopes have generally chosen grating cross-dispersion, in spite of a limited wavelength range and lower efficiency than a prism [39].

Highlights of the first observations of stars and quasars using UVES were presented by Sandro D'Odorico *et al.* in 2000 [40]. The results discussed include the detection of beryllium in two faint metal-poor stars using the Be doublet lines in the ultraviolet at 313 nm, determination of the lithium isotope ratio in a tenth magnitude halo star, and studies of the intergalactic medium using the Lyman α forest of absorption lines in high redshift ($z > 2$) quasars.

8.9 THE SLOAN DIGITAL SKY SURVEY SPECTROGRAPHS

The other major optical spectroscopic survey of faint galaxies and quasi-stellar objects, complementing the 2dF survey in the northern sky, is the Sloan Digital Sky Survey (SDSS). It is larger than the 2dF survey at the AAT, and started routine observations several years later (in 2000). It is interesting to compare the technical solutions to the challenging problems of large surveys that each project has adopted.

The SDSS uses a dedicated alt-az 2.5-m wide-field Ritchey–Chrétien Cassegrain telescope at Apache Point Observatory in New Mexico. The survey comprises both direct imaging (using a large 120-megapixel CCD camera) and spectroscopy. For the spectroscopy, two spectrographs are used, mounted on the telescope near the f/5 Cassegrain focus. They are fibre-fed multi-object instruments, each accepting 320 fibres (of length 2 metres). The resolving power is about 2000. In these

respects, the solutions are quite similar to 2dF, except for the choice of telescope focus.

The major differences in instrument design philosophy are, firstly, that SDSS uses a plug-plate for fibre placement. Plug plates are prepared in advance and during the day fibres are installed in holes drilled in the aluminium plate. In this respect it is less automated than the 2dF system. However, the fact that the SDSS telescope also is an imaging instrument makes this a practical solution, as object coordinates can be readily obtained from the direct images. As many as ten plug plates have to be prepared in advance for a night's observing, each of them with 640 fibres fitted into pre-drilled holes with a positioning accuracy of ±9 μm.

Secondly, the SDSS spectrographs employ dichroic filters, so that the blue (380–610 nm) and red (590–910 nm) beams are split and sent to two different dispersing elements and cameras. This simplifies the design of the f/1.3 dioptric (refracting) cameras, as each is optimized for a limited wavelength range. This solution compares with the Schmidt camera design adopted for 2dF, with the full wavelength range being received by the camera.

Thirdly, the SDSS spectrographs use grisms as the dispersing elements, not reflection gratings. The grisms are mounted close to the camera, and this leads to a smaller camera aperture and makes it easier to control off-axis aberrations. Each grism is a right-angled prism, with the transmission grating cemented to the hypotenuse. The optical layout is shown in Fig. 8.10.

The cameras of the SDSS spectrographs were designed by Harland Epps at the University of California in Santa Cruz. Each has eight lens elements, the first being a large (178 mm diameter) calcium fluoride singlet lens. One of the blue-arm cameras is shown in Fig. 8.11.

Each spectrograph of SDSS uses two 2048 × 2048-pixel CCD cameras (pixel size 24 μm), one for the blue arm and one for the red. The 180-μm fibres correspond to 3 arc seconds on the sky; the 2.5 times demagnification in the spectrograph results in the fibre image covering three pixels of the CCD.

The technical details of the SDSS instruments and programme are given by Donald York *et al.* [41].

The first SDSS survey (SDSS-I) was completed in 2005 and the data were released annually. The fifth

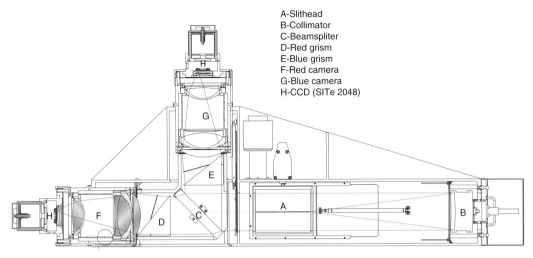

A-Slithead
B-Collimator
C-Beamspliter
D-Red grism
E-Blue grism
F-Red camera
G-Blue camera
H-CCD (SITe 2048)

Figure 8.10. Optical layout of one of the Sloan Digital Sky Survey (SDSS) spectrographs. Each spectrograph receives light from up to 320 optical fibres at A. Blue and red arms are separated by the dichroic filter.

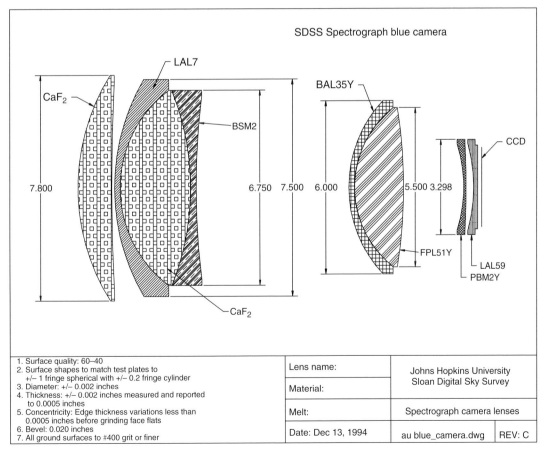

Figure 8.11. Lens elements of an SDSS blue-arm camera as designed by Harland Epps.

and final data release of SDSS-I (June 2005) includes redshifts for about 675 000 galaxies and over 90 000 quasars in an area of 5740 square degrees of sky. The red magnitude limit for the galaxy spectra is about 18 in these data [42].

8.10   THE HERCULES SPECTROGRAPH AT MT JOHN

Hercules (the High Efficiency and Resolution Canterbury University Large Echelle Spectrograph) was one of the first of the vacuum fibre-fed high resolution échelle spectrographs, designed for high precision radial-velocity observations, to come into service. It had first light in April 2001. Details of the instrument are reported by Hearnshaw *et al.* in [43].

Hercules is mounted on the 1-metre McLellan telescope at Mt John University Observatory in New Zealand. The design philosophy was to provide an instrument suitable for small telescopes (1 to 3 m) which has exceptional mechanical and thermal stability. The large vacuum tank, shown in Fig. 8.12, (4.5 m long; 1.2 m in diameter) was a key aspect of the design. In addition, the tank has no moving parts, so that every spectrum has precisely the same format, thereby promoting ease of reducing the data.

Hercules uses an R2 (31.6 gr/mm) 204 × 408-mm échelle. The cross-dispersion is from a large BK7 prism (height 276 mm; apex angle $\alpha = 49.5°$) in double pass, and placed in close proximity to the échelle. The prism

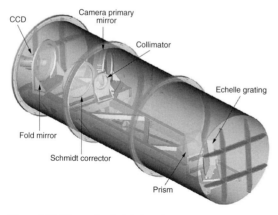

Figure 8.12. The mechanical design of the Hercules spectrograph at Mt John Observatory, New Zealand. It is mounted in a large vacuum tank and was commissioned in April 2001.

glass is of the highest quality, with refractive index variations of no more than $\pm 2 \times 10^{-6}$. The optical design is shown in Fig. 8.13. In some respects, the design resembles the Hamilton échelle at Lick Observatory [8], in that both instruments use an R2 échelle grating, a 20-cm beam size, prism cross-dispersion and a folded Schmidt camera. However, the Hamilton échelle employs two cross-dispersion prisms, mounted after the échelle in the optical path.

Three fibres are currently available to link the f/13.5 Cassegrain focus to the spectrograph. They give resolving powers of respectively 41 000, 70 000 and 82 000. The fibre length is 25 m. The lowest resolving power 100-μm fibre has an acceptance angle of 4.2 arc seconds on the sky, so essentially all the light in a stellar image enters the fibre, even in moderate seeing.

Hercules can deliver a wavelength range from 360 nm to 950 nm in a single exposure, using a Fairchild 486 4k × 4k CCD with 15-μm pixels. The format of the spectrograph focal plane is shown in Fig. 3.8. The Hercules room is heavily insulated and temperature controlled to maintain a temperature of $20.0 \pm 0.1\,°C$.

Hercules has a folded Schmidt camera. This gives outstanding sub-pixel spot diagrams for all wavelengths and field angles; the RMS spot sizes for a point source in the fibre exit are in all cases under 10 μm. The disadvantage of the folded Schmidt is some light loss (up to a maximum of 23 per cent) at some central green wavelengths, but progressively less than this in the blue and red orders and at the ends of the green orders.

A feature of Hercules is the small Littrow angle of $\theta = 3.0°$. This results in higher central order efficiency, but leads to a long spectrograph (4.5 m) for the incident and diffracted beams from the échelle to be sufficiently separated.

The evacuated environment in the tank leads to high stability, with the spectra being immune to ambient pressure and temperature changes (note that an air spectrograph, if subjected to a 1-mbar pressure change, sees the resulting wavelength change of the spectral lines as a spurious Doppler shift of -80 m s^{-1}, while a temperature change of just $+1\,°C$ mimics a velocity shift of $+270$ m s^{-1}). The tank is maintained at a pressure of between 2 and 4 torr, and is pumped about three or four times a year.

One vacuum échelle spectrograph that preceded Hercules is the Astronomy Research Consortium

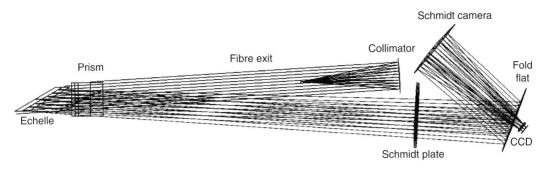

Figure 8.13. The optical layout of the Hercules spectrograph.

échelle spectrograph known as ARCES on the 3.5-m telescope at Apache Point Observatory, New Mexico [44]. This instrument came into service in 1999. Like Hercules it uses an R2 échelle grating and prism cross-dispersion. ARCES is at the Nasmyth focus and is not fibre-fed; high radial-velocity precision is not a feature of its performance.

The wavelength calibration of Hercules is by means of a Th–Ar hollow-cathode lamp in the Cassegrain fibre-feed module. The radial-velocity precision using a cross-correlation technique is about 15 m s^{-1} on late-type stars. A fibre double scrambler is planned in 2008, which should improve the precision by about an order of magnitude to near the photon-noise limit of radial-velocity precision. Hercules can deliver a signal-to-noise ratio of 100 to 1 on ninth magnitude stars in about one hour in good seeing conditions.

Hercules has been used for studies of high precision analysis of spectroscopic binary orbits, and for studies of asteroseismology of pulsating stars. An example of the results is in [45].

8.11 HARPS, THE HIGH ACCURACY RADIAL-VELOCITY PLANET SEARCHER

The HARPS instrument was designed to achieve the ultimate in radial-velocity precision. It has been installed on the European Southern Observatory's 3.6-metre telescope in Chile since 2003, and it is a joint project of ESO, the Geneva Observatory, the Universität Bern Physikalisches Institut, Haute-Provence Observatory and the Service d'Aeronomie of the CNRS. HARPS regularly achieves 1 m s^{-1} in radial-velocity precision on late-type slowly rotating stars. It uses many of the same principles as Hercules; the main difference affecting radial-velocity precision is a fibre double image scrambler in HARPS, which stabilizes the illumination of the fibre exit, regardless of the fluctuations in the input illumination of a fibre resulting from guiding variations or atmospheric scintillation.

HARPS uses a very large R4 échelle grating, measuring 837×208 mm. Cross-dispersion is with an FK5 grism (17 gr/mm) and the f/3.3 camera is dioptric, with six lens elements. This optical arrangement gives a resolving power of about 115 000. The tank pressure is about 10^{-2} torr. There is a facility for simultaneous Th–Ar and stellar spectra to be recorded, using two 70-μm fibres and interleaved échelle orders. The fibre core diameter corresponds to 1 arc second on the sky. The HARPS spectrograph is shown in Fig. 8.14.

The wavelength range of HARPS is 380 to 690 nm over 72 échelle orders, less than Hercules because of the higher cross-dispersion (which is needed for the simultaneous stellar and thorium exposures). Two butted $2k \times 4k$ CCD chips are used as the detector, giving an overall detector area of $60 \times 60 \text{ mm}^2$.

The technical details of HARPS are described by Francesco Pepe *et al.* [46] and by Michel Mayor *et al.* [47]. The principal observing programme of HARPS has been the detection of extrasolar planets by the radial-velocity method. By April 2007, HARPS had discovered 12 new extrasolar planets orbiting nine solar-type stars. Eight of these planets are listed in a paper by Christophe Lovis *et al.* reviewing the performance of HARPS [48].

Figure 8.14. The enclosed HARPS vacuum tank, seen here in the Geneva integration laboratory prior to installation on La Silla, Chile. The tank has a volume of about 2 m^3.

The high precision of HARPS velocities allows the detection of planets whose mass is of the order of ten times the Earth mass, the so-called rocky super-Earths.

REFERENCES

[1] Buzzoni, B., Delabre, B., Dekker, H. *et al.*, *European South. Observ. Messenger* **38**, 9 (1984)

[2] Dekker, H. and D'Odorico, S., *European South. Observ. Messenger* **37**, 7 (1984)

[3] Dupin, J. P., Fort, B., Mellier, Y. *et al.*, *European South. Observ. Messenger* **47**, 55 (1987)

[4] Giraud, E., *European South. Observ. Messenger* **51**, 37 (1988)

[5] Eckert, W., Hofstadt, D. and Melnick, J., *European South. Observ. Messenger* **57**, 66 (1989)

[6] Savaglio, S., Benetti, S. and Pasquini, L., *EFOSC1 Operating Manual*, ESO Garching, version #3, May 1997. See also www.eso.org

[7] Vogt, S. S. and Penrod, G. D., in *Instrumentation for Ground-based Optical Astronomy*, ed. L. B. Robinson, New York: Springer-Verlag, p. 68 (1988)

[8] Vogt, S. S., *Publ. astron. Soc. Pacific* **99**, 1214 (1987)

[9] Epps, H. W. and Vogt, S. S., *Appl. Optics* **32**, 6270 (1993)

[10] Epps, H. W., *Proc. Soc. Photo-instrumentation Engineers (SPIE)* **1235**, 550 (1990)

[11] Vogt, S. S., Allen, S., Bigelow, B. *et al.*, in 'Instrumentation in astronomy VIII', ed. D. L. Crawford and E. Craine, *Proc. Soc. Photo-instrumentation Engineers (SPIE)* **2198**, 362 (1994). See also Vogt, S. S., *HIRES User's Manual*, UCO/Lick Observatory Technical Report no. 67 (1994)

[12] Vogt, S. S., in 'Astronomical instrumentation and the birth and growth of astrophysics', ed. F. Bash and C. Sneden, *Astron. Soc. Pacific Conf. Ser.* **270**, 5 (2002)

[13] Oke, J. B., Cohen, J. G., Cromer, J. *et al.*, *Publ. astron. Soc. Pacific* **107**, 375 (1995)

[14] Weymann, R. J., *Astrophys. J.* **505**, L95 (1998)

[15] Reitzel, D. B. and Guharthakurta, P., *Astron. J.* **124**, 234 (2002)

[16] Brown, T., Gilliland, R., Noyes, R. and Ramsey, L. W., *Astrophys. J.* **368**, 599 (1991)

[17] Queloz, D., Mayor, M., Sivan, J. P. *et al.*, in 'Brown dwarfs and extrasolar planets' ed. R. Rebolo, E. L. Martín and M. R. Zapatero Osorio, *Astron. Soc. Pacific Conf. Ser.* **134**, 324 (1998)

[18] Baranne, A., Queloz, D., Mayor, M. *et al.*, *Astron. & Astrophys. Suppl.* **119**, 373 (1996)

[19] Mayor, M. and Queloz, D., *Nature* **378**, 355 (1995)

[20] Lewis, I. J., Cannon, R. D., Taylor, K. *et al.*, *Mon. Not. R. astron. Soc.* **333**, 279 (2002)

[21] Gray, P., Taylor, K., Parry, I., Lewis, I. and Sharples, R., *Astron. Soc. Pacific Conf. Ser.* **37**, 145 (1993)

[22] Cannon, R. D., *Publ. Astron. Soc. Australia* **12**, 258 (1995)

[23] Jones, D. J. A., *Appl. Optics* **33**, 7362 (1994)

[24] Taylor, K., Bailey, J., Wilkins, T., Shortridge, K. and Glazebrook, K., *Astron. Soc. Pacific Conf. Ser.* **101**, 195 (1996)

[25] Maddox, S., *Astron. Soc. Pacific Conf. Ser.* **200**, 63 (2000)

[26] Colless, M., Dalton, G., Maddox, S. *et al.*, *Mon. Not. R. astron. Soc.* **328**, 1039 (2002)

[27] Shanks, T., Boyle, B. J., Croom, S. M. *et al.*, *Astron. Soc. Pacific Conf. Ser.* **200**, 57 (2000)

[28] Croom, S. M., Smith, R. J., Boyle, B. J. *et al.*, *Mon. Not. R. astron. Soc.* **322**, L29 (2001)

[29] Bridges, T., *AAO Newsletter* **100**, 20 (2002)

[30] Timothy, J. G., *Publ. astron. Soc. Pacific* **95**, 810 (1983)

[31] Joseph, C. L., *Exper. Astron.* **6**, 97 (1995)

[32] Woodgate, B. E., Kimble, R. A., Bowers, C. W. *et al.*, *Publ. astron. Soc. Pacific* **110**, 1183 (1998)

[33] Woodgate, B. E. and Kimble, R. A., *Astron. Soc. Pacific Conf. Ser.* **164**, 166 (1999)

[34] Kimble, R. A., *Astrophys. J.* **492**, L83 (1998)

[35] Baum, S. A., Downes, R., Ferguson, H. C. *et al.*, in 'Conference on space telescopes and instruments, V', *Proc. Soc. Photo-instrumentation Engineers (SPIE)* **3356**, 271 (1998)

[36] Dekker, H., D'Odorico, S., Kaufer, A., Delabre, B. and Kotlowski, H., in 'Optical and infrared telescope instrumentation and detectors', ed. Masanori Iye and A. Moorwood, *Proc. Soc. Photo-instrumentation Engineers (SPIE)* **4008**, 534 (2000)

[37] Dekker, H. and D'Odorico, S., *European South. Observ. Messenger* **70**, 13 (1992)

[38] Dekker, H., *European South. Observ. Messenger* **80**, 11 (1995)

[39] Pilachowski, C., Dekker, H., Hinkle, K. *et al.*, *Publ. astron. Soc. Pacific* **107**, 983 (1995)

[40] D'Odorico, S., Cristiani, S., Dekker, H. *et al.*, in 'Discoveries and research prospects from 8- to 10-meter class telescopes', ed. J. Bergeron, *Proc. Soc. Photo-instrumentation Engineers (SPIE)* **4005**, 121 (2000)

[41] York, D. G. *et al.* (144 authors), *Astron. J.* **120**, 1579 (2000). See also www.sdss.org and www.astro. princeton.edu/PBOOK/

[42] Adelman-McCarthy, J. K. *et al.* (153 authors), *Astrophys. J. Suppl. ser.* 172, 634 (2007). See also www.sdss.org/dr5/

[43] Hearnshaw, J. B., Barnes, S. I., Kershaw, G. M. *et al.*, *Exper. Astron.* **13**, 59 (2002)

[44] Wang, S., Hildebrand, R. H., Hobbs, L. M. *et al.*, in 'Instrument design and performance for optical and infrared ground-based telescopes', ed. Masanori Iye and A. F. M. Moorwood, *Proc. Soc. Photo-instrumentation Engineers (SPIE)* **4841**, 1145 (2003)

[45] Skuljan, J., Ramm, D. J. and Hearnshaw, J. B., *Mon. Not. R. astron. Soc.* **352**, 975 (2004)

[46] Pepe, F., Mayor, M. and Rupprecht, G., with the collaboration of the HARPS team, *European South. Observ. Messenger* **110**, 9 (2002). Further details of HARPS can be found at http://www.ls.eso.org/lasilla/sciops/3p6/harps/instrument.html

[47] Mayor, M., Pepe, F., Queloz, D. *et al.*, *European South. Observ. Messenger* **114**, 20 (2003)

[48] Lovis, C., Pepe, F., Bouchy, F. *et al.*, *Proc. Soc. Photo-instrumentation Engineers (SPIE)* **6269** (2006)

Figure sources and acknowledgements

CHAPTER 1

Fig. no.	Source	Acknowledgements
1.1	[12]	
1.2	C. A. Young, *A Text Book of general astronomy* Ginn & Co., Boston (1888) Fig. 107	
1.3	J. Scheiner, *Die spectralanalyse der Gestirne* W. Engelmann Verlag, Leipzig (1890) Fig. 26	
1.4	J. N. Lockyer, *Solar Physics* MacMillan & Co., London (1874) Fig. 58	
1.5	J. N. Lockyer, *Solar Physics* MacMillan & Co., London (1874) Fig. 48	
1.6	[37] Fig. 1	
1.7	[37] Fig. 2	
1.8	[43] Plate 38	
1.9	H. Grubb, *Catalogue of Astronomical Instruments* Dublin (1885)	Courtesy I. S. Glass
1.10	A. Secchi, *le Soleil* (part 1, 2nd edn.) Gauthiers-Villars, Paris (1875)	
1.11	J. N. Lockyer, *Solar Physics* MacMillan & Co., London (1874) Fig. 49	
1.12	[84] Plate XXXIV	
1.13	[89]	
1.14	[101] Plate I	
1.15	G. Eberhard *Handbuch der Astrophysik* 1, 299 (1933) Chapter 4, Fig. 17	
1.16	[93] Plate II	Reproduced by permission of the AAS
1.17	[93] Plate I	Reproduced by permission of the AAS
1.18	J. S. Plaskett, Report of the Chief Astronomer Dominion Observ., Ottawa, 1909, p. 143, Figs. 2 and 4	
1.19	[101] Plate III	
1.20	[101] Plate VI	
1.21	H. Kayser, *Handbuch der Spectroscopie* vol. I, Verlag S. Hirzel, Leipzig (1900) Fig. 120	
1.22	[113]	
1.23	[169] Fig. 1	Reproduced by permission of the AAS
1.24	[171] Fig. 8, Plate VII	Reproduced by permission of the AAS

Fig. no.	Source	Acknowledgements
1.25	[176] Fig. 2	Courtesy of W. Liller and of the Optical Society of America
1.26	J. Scheiner, *Die Spectralanalyse der Gestirne* W. Engelmann Verlag, Leipzig (1890) Fig. 41	
1.27	J. Scheiner, *Die Spectralanalyse der Gestirne* W. Engelmann Verlag, Leipzig (1890)	
1.28	[152] Plate V	
1.29	[197] Plate I	
1.30	[123]	Reproduced by permission of the AAS
1.31	[201]	Courtesy of the Astronomical Society of the Pacific
1.32	[223] Plate X	Reproduced by permission of the AAS
1.33	[216] Plate XIII	Reproduced by permission of the AAS
1.34	[226] Fig. 2	

CHAPTER 2

Fig. no.	Source	Acknowledgements
2.1	Drawn by author	
2.2	Plotted by author	
2.3	Drawn by author	
2.4	Drawn by author	
2.5	Drawn by author	
2.6	Drawn by author	
2.7	Plotted by author	
2.8	Drawn by author	
2.9	Drawn by author	
2.10	Drawn by author	
2.11	[36] Fig. 4	Courtesy D. D. Walker and F. Diego and of Blackwell Publishing
2.12	Drawn by author	
2.13	[52]	Courtesy of A. Baranne

CHAPTER 3

Fig. no.	Source	Acknowledgements
3.1	[6]	
3.2	[1]	Courtesy of Elsevier
3.3	[26] Fig. 1	
3.4	[21] Fig. 117	
3.5	[19] Fig. 1	
3.6	[43] Fig. 2	Reproduced by permission of the AAS
3.7	[51] Fig. N–21	
3.8	Adapted from [43] Fig. 4	Courtesy of S. I. Barnes

Fig. no.	Source	Acknowledgements
3.9	[58] Fig. 6	Courtesy of D. D. Walker and F. Diego and of Blackwell Publishing
3.10	Supplied by the author	
3.11	[90] Fig. 1	
	Supplied by the author	
3.12	Exposed by the author	

CHAPTER 4

Fig. no.	Source	Acknowledgements
4.1	[8] Fig. 9	
4.2	A. Secchi, *le Soleil* (part 2, 2nd edn.) Gauthiers-Villars, Paris (1875) Fig. 151	
4.3	A. Secchi, *le Soleil* (part 2, 2nd edn.) Gauthiers-Villars, Paris (1875) Fig. 152	
4.4	A. Secchi, *le Soleil* (part 2, 2nd edn.) Gauthiers-Villars, Paris (1875) Fig. 153	
4.5	H. Kayser, *Handbuch der Spectroscopie* vol. I, Verlag S. Hirzel, Leipzig (1900)	
4.6	[34]	
4.7	[56]	
4.8	[61] Plate I	
4.9	[61] Plate III	
4.10	G. Millochau & M. Stefánik, *Astrophys. J.* **24**, 42 (1906)	Reproduced by permission of the AAS
4.11	[62]	
4.12	[75] Plate 22(1)	
4.13	[75] Plate 46(1)	
4.14	[75] Plate 23	
4.15	[73] Plate 5	With kind permission of Springer Science + Business Media
4.16	[82] Fig. 1	Courtesy of Elsevier
4.17	G. E. Hale, *Contrib. Mt Wilson Solar Observ.* no. 10 (1906) Plate XIV, also in *Astrophys. J.* 24, 61 (1906)	Reproduced by permission of the AAS
4.18	G. E. Hale, *Contrib. Mt Wilson Solar Observ.* no. 7 (1906) Plate XII, also in *Astrophys. J.* 23, 54 (1906)	Reproduced by permission of the AAS
4.19	G. E. Hale, *Contrib. Mt Wilson Solar Observ.* no. **23** (1908) Plate XXIV, also in *Astrophys. J.* **27**, 204 (1908)	Reproduced by permission of the AAS
4.20	[90] Fig. 2	Reproduced by permission of the AAS
4.21	[65] Plate XXV	Reproduced by permission of the AAS
4.22	C. E. St John, *Astrophys. J.* **67**, 1 (1928) Plate II	Reproduced by permission of the AAS
4.23	C. E. St John, *Astrophys. J.* **67**, 1 (1928) Plate I	Reproduced by permission of the AAS
4.24	[105] Chapter 2, Fig. 15	
4.25	[105] Chapter 2, Fig. 13	
4.26	[95] Fig. 1	

Fig. no.	Source	Acknowledgements
4.27	[105] Chapter 2, Fig. 27	
4.28	[113] Plate I	
4.29	[113] Fig. 1	
4.30	R. R. McMath in *The Sun* ed. G. P. Kuiper, p. 605, University of Chicago Press (1953) Fig. 10	Reproduced by kind permission of the University of Chicago Press
4.31	[115] Plate I	Reproduced by kind permission of the University of Michigan, Department of Astronomy
4.32	R. R. McMath in *The Sun* ed. G. P. Kuiper, p. 605, University of Chicago Press (1953) Fig. 9	Reproduced by kind permission of the University of Chicago Press
4.33	R. R. McMath in *The Sun* ed. G. P. Kuiper, p. 605, University of Chicago Press (1953) Fig. 15	Reproduced by kind permission of the University of Chicago Press
4.34	[126]	Reproduced by kind permission of the University of Michigan Press
4.35	R. R. McMath in *The Sun* ed. G. P. Kuiper, p. 605, University of Chicago Press (1953) Fig. 13	Reproduced by kind permission of the University of Chicago Press
4.36	[133] Fig. 4	Courtesy of the Optical Society of America
4.37	[150] Fig. 2	Courtesy of the Optical Society of America
4.38	[150] Fig. 4	Courtesy of the Optical Society of America

CHAPTER 5

Fig. no.	Source	Acknowledgements
5.1	[12]	
5.2	E. C. Pickering, *Astron. & Astrophys.* **11**, 199 (1892)	
5.3	F. McLean *Spectra of Southern Stars* E. Stamford, London (1898)	
5.4	[90] Plate XI	Reproduced by permission of the AAS
5.5	[35]	
5.6	[44] frontispiece to part 1	
5.7	[52] Fig. 1	Reproduced by permission of the AAS
5.8	[70] Fig. 15	With kind permission of Springer Science + Business Media
5.9	[73] Fig. 1	With kind permission of Springer Science + Business Media
5.10	[74] Fig. 1	With kind permission of Springer Science + Business Media
5.11	[90] Plate X Fig. 2	Reproduced by permission of the AAS
5.12	[96] Fig. 5	With kind permission of Springer Science + Business Media
5.13	[102] Fig. 4	Courtesy of Elsevier
5.14	D. A. MacRae, *Sky and Tel.* **30**, 7 (1965)	

CHAPTER 6

Fig. no.	Source	Acknowledgements
6.1	[3]	
6.2	[5]	Reproduced by permission of the AAS
6.3	[9] Fig. 1	Reproduced by permission of the AAS
6.4	[9] Fig. 2	Reproduced by permission of the AAS
6.5	[11] Fig. 1	
6.6	[11] Fig. 5	
6.7	[20]	Reproduced by permission of the AAS
6.8	[22]	Reproduced by permission of the AAS
6.9	[22]	Reproduced by permission of the AAS
6.10	[26]	With kind permission of Springer Science + Business Media
6.11	[26]	With kind permission of Springer Science + Business Media
6.12	[28]	Courtesy of G. R. Lemaitre
6.13	[32] Fig. 2	Reproduced by permission of the AAS
6.14	[35] Fig. 1	Courtesy of T. P. Stecher
6.15	[39] Fig. 2	Reproduced by permission of the AAS Courtesy of D. C. Morton
6.16	[39] Fig. 3a	Reproduced by permission of the AAS Courtesy of D. C. Morton
6.17	[41]	Reproduced by permission of the AAS Courtesy of G. Carruthers
6.18	[49]	Reproduced by permission of the AAS Courtesy of Y. Kondo
6.19	[51]	Reproduced by permission of the AAS Reproduced by permission of the Optical Society of America
6.20	[52] Fig. 2	Courtesy of A. D. Code
6.21	[52] Fig. 6	Courtesy of A. D. Code
6.22	[54] Fig. 1	Courtesy of J. B. Rogerson Reproduced by permission of the AAS
6.23	[58]	Courtesy of K. A. van der Hucht
6.24	[59] Fig. 1	Courtesy of A. Boggess
6.25	[59] Fig. 2	Courtesy of A. Boggess
6.26	[65]	Courtesy of G. A. Gurzadyan
6.27	[68]	Courtesy of G. A. Gurzadyan

CHAPTER 7

Fig. no.	Source	Acknowledgements
7.1	[3] Fig. 1	Courtesy of H. R. Butcher Reproduced by permission of SPIE
7.2	[3] Fig. 2	Courtesy of H. R. Butcher Reproduced by permission of SPIE

Fig. no.	Source	Acknowledgements
7.3	[3] Fig. 3	Courtesy of H. R. Butcher Reproduced by permission of SPIE
7.4	[15] Fig. 3	Courtesy of J. R. P. Angel Reproduced by permission of the AAS
7.5	[19] Fig. 1	Courtesy of J. M. Hill Reproduced by permission of SPIE
7.6	[18] Plate L4, Fig. 1	Courtesy of J. M. Hill Reproduced by permission of the AAS
7.7	[21] Fig. 11	Courtesy of J. M. Hill Reproduced by permission of SPIE
7.8	[2] Fig. 3	Courtesy of J. M. Hill

CHAPTER 8

Fig. no.	Source	Acknowledgements
8.1	[1] Fig. 1	Courtesy of H. Dekker
8.2	[1] Fig. 2	Courtesy of H. Dekker
8.3	[9] Fig. 1	Courtesy of S. S. Vogt and of the Optical Society of America
8.4	[9] Fig. 8	Courtesy of H. W. Epps and of the Optical Society of America
8.5	[13] Fig. 1	Courtesy of J. B. Oke and J. G. Cohen
8.6	[18] Fig. 1	Courtesy of A. Baranne and D. Queloz Reproduced by permission of ESO
8.7	[21] Fig. 5	Courtesy of P. M. Gray
8.8	[32] Fig. 2	Courtesy of B. E. Woodgate
8.9	[37] Fig. 3	Courtesy of H. Dekker
8.10	see note	Courtesy of M. A. Strauss, D. H. Weinberg and A. Uomoto
8.11	see note	Courtesy of M. A. Strauss, D. H. Weinberg and A. Uomoto
8.12	[43] Fig. 6	Courtesy of G. M. Kershaw
8.13	[43] Fig. 2	Courtesy of G. M. Kershaw
8.14	[46] Fig. 2	Courtesy of F. Pepe

Note:
Figures 8.10 and 8.11 appear in the SDSS Project Book website at
www.astro.princeton.edu/PBOOK/spectro/spectro.htm,
being Figs. 7.6 and 7.7.

Name index

Page numbers in bold indicate that the dates of the person referenced are given on that page.

Subject index